Essentials of
Engineering
Thermodynamics

About the Author

Clement Kleinstreuer is a professor in the Mechanical and Aerospace Engineering Department at North Carolina State University in Raleigh, North Carolina. He received his Dipl.-Ing. degree at the Technical University Munich, his MS at Stanford University, and his PhD at Vanderbilt University. While his previous textbooks are for graduate students and professionals, this is his first text for undergraduate students.

Essentials of Engineering Thermodynamics

Principles and Applications

Clement Kleinstreuer, PhD

New York   Chicago   San Francisco
Athens   London   Madrid
Mexico City   Milan   New Delhi
Singapore   Sydney   Toronto

Library of Congress Control Number: 2020952346

Essentials of Engineering Thermodynamics: Principles and Applications

ISBN 978-1-260-46780-2
MHID 1-260-46780-5

This book is printed on acid-free paper.

Sponsoring Editor
    Robert Argentieri

Copy Editor
    Heather Mann and Fred Dahl

Editing Supervisor
    Stephen M. Smith

Proofreader
    Nikhil Roshan, MPS Limited

Production Supervisor
    Lynn M. Messina

Indexer
    Mary Kidd

Acquisitions Coordinator
    Elizabeth M. Houde

Art Director, Cover
    Jeff Weeks

Project Manager
    Rishabh Gupta, MPS Limited

Composition
    MPS Limited

To my family,
Christin, Joshua, Nicole, Andrew, and Gavin

Contents

Preface

The main objective here is to develop an engaging textbook that improves on existing undergraduate thermo-books in several categories. Ultimately, it will result in students' higher success, i.e., a more substantial knowledge base and independent problem-solving skills. This is achieved by stressing the fundamentals, i.e., salient concepts and physical insight, via a novel learning approach, specifically, discussions of detailed sample-problem solutions, visual learning experiences, and two-part homework sets, tests, and exams, featuring both insight questions and engineering problems. Course projects, their associated reports, and result presentations are highly recommended. The novel instructional strategy, concerning thermal systems analyses, is a *deductive* approach, i.e., from the general to the specific. After a brief review of some Physics 101 (or Engineering 101) essentials, the fundamentals are first presented in general form, followed by discussions of specific applications that are special cases of the generalized system. While following the deductive approach, the essentials of thermodynamics are often repeated, because the process of understanding engineering thermodynamics is quite similar to learning a new language.

Although the mathematical needs in introductory engineering thermodynamics are rather minimal, the subject can be for some students overwhelming in terms of the amount of new material, definitions, and concepts. Thus, salient topics in this book are presented in a concise and illustrative form. Nevertheless, for many students most challenging are the requirements for *logical thinking and multi-tasking*. Here, the deductive learning approach, illustrated with detailed answers to concept questions and solutions to representative problems, should help the reader to develop a *fundamental knowledge base* and to learn the skills for *independent problem solving*.

Engineering thermodynamics is an ideal subject to demonstrate the merits of this novel learning approach, especially when embedded in the e-book format. Specifically, Chap. 1 conveys the course foundation by introducing basic definitions, concepts, and laws as well as thermal systems/devices, working fluids, energy forms, and their transfer processes. Chapters 2 and 3, relying on the Chap. 1 material, introduce the 1st and 2nd Laws of Thermodynamics with applications to all types of systems, devices, and basic cycles. Chapter 4 evaluates exergy, i.e., ways to maximize the work potential of energy, while Chap. 5 introduces somewhat more advanced closed and open power cycles. At some universities, most of Chap. 4 and all of Chap. 5 are typically not part of a first course in thermodynamics. However, that added material is quite important for all engineering students. It can be readily covered in one semester via the deductive learning approach; furthermore, both chapters are fine sources for course projects plus presentations. In fact, Chaps. 4 and 5 help to eliminate a second engineering thermodynamics course or turn it into an elective. Specifically, the basics of psychrometry, chemical and phase equilibrium, and gas mixtures and their applications can be discussed in follow-up courses, such as combustion, HVAC, and heat-transfer system design.

Questions and problems for homework assignments and tests are grouped into (i) Part A, *Insight,* i.e., concept questions, definitions, derivations of basic correlations, etc. (worth 20% to 30%), and (ii) Part B, *Problems,* i.e., basic applications (worth 70% to 80%). All Part B solutions follow a strict format of system *Sketch,* salient *Assumptions,* and basic *Method,* i.e., system identification, type of fluid/process, reduced mass/energy balances, correlations, etc. *These three interconnected preliminaries fully encapsulate the problem statement and are the gateway to successful problem solving.* After Chap. 4, suggestions for course projects are listed as optional assignments to tie the material together and learn how to write project reports. Students achieve a further enhanced learning experience via PowerPoint presentations and via group solutions of mock tests in a "flipped-classroom mode." Additionally, typically very similar problems may be found on the Web or in introductory books. Examples include Çengel and Boles (the required text at North Carolina State University for decades), as well as Moran and Shapiro, Borgnakke and Sonntag, and Turns and Pauley, which are (on and off) recommended texts at NCSU. They naturally have all influenced the author's teaching and hence the writing of this textbook.

Appendix A contains *Equation Sheets* that are useful for homework assignments and closed-book tests. *Property Tables* in SI units as well as conversion factors are provided in App. B. Appendix C is *Guidelines for Report Writing,* and *typical two-part tests and exams* can be found in App. D. (Appendices A and B may be downloaded to use for homework, tests, and exams from www.mhprofessional.com/thermo.)

A detailed solutions manual, authored by Ryan DeBoskey, will be available for instructors adopting this textbook.

The expert assistance of Victoria Grace Augoustides and Karthik Sridhar in text/equation typing, figure drawing, format checking, and material researching, as well as the professional support of McGraw Hill's Senior Editor Robert Argentieri, MPS's Project Manager Rishabh Gupta, and their staffs, are gratefully acknowledged.

Clement Kleinstreuer, PhD

Essentials of Engineering Thermodynamics

CHAPTER 1

Background Information

This chapter reviews Physics 101 basics and, in doing so, provides a common knowledge base on which subsequent chapters will build. *Essentials of Engineering Thermodynamics* describes on the macroscopic level *quasi-equilibrium processes* of *simple substances (or working fluids)*, as well as interactions between *systems (or devices)* and their *surroundings*. Typical *quasi-equilibrium* processes include a high-pressure supply line filling an air tank, steam expansion in a turbine, compression of a gas, and heat/work transfer to a piston–cylinder device. For such complex activities, it is assumed that the temporal changes of the fluid state occur very slowly, while spatial gradients vanish immediately, for example, via the use of a stir for rapid mixing. Recall that a spatial gradient is the rate of change of a physical scalar quantity with respect to its position in space. The compressible working fluids are "simple" because all their properties can be determined with just two (independent intensive) quantities by using equations of state, graphs/diagrams, and/or property tables. Idealized process/device examples from everyday life include:

- Pumping up a bicycle tire, where the work input produces increased air pressure and hence tube volume as well as wasted heat.

- Mixing a very hot water stream with cold water for a temperature-controlled shower.

- Producing electricity via high-temperature, high-pressure steam entering a turbine that will generate power for multitudinous uses (such as powering your favorite gaming station).

- Enjoying seasonally cool or warm air in the home or workplace via a heat pump.

- The cyclic process of cooling a refrigerator compartment so beverages stay cold and food stays fresh.

- The working of components of power cycles, especially those named after Carnot, Rankine, and Brayton.

To get started, we must refresh our knowledge of some preliminary physics concepts. First, we review the differences between closed and open systems, dimensions and units, gage and absolute pressures, and local and thermodynamic equilibrium. Second, we provide a more technical definition of energy forms and energy transfer mechanisms and we state the associated laws of thermodynamics. Finally, we discuss some simple applications to systems, devices, processes, and working fluids.

1.1   Review of Basic Concepts in Thermodynamics

Thermodynamics deals with *energy transfer* in *systems* or *devices* that are in contact with their *surroundings*, using different *fluids* that are undergoing different physical or bio-chemical *processes*. Before launching into the meaning and implications of these key words, we review some preliminary concepts.

1.1.1   Temperature, Pressure, and Processes

Thermodynamic systems range from aircraft components (e.g., compressor, combustion chamber, turbine, nozzle) to heating-cooling-ventilation devices and refrigerators to power plants. Some are closed systems (e.g., a piston–cylinder device where m_{fluid} = const.). Others are open systems, either operating under steady-state conditions (e.g., turbines and compressors) or being transient, such as filling a tank via high-pressure fluid-supply lines (see Sec. 1.2). Each system (or device) interacts with its surroundings from which (or to which) heat, work, and, for open systems, energy-carrying mass flow are being supplied. *It should be noted that for analysis purposes, the terms* system, device, process, *and* fluid *are often equivalent.*

Thermodynamic properties, such as a fluid's volume, mass, and energy, are *extensive* properties (i.e., mass dependent), while temperature, pressure, and density are *intensive* properties (i.e., independent of the size or mass of a fluid or solid). As mentioned, simple working substances considered in this book are fully determined by only two property values, and this condition is known as the *state postulate.* For example, knowing the steam temperature T (°C or K) and its specific volume $v = V/m$ (m³/kg), all other properties, such as steam pressure and quality as well as internal energy, enthalpy, and entropy, can be readily determined via charts (see the Property Tables in App. B) or correlations.

As already indicated, the concept of *thermodynamic equilibrium* is fundamental. While something in mechanical equilibrium only requires that all opposing forces are the same, thermodynamic equilibrium implies mechanical, thermal, phase, and chemical equilibrium. Practically, all forces, temperatures, phases, and concentrations are balanced. While this is not the case for real processes, even for "simple substances," a key assumption is necessary to make calculations manageable. Specifically, when processing, say, a fluid from its initial state to its final state, everything has to be in *quasi-equilibrium condition* at any point in time and space. This is achieved when the process advances very slowly so that every variable has time to equilibrate. That way, a simple substance is, at inlets/outlets and initial and final states, in equilibrium; however, the actual nonequilibrium effects, such as turbulence, friction, and sudden expansion/compression, are captured via the unavoidable increase in the system's *entropy* (see Sec. 1.2.3).

Very simple concepts include *dimensions* and *units.* A *dimension* is a measure of a physical quantity, such as length L, time t, mass M, temperature T, and current A. In contrast, a *unit* assigns a measured or calculated number to the dimension, say, in meters, seconds, kilograms, kelvins, and amperes. For example, the force in newtons, N, is an M-L-t derivative, using mass and acceleration: $1\,N = 1\,\text{kg} \cdot 1\,\text{m/s}^2$ ($M \cdot L/t^2$). The equivalence of work and energy appears unit-wise in $1\,kN = 1\,kJ$, where kJ is short for kilojoule (see the Conversion Factors with the Property Tables of App. B). Furthermore, energy forms are often expressed per unit time; for example, energy flow rate $\dot{E} = E/t$, or heat flow rate $\dot{Q} = Q/t$, or power $P \equiv \dot{W} = W/t$, all having the units $\text{kJ/s} \equiv \text{kW}$

(i.e., kilowatt). Some of the energy or energy-rate units can be unusual (see the Conversion Factors with the Property Tables):

- $1 \text{ kJ} = 1000 \text{ N} \cdot \text{m} = 1 \text{ kPa} \cdot \text{m}^3$
- $1 \text{ kWh} = 3600 \text{ kJ}$
- $1 \text{ kJ/kg} = 1000 \text{ m}^2/\text{s}^2$
- $1 \text{ kW} = 1 \text{ kJ/s} = 1.341 \text{ hp}$

Another application is *specific properties*, which are calculated per unit mass ($m = \rho V$), such as specific volume $v = V/m$ (m^3/kg), specific internal energy $u = U/m$ (kJ/kg), specific enthalpy $h = H/m$ (kJ/kg), as well as a combination of internal energy u, plus flow work pv, and specific entropy $s = S/m$ (kJ/K · kg).

Temperature and pressure are two of the most important state variables in engineering thermodynamics. In SI units, temperature T is expressed in Celsius or in kelvins, where $T(°C) = T(K) - 273.15$; the conversion to Fahrenheit reads $T(°F) = 1.8T(°C) + 32$. Pressure units use pascals (Pa), where $1 \text{ kPa} = 10^3 \text{ N/m}^2$ or $1 \text{ MPa} = 10^6 \text{ N/m}^2$, while 1 bar $= 10^5 \text{ N/m}^2$ and one standard atmospheric pressure (i.e., 1 atm = 101.325 kPa). Recalling that the *gage pressure* is a manometer-pressure reading, we have $p_{gage} = p_{absolute} - p_{atm}$. Note that given values are absolute pressures.

As mentioned, engineering thermodynamics describes energy transfer via fluids or solids to/from devices or systems. During such processes, which naturally interact with their surroundings, the focus is typically only on:

- The *initial and final states* of the working substance for transient processes; say, gas compression in a piston–cylinder device (see Sec. A.10.3 in App. A).
- Or the state of the fluid at the inlet and outlet of the device in case of steady flow processes, such as superheated, high-pressure steam expansion in a turbine (see Sec. A.10.4).

Thus, what happens to the fluid during the quasi-equilibrium process from the beginning to the end, or transport phenomena occurring inside a device/system between inlets and outlets, is immaterial. This is because we assume that all initial/final or inlet/outlet end-point conditions of the simple working fluid are *states of thermodynamic equilibrium.* In reality, the development toward the state of equilibrium is due to irreversible processes, such as heat conduction, species-mass diffusion, turbulence, and chemical reaction. The *process direction* is defined by an increase in total entropy [i.e., an increase in the degree of molecular and structural disorder [see Eq. (1.5)].

1.1.2 Forms of Energy and Associated Transfer Mechanisms

Heat and work are the energy forms transferred to/from closed systems, such as piston–cylinder devices. For open systems (e.g., compressors, turbines, throttles, nozzles, diffusers, and heat exchangers; see the Equation Sheets in App. A), steady, uniform fluid streams carry different energy forms into (or out of) a device as well. Therefore, the most important energy forms are as follows.

    i. Heat, $Q \sim \Delta T$, measured in kilojoules (kJ), is caused by temperature differences and is typically supplied to or removed from a system/device via thermal conduction, convection, or radiation.

ii. Work, $W = \int F ds$ (kJ), is done by the surrounding onto a system (positive W_{in}) or performed by a device (negative W_{out}). While boundary work $W_b = \int_1^2 p(\forall) d\forall$ appears most frequently, additional work forms are listed in Table 1.1.

iii. $E_{kinetic} = m \cdot v^2/2$ (kJ) is especially important when dealing with relatively high-speed, typically steady uniform open flow devices, such as nozzles and diffusers.

iv. Enthalpy, $H \equiv U + p\forall$ (kJ) is carried by *mass flow*. It encapsulates internal energy, U, plus flow work, $p\forall$. *Internal energy* U (kJ) is the sum of all molecular interactions and is mainly a function of temperature. Flow work is due to net pressure moving fluid volume $\forall$ into or out of a system/device.

Type	Differential Form	Integral Form	Application
Mechanical Forms of Work			
Boundary Work	$\delta W_b = F * ds$ $F = pA$ $A ds = d\forall$ $\therefore\ \delta W_b = p d\forall$	$W_b = \int_1^2 p(\forall) d\forall$ *Note:* Need $p(\forall)$ function! For example, $p\forall = m R_{gas} T$ for ideal gases.	Piston–cylinder device Balloon Any deforming Control Volume
Spring Work	$\delta W_{sp} = F * ds;\ F = ks$ *Note:* Spring "constant" k can be a function of s	$W_{sp} = \int_1^2 k s\, ds = \dfrac{k}{2}\left(s_2^2 - s_1^2\right)$	Spring-loaded piston–cylinder device: $W = W_b + W_{sp}$
Shaft Work	$\delta W_{sh} = 2\pi n dT$	$W_{sh} = 2\pi n T$, where n is the shaft revolution number and T is the torque	Fluid mixing: W_{stir}
Elastic Solid Bar Work	$\delta W_{elastic} = F * dx;\ F = \sigma_n A$ σ_n = normal stress A = cross-sectional area	$W_{elastic} = \int_1^2 F\, dx = \int_1^2 \sigma_n A\ dx$	Cantilever beam
Surface Tension Work	$\delta W_{sur} = F * dx;\ F = \sigma_s 2b$	$W_{sur} = \int_1^2 \sigma_s 2b\, dx$	Bubbles
Nonmechanical Forms of Work			
Electrical Work	$\delta W_{el} = VI dt$	$W_{el} = \int_2^1 VI dt = VI\Delta t$ *Note:* For constant voltage V and current I	Resistance Heating

TABLE 1.1   Mechanical and Nonmechanical Work Forms

Notes

- *Enthalpy is a constructed energy form.* It is very convenient to describe energy balances for *open systems* because $H = U + p\forall$ is directly associated with mass transfer in or out of a device/system.

- In contrast, heat transferred, Q, and work performed, W, relate to both *open and closed* systems (see Sec. 1.4.3).

- *Internal energy U* describes the energy content of a simple substance at rest; its *changes during process time Δt occur inside transient closed or transient open systems.*

- The state of a simple substance is completely determined by just two intensive properties; for example, when knowing T and p, all *specific properties v, u, h, and s* can be calculated or obtained from the Property Tables in App. B. Here, "specific" implies "per-unit-mass" (e.g., $v = \forall/m$, $h = H/m$, $u = U/m$, etc.).

Note Because work W (and heat Q) are *process path–dependent quantities*, minute changes are indicated by inexact differentials δW (or δQ), in contrast to infinitesimal changes in *state variables* like dT, dp, dV, dU, dH, and dS, which are *path-independent quantities*.

Example 1.1 Consider heating water in a tall cylinder closed by a movable piston with negligible wall friction.

a. Sketch the system (i.e., the piston–cylinder device): (i) before the water has boiled; (ii) after 50% of the water has turned into steam; and (iii) when all the water has turned into high-temperature steam and an external force is exerted on the piston to compress the steam.

b. Does the pressure change inside the cylinder during this initial heating and phase-change process [Cases (i) and (ii)]?

c. List all energy transfer mechanisms and the directions of energy flow.

d. For Case (iii), the superheated steam is being compressed. Find W_b.

Solution

a.

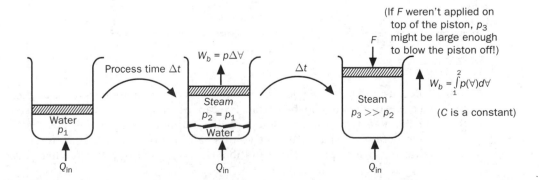

b. As long as there is no external force on the piston, the pressure exerted on the fluid is $p = W/A$, where W is the weight of the piston and A is its cross-sectional area. As the piston moves frictionless upward when the water boils and steam is generated, the pressure imposed on the fluid never changes, i.e., $p = $ const.

c. Most importantly, heat is being transferred at a constant rate to the p-c device (e.g., via conduction when sitting on a hot plate), and some heat may get lost through the system's walls. Clearly, from (1.1b), boundary work (see Table 1.1) is being performed by the system onto the surrounding.

d. From Table 1.1, $\delta W_b = Fds$ or $W_b = \int_1^2 p(\forall)d\forall$, where $p(\forall)$ has to be known.

Example 1.2 Consider air in an insulated, spring-loaded piston–cylinder device with an electric resistance heater. At an initial time, $t = 0$, the frictionless piston just touches the spring with parameter $k = cz^{1/2}$ and the heater is turned on. The air-expansion process ends later at t_{final}.

a. Sketch the thermodynamic State 1 of the air at $t = t_{initial}$ and then State 2 at $t = t_{final}$.

b. List all energy transfer mechanisms.

c. In symbolic math, provide expressions for the energy forms transferred in and out of the system.

Solution
a.

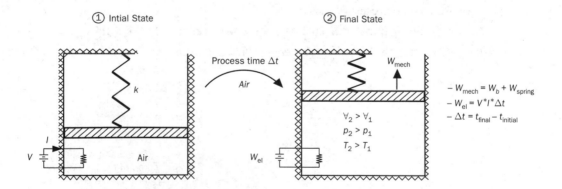

b. Because the p-c device is insulated (i.e., adiabatic), no conduction/convection/radiation heat transfer can occur into or out of the system. However, the electric work is being supplied, which elevates the air temperature, expands the air volume, and hence raises the piston. During the expansion process of duration $\Delta t = t_{final} - t_{initial}$, the moving piston compresses the nonlinear spring, and simultaneously, boundary work is performed by the system onto the surrounding. The general, closed-system energy balance $\Sigma E_{in} - \Sigma E_{out} = E_{final} - E_{initial}$ can be reduced to:

$$W_{el} - (W_b + W_{sp}) = [\Delta E]_{\text{p-c device}}$$

where the work forms are listed in Table 1.1 and $[\Delta E]_{p-c}$ is the total energy change inside the system during $\Delta t = t_{final} - t_{initial}$. The change in total internal energy, $\Delta E = \Delta kE + \Delta pE + \Delta U$, is typically approximated by the fluid's change in internal energy; thus, $[\Delta E]_{p-c} \approx \Delta U_{p-c}$.

c. Based on the answer to (1.2b), three work forms have to be considered. From Table 1.1:

- $P_{el} \equiv \dot{W}_{el} = W_{el}/\Delta t$, where $W_{el} = VI\Delta t$ for constant electric current and applied voltage.

- $W_b = p\Delta\forall$, as $p = const$.

- $W_{sp} = \int_1^2 k(z) \cdot z\,dz = c\int_1^2 \sqrt{z} \cdot z\,dz = \frac{2c}{5}(z_2^{5/2} - z_1^{5/2})$.

Discussion

As indicated with Examples 1.1 and 1.2, associated with different energy transfer mechanisms are different *forms of energy*, which either cross a system's boundary (here Q and W_b) or reside inside the system (actually E_{total}, but just U captures >95% of all internal energy forms).

Energy forms we are dealing with in engineering thermodynamics include kinetic, potential, internal, flow, heat, electrical, and mechanical energies. These energies can be converted from one form into another (e.g., potential into kinetic energy), or supplied heat into internal energy, or electric energy into heat, etc. However, the sum of all energy forms, that is, $E_{system, total}$, is always constant.

Macro-Scale Sum of Energies The *conservation of energy principle* states that energy can take on different forms, but the total sum is always a constant (see a simple example in Fig. 1.1):

$$E_{total} = E_{pot} + E_{kin} + E_{internal} + E_{flow-work} + E_{heat-supply} + E_{electric}$$
$$+ E_{mechanical} + E_{electro-magnetic} + \cdots = \cent \tag{1.1}$$

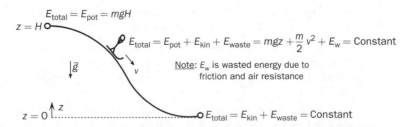

FIGURE 1.1 The principle of energy conservation demonstrated via a skier zipping down a hill.

As already mentioned, in Eq. (1.1) the most important energy forms are:

- $E_{\text{heat-transfer}} \equiv Q$, i.e., the heat transferred to or from the system/device.
- $E_{\text{internal}} \equiv U$, which is equivalent to the total energy content of the fluid residing inside a device/system.
- $E_{\text{flow}} \equiv H$, which mainly represents a fluid stream's temperature and pressure.
- $E_{\text{mechanical}} \equiv W_{\text{mech}}$ could be boundary, spring, or shaft work (see Table 1.1).
- $E_{\text{electric}} \equiv W_{\text{el}}$ is a function of applied voltage V and electric current I.
- $E_{\text{kinetic}} = m \cdot v^2/2$, which is important when dealing with nozzles and diffusers.

Note   There are three additional laws of thermodynamics, as discussed in Sec. 1.3.

Internal Energy vs. Enthalpy   As implied earlier, it is important to distinguish between internal energy U and enthalpy H:

- U [kJ] (or u [kJ/kg]) encapsulates energies due to molecular interactions, such as translation, collision, rotation, spin, and vibration of a *fluid at rest*, i.e., inside the system or device, which is also known as the *control volume* (C$\forall$). Thus, the internal energy is mainly a function of temperature. In fact, for *ideal gases U* (or *u*) is a function of temperature only!

- H [kJ] (or h [kJ/kg]) is *a made-up energy form*, representing both internal energy U and flow work ($p\forall$) of a *fluid in motion*. Thus, the enthalpy is determined by a fluid's temperature and pressure levels, as they are most crucial for two-phase flow devices, e.g., steam turbines. Interestingly, as shown later, for *ideal gases*, $h = h(T)$ only as well.

1.2   Hierarchy of Thermodynamic Systems

The *systems, devices,* or *compartments* in which energy transfer, exchange, and conversion occur include (see Figs. 1.2 and 1.3):

- *Transient* (i.e., unsteady, time-dependent) *open* systems, such as filling of a tank from a high-pressure fluid-supply line or fluid removal or fluid loss from a system, where in both cases, the fluid mass and temperature inside the system change with time. A fire extinguisher in operation is another example (http://www.youtube.com/watch?v=jR8vZHtOLdA—see this video for some information on closed, open, and isolated systems).

- *Closed* systems, such as tanks, compartments, or piston–cylinder devices (when the valves are closed), are transient systems as well.

  - *Open devices that operate under steady uniform flow conditions*, such as turbines and compressors, nozzles, and diffusors, as well as heat exchangers and mixing chambers. (See the latest YouTube video for a basic idea of how a gas turbine works)

- A *cyclic* system is a ring of *open steady uniform flow devices*, where a fluid, typically undergoing phase change (say, H_2O or R-134a), is continuously recycled inside the system. A steam power plant is a good example: A boiler supplies high-enthalpy steam to a turbine from which the exhausted steam is turned into water inside a condenser (form of a heat exchanger) from which the water is pumped back to the boiler to produce steam. Other cyclic systems are heat pumps, refrigerators, and air conditioners. (See the latest YouTube video for an idea of how a heat pump works; such website has really good animations of thermodynamic systems.)

The word "system" may describe all kinds of things. For example, "thermodynamic system" is often generically used for a device, or a compartment, or a working fluid/substance/material/body or a process, focusing on what is inside the system and how it interacts with its surroundings. Essentially, these systems can be all regarded as *control volumes (C∀s)*, where the *control surface* (CS) represents the system's solid walls and inlet/outlet ports—the latter in the case of an open system.

Control volumes encompass:

- Steady or transient open, uniform flow devices with fluid mass transfer (see Figs. 1.2a and 1.3).

- Transient closed compartments or devices where fluid mass m is constant (see Fig. 1.2b).

Mass and Energy Balances   Recalling from Engineering 101 (or Physics 101) and as outlined in the Equation Sheets (see App. A), the conservation laws on the macro scale can be written as follows.

For transient uniform fluid flow:
$$\sum \dot{m}_{in} - \sum \dot{m}_{out} = \left(\frac{\Delta m}{\Delta t}\right)_{c.v.} \tag{1.2a}$$

Or during process time Δt:
$$\boxed{\sum m_{in} - \sum m_{out} = (\Delta m)_{c.v.} = (m_f - m_i)} \tag{1.2b}$$

where m_f is the final and m_i is the initial amount of mass in the C∀.

For transient energy transfer:
$$\sum \dot{E}_{in} - \sum \dot{E}_{out} = \left(\frac{\Delta E}{\Delta t}\right)_{c.v.} \approx \left(\frac{\Delta U}{\Delta t}\right)_{c.v.} \tag{1.3a}$$

Or during process time Δt:
$$\boxed{\sum E_{in} - \sum E_{out} \approx (\Delta U)_{c.v.} = \left[(mu)_f - (mu)_i\right]_{c.v.}} \tag{1.3b}$$

In Eq. (1.3a) $\sum \dot{E}_{in/out} = \sum \dot{Q} + \sum \dot{W} + \sum \dot{H} + \sum \dot{E}_{kin} + \sum \dot{E}_{pot}$ in kJ/s = kW. In general, $\dot{X} = \dot{m}x$, with x being a specific property, e.g., u or h; $\dot{E}_{kin} = \frac{1}{2}\dot{m}v^2$; $\dot{E}_{pot} = \dot{m}gz$;

$\dot{H} = \dot{m}h = \dot{m}(u + pv)$

Notes

- For *steady* flow processes, the right-hand sides of Eqs. (1.2a) and (1.3a) are zero, as there are no changes over time inside the system or device (see Fig. 1.3).

- The differential change in specific enthalpy is defined as:
$$dh = du + d(pv) \tag{1.4}$$

1.2.1   Transient Systems and Devices

Figures 1.2a and b depict the schematics of two basic transient systems. The first one is an *open system* as fluid mass (carrying enthalpy and possibly kinetic energy) crosses the control surface. It is also *transient* because mass and energy changes occur over time *inside the control volume.* The second one is a *closed system,* e.g., a tank, compartment, or piston–cylinder device where m_{system} = constant. Closed systems are generally *transient* because of internal energy accumulation (or depletion) over process time Δt. Recall that there are *no spatial variations in fluid properties* because a quasi-equilibrium process with perfect mixing is always assumed.

Note As implied in the previous conservation equations, all devices to be discussed from here on are *special cases of the transient open uniform flow system* of Fig. 1.2a.

Notes

- The piston–cylinder (p-c) device is the best-known example of a closed (transient) system.
- An isolated system is totally shut off from its surroundings, i.e., no energy/mass transfer.

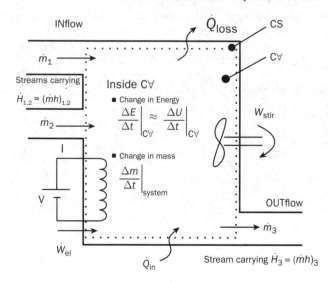

(a)

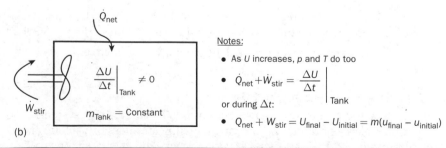

(b)

Figure 1.2 Forms of energy and energy transfer mechanisms in transient systems: (a) open and (b) closed.

1.2.2 Steady Open Uniform Flow Devices

In industrial processes and energy production, steady-state operation is preferred. Thus, most devices operate under steady flow conditions. Figure 1.3 depicts some of the common systems with basic mass and energy balances [see Eqs. (1.2b) and (1.3b) with the right-hand sides set to zero].

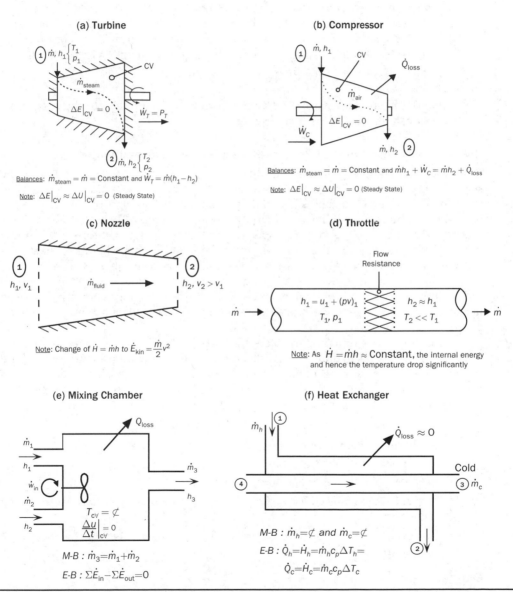

FIGURE 1.3 Schematics of common steady uniform flow devices.

Example 1.3   Considering the transient open uniform flow device of Fig. 1.2(a), what geometric design and energy transfer changes do you have to implement to turn it into a closed steady-state system?

Solution   First, the three ports (two inlets, one outlet) have to be blocked off in order to have a "closed system"; hence, there is no mass and enthalpy flow into and out of the system. To force the closed system to operate under steady-state conditions, all energies transferred IN have to be balanced by all energies going OUT, i.e., $\Sigma\dot{E}_{IN} = \Sigma\dot{E}_{OUT}$. If such an IN = OUT balance is not achieved, the *energy level inside the system (or control volume) would either increase or decrease with time*—the hallmark of a *transient* system/device. Applied to the present case, we require on a rate basis for all times:

$$\dot{W}_{el} + \dot{W}_{stir} + \dot{Q}_{in} = \dot{Q}_{loss}$$

Thus, all the constant work rates and heat-flow rates supplied to the system are being constantly lost to the surrounding.

1.2.3   Fluid Flow, Types of Processes, and System Irreversibilities

In a system of compressible working fluids (i.e., gases and liquids), only two intensive properties are required to fully describe it. The gases and fluids in this system are known as simple substances, and the definition of the system from two properties is known as the state postulate. Common gases, such as air, CO_2, N_2, He, etc., can be treated as ideal gases ($p\forall = mRT$) at moderate pressures and temperatures. Concerning liquids, water is typically used in heat exchangers and mixing chambers. Examples of two-phase fluids are H_2O and the *refrigerant R-132a*, which can appear in the compressed liquid, saturated mixture (i.e., liquid + vapor), and pure vapor states depending on the temperature *and* pressure levels. Such phase-changing fluids are employed in cyclic processes to generate, say, a constant torque with a steam turbine or low temperatures in a freezer.

A *process* moves a fluid and transfers energy from its inlet state to its exit state; for example, in steady open, uniform flow devices. Alternatively, a process conveys a fluid and transfers energy from its initial state to its final state, as in transient open or closed systems. Another application is a two-phase fluid being processed in a steady cycle. Processes may occur under fixed conditions; for example:

- Isobaric (pressure $p = \mathcal{C}$) as in ideal p-c devices
- Isothermal (temperature $T = \mathcal{C}$) when work-input and heat-loss balance are steady, i.e., at all times
- Isochoric (volume $\forall = \mathcal{C}$) as in solid tanks
- Isenthalpic (enthalpy $H = \mathcal{C}$), i.e., no change in enthalpy as in ideal throttling devices
- Isentropic (entropy $S = \mathcal{C}$), i.e., reversible (no waste) *and* adiabatic ($Q = 0$) conditions

Entropy

In actual processes, even when using ideal gases, some of the energy input is wasted due to *irreversibilities* such as friction, heat exchange, turbulence, energy losses due to sudden fluid expansion/compression, etc. These internal and external process irreversibilities have a net accumulation that is quantitatively captured on the macroscale as "entropy generation": S_{gen}. It is caused by a system's entropy change plus that of its surroundings, i.e., $S_{gen} = \Delta S_{system} + \Delta S_{surrounding}$. Simply put, entropy S [kJ/K] is a measure of molecular disorder, i.e., the level of randomness of the material's molecules. For example, a 1-kg block of ice has a very low entropy; when heated up and turned into water, that 1 kg has a significant S-value; after further heat transfer, the water turns into steam, and as a result: $S_{vapor} \gg S_{water} \gg S_{ice}$. Clearly, $\Delta S \sim Q$ and for all *real* processes $S_{gen} \equiv \Delta S_{total} = \Delta S_{system} + \Delta S_{surrounding} > 0$, *which is the 2nd Law of Thermodynamics.* Specifically, wasteful mechanical (e.g., friction) and thermal (heat transfer) effects cannot be avoided in the real world. Therefore, the *"increase-of-entropy" principle* reads:

$$S_{gen} \equiv \Delta S_{total} = \Delta S_{system} + \Delta S_{surrounding} > 0 \tag{1.5}$$

http://www.youtube.com/watch?v=heP5NX7bDG8—an interesting video that relates entropy and the arrow of time.

Equation (1.5) is the *Second Law of Thermodynamics*, written on the macro scale. Practically (see Fig. 1.4), a process cannot proceed unless *both the 1st and 2nd Laws* are fulfilled.

As should be transparent by now, temperature and pressure are the two most important properties determining the state of a simple substance. As mentioned, *temperature*, measured in degrees Celsius [°C] or Kelvin [K], where T [K] $= T$ [°C] $+ 273.15$, commonly expresses how hot or cold something is. For ideal gases, the temperature is directly proportional to the mean translational kinetic energy of its molecules, which is part of the fluid's internal energy. In fact, it can be shown that for ideal gases internal energy $U = U(T)$ only. *Pressure*, measured in pascals [Pa $=$ N/m²], is the normal force per unit area, i.e., scalar $p = dF_n/dA$. Gauge pressure is recorded relative to the atmospheric pressure, while the absolute pressure $p_{abs} \equiv p = p_{ambient} + p_{gage}$. For an ideal gas in a closed container, pressure $p = mRT/\forall$ is exerted by collisions of molecules onto a surface A. It should be recalled that the atmospheric pressure 1 atm $= 101.3$ kPa, while 1 bar $= 100$ kPa. *All units in this book, i.e., magnitudes to the basic mass-length-time-temperature dimensions (M-L-t-T), are SI units, and the pressures are absolute pressures.*

Notes   All substances have a *mass*:

$$m = \rho \forall \; [\text{kg}] \tag{1.6}$$

or appear as a fluid stream with *mass flow rate*:

$$\dot{m} = \text{v} \cdot A \cdot \rho \; [\text{kg}] \tag{1.7}$$

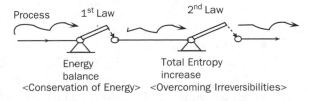

FIGURE 1.4   Real processes should obey both laws to proceed (adapted from Cengel & Boles, 2015).

where fluid streams have a temperature and pressure and hence carry enthalpy, kinetic energy, and entropy. Here v is the specific volume, v is the average velocity of the fluid stream, and A is the cross-sectional area of the device inlet/outlet pipes, while v·A is the *volumetric flow rate:*

$$\dot{\forall} = v{\cdot}A \ \left[\text{m}^3/\text{s}\right] \tag{1.8}$$

As implied, the inverse of density ρ is the *specific volume:*

$$v = \forall/m \ \left[\text{m}^3/\text{kg}\right] \tag{1.9}$$

Outside Einstein's famous energy–mass relation for subatomic particles, i.e., $E = mc^2$, *the principle of mass conservation* can now be applied to any open system with more insight (see Sec. 2.1).

Temporal vs. Spatial Changes

As indicated, some systems or devices operate under transient conditions (see Fig. 1.2), while others execute processes steadily (see Fig. 1.3). For the first case it is assumed that *property changes only occur with time, as everything is perfectly mixed inside the system (or control volume).* This is practically achieved with mixers via work input W_{stir}. In contrast, *during steady-state operations, property changes occur between inlets and outlets, where at any location inside the device nothing ever changes with time, i.e.,* $\dfrac{dm}{dt} = 0$ and $\dfrac{dU}{dt} = 0$.

Thus, as the very first assumption of a problem, one has to determine if the process, system, or device is transient or steady. Realizing this and identifying the type of working fluid are the key assumptions for a proper setup of the method prelim and subsequent calculations in the solution.

> **Example 1.4**   Classify the following systems and devices, as well as fluids and substances with brief descriptions of the energy and entropy processes: (a) Air is being compressed in an idealized piston–cylinder device; (b) wet steam is being released from a tank; (c) a compressed water stream is being boiled in a pipe; and (d) two gases of different temperatures are being mixed in a compartment.
>
> **Solution**

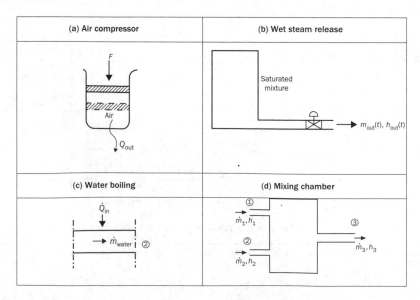

a. An air compressor is a *closed (transient) system* with a deforming control volume, where *air* can be assumed to behave like an *ideal gas. Compression, in the form of boundary work, generates heat,* which is rejected to the surrounding environment. Compression reduces the air's entropy, i.e., $\Delta S_{system} < 0$, as molecules move less randomly in denser gases, but heat loss to the surrounding environment produces more entropy so that the total change is greater than zero. Hence, $S_{gen} \equiv \Delta S_{total} = \Delta S_{system} + \Delta S_{surrounding} > 0$!

b. "Wet steam" implies a *mixture of H_2O vapor and liquid water,* i.e., the mixture quality is initially somewhere in the range of $0 < x < 1$, where the mixture quality is defined as $x = m_{vapor}/m_{total}$. Clearly, this is a *transient open uniform flow system* with a reduction in mass and energy inside the tank over time (due to heat loss and enthalpy leaving with the stream). Concerning entropy and the 2nd Law of Thermodynamics, $\Delta S_{total} > 0$ comes about due to entropy changes via the exiting mass flow, heat loss to the surroundings, irreversibilities (i.e., S_{gen} due to friction and heat transfer) inside the tank, and net changes inside the tank.

c. Phase change of water in a pipe (actually, an array of boiler tubes) is an essential part of a steam power plant. It is a *steady open uniform flow device* operating with a constant mass flow rate in cyclic fashion. Ignoring heat loss and pressure drop over tube length, the change in entropy of the inlet/outlet stream and heat input generate $S_{gen} \equiv \Delta S_{total} > 0$.

d. This is a *mixing chamber* operating under *steady open uniform flow* conditions. The fluids are assumed to be ideal gases, where the outlet gas mixture has a temperature somewhere between the ones of the hot and cold gas streams. It is typically assumed that there are no heat losses, and hence the energy (in terms of enthalpy) given up by the hot stream is 100% received by the cold stream, forming a third stream. The irreversibilities of S_{gen} are solely due to the mixing process, as adiabatic chamber walls are assumed.

1.3 Laws of Thermodynamics and Math Descriptions

Historically four Laws of Thermodynamics have been established, although the 1st and 2nd laws take center stage in this text. In addition to these two foundations, the conservation of mass principle is needed to describe real-world systems, devices, and processes. Following our deductive approach, these three laws can be mathematically described by just one rate equation (see Sec. 1.3.1 and the Equation Sheets in App. A).

1.3.1 The Four Laws of Thermodynamics

In addition to the 1st Law of Thermodynamics, which encompasses the fundamental principle of energy conservation, and the 2nd Law of Thermodynamics, which indicates the direction of real processes (see Fig. 1.4), two additional laws deal with the temperature. In summary,

- *Zeroth Law*: Two bodies are in thermal equilibrium if both have the same temperature, regardless if they are in contact or not.
- *First Law*: E_{total} = constant [see Eq. (1.1)], or applied to a closed system such as a piston–cylinder device: $Q_{in} - W_{out} = \Delta E_{total} \approx [\Delta U]_{system}$
- *Second Law*: Process irreversibilities are unavoidable; hence, $S_{gen} \equiv \Delta S_{total} > 0$ [see Eq. (1.5)].

- *Third Law:* The entropy of a perfect crystal at 0 K is zero, although zero Kelvin is not steadily attainable because temperature fluctuates temporally and spatially around average values. Nevertheless, it establishes the *absolute temperature scale*, starting with 0 K = −273.15°C.

Clearly, the 1st and 2nd Laws were established before R. H. Fowler came up with the "thermal-equilibrium" law in 1931, which logically had to be placed before the 1st Law. *The assumption that the total energy changes inside a transient system or device, ΔE_{total}, can be approximated by ΔU_{system} is realistic because temporal changes of other energy forms (e.g., ΔE_{pot} and ΔE_{kin}) are negligible inside most systems.*

1.3.2   Mathematical Descriptions

It is interesting to note that the processes in the devices presented so far and throughout the book can be described by just one *generalized first-order rate equation*, encompassing mass conservation and energy conservation, as well as entropy and exergy balances, as discussed in the upcoming chapters:

$$\sum \dot{X}_{in} - \sum \dot{X}_{out} \pm \dot{X}_{source/sink} = \left(\frac{dX}{dt} \right)_{system} \approx \left(\frac{\Delta X}{\Delta t} \right)_{system} \qquad (1.10a, b)$$

In Eq. (1.10) $\dot{X}$ may represent:

i. The mass flow rate [see Eq. (1.7)] of the simple substance (or working fluid).

ii. Energy rate forms such as heat, work, and/or enthalpy per unit time.

iii. Entropy per unit time, including the source term $\dot{X}_{source} \equiv \dot{S}_{generation}$.

iv. Exergy per unit time, i.e., rate of useful energy, including the sink term.

$$\dot{X}_{sink} \equiv \dot{X}_{destruction} = \dot{S}_{generation} \cdot T_{ambient}$$

Multiplying Eq. (1.10b) by Δt produces the general balance equation for the observation interval (or process duration) $\Delta t = t_{final} - t_{initial}$:

$$\sum X_{in} - \sum X_{out} \pm X_{source/sink} = \Delta X_{CV} = \left[X_{final} - X_{initial} \right]_{CV} \qquad (1.10c)$$

In Eq. (1.10c) we replaced the generic "system" subscript with the more specific "control volume" (see Fig. 1.2 depicting C.∀.). Clearly, in both the mass and energy balances, the $\pm X_{source/sink}$ terms are zero because of the conservation laws. As implied, $X_{source} = S_{gen}$ in the entropy balance. Not surprisingly, $X_{sink} = T_{ambient} S_{gen}$, i.e., energy is wasted/destroyed because of irreversibilities due to frictional and heat transfer effects (see Chap. 4). In energy balances $\left[X_{final} - X_{initial} \right]_{C.V.}$ can be approximated by $\left[U_{final} - U_{initial} \right]_{CV}$ as other energy forms inside the control volumes are negligible.

It should be noted that a solution of Eq. (1.10a) provides $X(t)$ for all times, whereas solutions of Eqs. (1.10b and 1.10c) give only thermodynamic state values for X at process beginning $X(t = t_{initial})$ and at process end, i.e., $X(t = t_{final})$. As an example, the general mass and energy balances for *transient* systems in Fig. 1.2 would read:

$$\sum X_{in} - \sum X_{out} = \left[X_{final} - X_{initial} \right]_{CV} \qquad (1.11)$$

In contrast, for *steady* flow devices shown in Fig. 1.3:

$$\sum \dot{X}_{in} - \sum \dot{X}_{out} = 0 \text{ or } \sum \dot{X}_{in} = \sum \dot{X}_{out} \qquad (1.12a, b)$$

It is interesting to note that for steady, uniform flow devices, energy transfer is described in terms of constant energy rates, $\dot{X}$; while for transient systems, processes are evaluated over a fixed observation time.

Sign Convention   In any energy balance, *energies entering the system/device appear as positive ("+") terms and leaving energies as negative ("−") terms.* While this sign convention is easy for heat Q, enthalpy H, and most work forms (say, $W_{electric}$ and W_{shaft}), it appears to be problematic for $W_{boundary}$. Specifically, when the system (i.e., the control volume) expands, work is done by the system onto the surrounding environment, and we have "−W," while for contracting systems (or shrinking C∀s) work is performed onto the system, which requires a "+W" in the energy balance.

1.4   Energy Transfer Modes

Section 1.1 and the subsections therein have already alluded to different forms of energy and how those energies may be transferred from the surroundings to a system/device and *vice versa*. Clearly, Eq. (1.10c) with $X \equiv E$ provided the energy balance for any system or device. In this section, the physics of energy forms and transfer modes are discussed and the reader's understanding is interactively checked. *Heat transferred and work done exhibit a few similarities in being both path-dependent functions and both being able to cross solid boundaries.* Kinetic energy changes are usually only significant in (steady-state) nozzles, diffusors, and throttling devices, while inside *transient* systems or control volumes, ΔE_{kin} is negligible, so that for all practical reasons $\Delta E_{total} \approx [\Delta U]_{system}$.

1.4.1   Heat Flow

Heat, i.e., *thermal energy*, is transferred to or from a system/device via conduction (e.g., a hot plate or ice pack), convection (e.g., a hot or cold fluid stream), and/or radiation (e.g., from a very hot heat source). Details of heat transfer are discussed in separate undergraduate and graduate courses. In this introductory e-book, the amount of heat transferred, Q [kJ], or heat-flow rate delivered, $\dot{Q}$ [kJ/s], is typically a given (or an unknown) constant value. Fundamentally, heat supplied increases the fluid's temperature (plus the pressure in tanks) and hence its internal energy, i.e., the sum of all forms of microscopic energies. As outlined, molecular activities occur in terms of kinetic energies due to molecule translation, rotation, vibration, and spin, as well as potential energies because of molecular binding forces.

In general, heat can cross walls or spaces driven by a temperature difference between the surrounding and the fluid inside the system or between two bodies, containers, fluid streams, etc. A process without any heat transfer is an *adiabatic process*; in practical terms, it occurs in systems with perfectly insulated walls. Like work, *heat is a path-dependent quantity*, i.e., differential changes are expressed as δQ (rather than dQ) because there are endless ways supplying heat in order to reach the same final (or outlet) state. Thus, $\int_1^2 \delta Q = Q$ and not ΔQ. Specifically, for steady, no-loss heat exchange between two bodies or fluid streams:

$$Q = m \cdot c(T_2 - T_1) \tag{1.13}$$

where m is the fluid/body mass, c is the specific heat capacity of the fluid (or solid body), and $\Delta T = T_2 - T_1$ is the temperature difference between the two fluids/bodies. Equation (1.13) is most useful when analyzing heat exchangers and mixing chambers, where $Q_{hot\ stream} = Q_{cold\ stream}$ (which is equivalent to the enthalpy exchange), implying that there are no heat losses to the surroundings.

1.4.2 Work Forms

The various work forms of interest are listed in Table 1.1. Like heat, work can cross the boundaries of a closed system, e.g., stirring to mix the fluid in a tank to eliminate spatial temperature gradients, or electric work to heat up a fluid in a compartment via an electric resistance heater, or piston displacement caused by heat input in a piston–cylinder device. As heat, *work is path dependent*. For example, in a piston–cylinder device

$$W_b \equiv W_{1\text{-}2} = \int_1^2 pd\forall$$ (see Table 1.1). Thus, in order to carry out the integration, one of

many possible $p(\forall)$-functions can be inserted to proceed from State 1 to State 2, clearly demonstrating the path dependence of work. Such $p(\forall)$-functions include:

$$p = mRT/\forall \text{ for } ideal \text{ gases; and more generally}$$ (1.14)

$$p = C/\forall^n \text{ for } polytropic\ gas \text{ expansion or compression}$$ (1.15)

While the range of the exponent n is theoretically $-\infty < n < \infty$, it practically varies between $n = 1$ for isothermal gas processes and $n = \kappa = c_p/c_v$ for isentropic processes, as discussed in Sec. 3.2. The constant $C = p_1 \cdot \forall_1^n = p_2 \cdot \forall_2^n = \text{¢}$.

There are two distinct work forms for which the generalized gas equation of state [Eq. (1.15)], $p(\forall) = C/\forall^n$, can be applied:

- Boundary work, characterized by volume changes, i.e., $W_b = \int_1^2 p(\forall)\,d\forall$, as in a piston–cylinder device.

- Compressor (or flow) work characterized by pressure changes to obtain $W_c = \int_1^2 \forall(p)\,dp$, using $\forall(p) = (C/p)^{1/n}$, where C is a constant.

As mentioned, work per unit time is power, i.e., $P \equiv \dot{W} = W/t$, being of interest when dealing with steady-state turbines and power cycles. Also worth mentioning is the fact that work does not carry entropy, but work performed can indirectly cause entropy changes in the substance and hence a system or device (see Chap. 3).

1.4.3 Enthalpy Transfer

As alluded to earlier, enthalpy is actually a combination property, which is conveyed via *mass transfer*, the third energy transfer mechanism, as mentioned in Sec. 1.1. Specifically, any fluid stream entering or exiting a control volume carries energy in the form of internal energy U as well as flow energy in terms of $p\forall$, pushing the fluid volume into/out of a device. For example, having steam at just a very high temperature but atmospheric pressure is useful for heat exchangers but inappropriate for running steam turbines where both high inlet temperatures *and* pressures (hence, high H_{in} values) are needed for efficient operation. So, for open systems a new energy form, which encapsulates temperature and pressure levels, was introduced as:

$$H = U + p\forall \tag{1.16}$$

or on a differential basis per unit mass,

$$dh + du + d(pv) \tag{1.17}$$

For *setting up energy balances,* it is important to note that enthalpy is the microscopic energy of a *moving* fluid, i.e., inlet/outlet streams. In contrast, internal energy is the microscopic energy form of a *stationary* fluid, i.e., changing inside a transient system.

1.5   Sources for Property Values

As mentioned, the properties of simple substances are determined by just two values; for example, knowing the temperature and pressure (or specific volume, or specific enthalpy) allows us to find all other property values in the Property Tables (see App. B). Homogeneous solids and liquids, being basically incompressible, are the easiest to analyze thermally. For ideal gases (IGs), numerous equations and basic physical explanations are available. More challenging are fluids, which change their phase with changing temperature and/or pressure. Key examples are water steam H_2O and refrigerant R-134a, which both may switch from compressed liquids via saturated liquid—vapor mixtures to superheated vapors and back (see Sec. 1.5.1). In any case, the sources to obtain property values include the following.

Equations of State

For IGs Eq. (1.14) holds, as well as the relations:

$$du = c_v \cdot dT \ \text{ and } \ dh = c_p \cdot dT \tag{1.18a, b}$$

The relations for du and dh are based on the fact that for IGs u and h are functions of temperature only (see the Property Tables for air, CO_2, He, etc.), in conjunction with the underlying general definitions for the specific heats:

$$c_v = \left(\frac{\partial u}{\partial T}\right)_{v=c} \ \text{ and } \ c_p = \left(\frac{\partial h}{\partial T}\right)_{p=c} \tag{1.19a, b}$$

Although the specific heats are temperature dependent, i.e., $c = c(T)$ (see App. B), for most practical applications *average c-values* at mean fluid temperatures are being used, even when additional volume and pressure changes occur.

For liquids and solids, we simply have:

$$dh = c \cdot dT \tag{1.20}$$

Property Tables

Given two property values of a simple substance, typically T and p, all other values, usually v, h, u, and s, can be found in the Property Tables, as explained in Sec. 1.5.1.

Example 1.5   Consider two cases of a high-pressure, high-temperature gas supply line filling up an insulated tank, which is initially evacuated, i.e., $m_{initial} = 0$ and $p_{initial} = 0$. Evaluate the final gas temperatures in the tank. *Case 1:* Supply helium at 120°C and 200 kPa until the tank pressure has also reached 200 kPa. *Case 2:* Supply air at 17°C and 95 kPa until the tank pressure has also reached 95 kPa.

Sketch

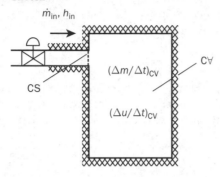

Assumptions

- Clearly, these are *transient open uniform flow systems*, as mass crosses the C.S., while *mass and (internal) energy content inside the tank change with time.*
- Helium and air are assumed to be IGs with constant specific heats.
- Potential and kinetic energy changes are neglected, $Q = 0$, and no work is being performed.

Method

- Reduced transient energy and mass balances with the IG law as well as expressions for u and h are needed. Specifically, $H_{in} \approx \Delta U_{CV} = (mu)_{final} - (mu)_{initial}$ and $m_{in} = m_{final}$.
- Note that there is no heat transfer, Q, as the tanks are insulated, but enthalpy enters with the mass flow, while the internal energy inside the tank, i.e., inside the C∀, increases.
- Property Tables for parameter values.

Solution   Considering the mass and energy balances for process time $\Delta t = t_{final} - t_{initial}$ we have with Eqs. (1.11), (1.14), and (1.18):

- *Mass Balance:* $m_{in} - 0 = m_{final} - 0$ or $m_{in} = m_{final} = m$, as nothing flows out and $m_{initial}$ was zero.
- *Energy Balance:* $mh_{in} - 0 = mu_{final} - 0$ or $h_{in} = u_{final}$, where $h = c_p T$ and $u = c_v T$.
- $pv = RT$ is equal to w_{flow}, which is part of the specific enthalpy $h = u + pv$, implying that with $u = u(T)$ only for IGs, $h = h(T)$ only and as always $h > u$.
- $k = c_p/c_v$.

With assumed *constant* values from the Property Tables in App. B, the results are as follows.

Case 1: From the energy balance we have per-unit mass,

$$c_p T_{supply-line} = c_v T_{tank}, \text{ so that } T_{tank} = \kappa_{helium} T_{supply-line} = 1.667\,(120 + 273\text{ K}) = 655\text{ K}$$

Case 2: Similarly, $T_{tank} = \kappa_{air} T_{supply-line} = 1.4\,(17 + 273\text{ K}) = 406\text{ K}$

Comments

- Clearly, $h_{supplied} > u_{accumulated}$ (because of the specific flow work) leads to a temperature increase of the gas inside the tank so that $T_{tank} > T_{supply-line}$.

- For fluid *discharge* problems, one has to consider *variable* enthalpies and pressures of the existing streams as $h = h(t)$ and $p = p(t)$.
- Recall that all temperature calculations dealing with ideal gases are typically in kelvins.

1.5.1 Phase Diagrams and Property Tables

Consider the experiment of heating 1 kg of liquid water in a piston–cylinder device where the weight of the frictionless piston exerts $p = 1$ atm $= 101.3$ kPa. Starting at room temperature (i.e., $T = 24°C$), the water volume will hardly increase until $T = 100°C$, the boiling point at atmospheric pressure. Then a remarkable thing happens until all water has evaporated during this isobaric ($p = ¢$) process: While the volume of the water–steam mixture increases greatly, the temperature does not rise during phase change, i.e., when advancing from the *saturated liquid point* ($x = 0$) to the *saturated vapor point* ($x = 1$) (see Fig. 1.5). The reason is that all the constantly supplied heat is used to break up the liquid molecular H_2O bonds and hence convert water to vapor. Specifically, the energy needed for phase conversion is known as the *latent heat of vaporization* (e.g., $h_{vap} \equiv h_{fg} = h_g - h_f = 2256.5$ kJ/kg for water at 1 atm), as listed in the Property Tables in App. B. Once the saturated vapor point has been reached, i.e., $x = 1$, further heat transfer results in superheated vapor, i.e., steam. This should not be confused with *wet steam*, which is a saturated mixture, i.e., composed of both saturated liquid and saturated vapor.

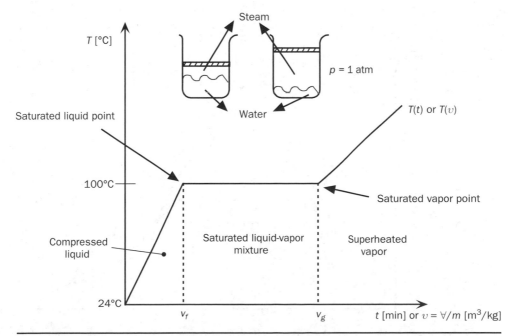

Figure 1.5 Heating H_2O at constant pressure and plotting temperature vs. specific volume.

Repeating the experiment at different pressures and connecting the saturated liquid points as well as the saturated vapor points forms a "dome" that separates three regions:

- Compressed liquid region, where properties are evaluated with the temperature at the saturated liquid point, as the pressure has only a minor impact on $v, u, h,$ and s.

- Saturated mixture region, where the quality $x \equiv m_{vapor}/m_{total}$ ranges from $x = 0$ to $x = 1$.

- Superheated vapor region, where temperature and volume (and hence enthalpy) greatly increase.

The resulting T-v *phase diagram for water* is depicted in Fig. 1.6, where the isobars are ascending, i.e., very steeply and crowded in the compressed liquid region. The *saturated liquid line* and the *saturated vapor line* merge at the critical point ($T_{critical} = 373.95°C, p_{cr} = 22.06$ MPa, and $v_{cr} = 0.003106$ m^3/kg) where both thermodynamic states are identical. However, above the critical point, i.e., for $T > T_{cr}$, the substance is referred to as superheated vapor.

A similar experiment where the piston is subjected to different loads (generating different pressures) generates a p-v diagram with isotherms now descending (see Fig. 1.7).

Concerning Figs. 1.6 and 1.7, as well as the associated Property Tables, the following should be noted:

- The property values on the saturated liquid lines in the T-v and p-v diagrams and columns of the Property Tables often carry the subscript "f," indicating "*flüssig*," which is German for "liquid."

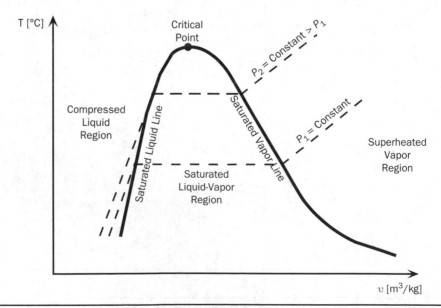

FIGURE 1.6 The three regions in a T-v diagram for H$_2$O.

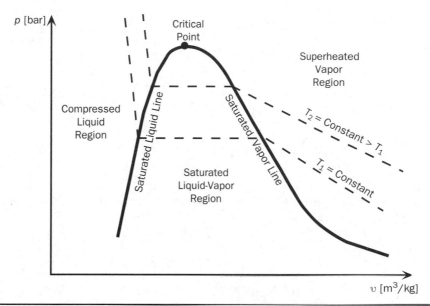

FIGURE 1.7   The three dual-phase regions in a p-v diagram for H_2O.

- The property values on the saturated vapor lines in both the T-v and p-v diagrams and columns of the Property Tables carry the subscript "g," indicating "*gasförmig,*" which is German for "gaseous."

- Differences in property values between the gaseous and liquid phases are denoted with subscript "fg"; for example, the heat necessary to vaporize 1 kg of water is the specific enthalpy $h_{evap} \equiv h_{fg} = h_g - h_f$, being 2256.5 kJ/kg at $p = 1$ atm. *Alternatively, $f \equiv l$, and hence $h_{lg} = h_g - h_l$.*

- During phase change at any given pressure, the saturation temperature and saturation pressure remain constant; the resulting liquid–vapor saturation curves, i.e., the nonlinear relations $T_{sat} = fct.\ (p_{sat})$ for different substances, have important implications in boiling, cooking, cooling, and freezing (see Fig. 1.8).

- Property values of "compressed liquids" are determined as a function of the fluid's temperature—not the pressure—because pressure changes have hardly any effect on the specific volume, internal energy, enthalpy, and entropy.

- Property values in the *saturated mixture region,* i.e., the space under the dome, are computed with the definition of the *mixture quality*:

$$x = \frac{m_{vapor}}{m_{total}}; m_{total} = m_{liquid} + m_{vapor} \qquad (1.21a, b)$$

1.5.2 Property Evaluation in the Mixture Region

Consider a tank of volume $\forall$ filled with saturated liquid and vapor, so that the specific volume of the saturated mixture is:

$$v = \left(\forall / m \right)_{total} = \forall_{liquid} / m + \forall_{vapor} / m \qquad (1.22a, b)$$

or with $\forall_{\text{liquid}} = m_{\text{liquid}} \, v_f$ and $\forall_{\text{vapor}} = m_{\text{vapor}} \, v_g$ we have:

$$v = \frac{m_{\text{liquid}}}{m_{\text{vapor}}} v_f + \frac{m_{\text{vapor}}}{m_{\text{total}}} v_g = (1 - x) \cdot v_f + x \cdot v_g \tag{1.23a, b}$$

or

$$v = v_f + x \cdot \left(v_g - v_f \right) \equiv v_f + x \cdot v_{\text{fg}} \tag{1.24a, b}$$

In general, for any specific property of a simple substance, i.e., $y \equiv v, u, h$ or s:

$$y = y_f + x \cdot y_{\text{fg}} \tag{1.25}$$

If a specific mixture property is known, the quality can be computed from Eq. (1.26) as:

$$x = \frac{y - y_f}{y_{\text{fg}}} \tag{1.26}$$

where the mixture range for quality x is $0 \le x \le 1.0$, as graphically depicted in the Equation Sheets (see App. A). Clearly, $x = 0$ indicates that we are on the saturated liquid line, and $x = 1$ (or 100%) covers the saturated vapor line—in both cases all the way up to the critical point (Figs. 1.6 and 1.7). The definition of x implies that it has no meaning in either the compressed liquid or the superheated vapor region.

Given two property values, one of them is usually either the temperature or pressure. The relation of the fluid's saturation pressure and saturation temperature (see Fig. 1.8) determine in which of the three regions the state of the fluid falls. For example, when $T > T_{\text{sat}}$ the fluid is in the superheated vapor region. In general, it is always advisable to check the associated p_{sat} for a given temperature relative to the given pressure (and vice versa) in order to determine in which region the given thermodynamic state resides. Figure 1.8 indicates the well-known fact that water boils faster (i.e., at a lower temperature) on top of a high mountain.

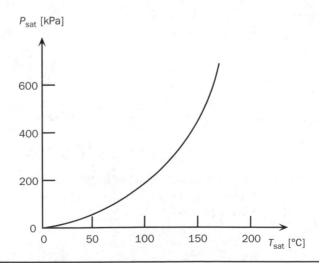

FIGURE 1.8 Liquid–vapor saturation curve for H_2O.

In summary, the three major ways to find the thermodynamic state values of a working fluid/system in the Property Tables in App. B are as follows:

- *Case I:* Compressed/saturated liquid region (see Fig. 1.6)

   a. Given T and p, where $T < T_{sat}$, use the temperature to find $y = v, u, h$, and s, as the pressure has no significant impact on the y quantities.

   b. If the given $T = T_{sat}$, y-values may be found on the saturated liquid line ($x = 0$, or on the saturated vapor line $x = 1.0$, or in the saturated mixture region).

- *Case II:* Saturated mixture region (see Fig. 1.6)
   Given T (or p) plus another property $y = v, u, h$, or s, all other property values are on the saturated liquid line, i.e., $y = y_f$, or on the saturated vapor line, i.e., $y = y_g$, or being in the mixture region, $y_f < y < y_g$; thus, they have to be evaluated with quality x: $y = y_f + x\, y_{fg}$.

- *Case III:* Superheated vapor region (see Fig. 1.6)
   Here, $T > T_{sat}$ and $p < p_{sat}$ while $y > y_g$!

Example 1.6   Given two property values for H_2O, determine the phase region and find the other property values for States 1–3.

- *State 1:* $p_1 = 200$ kPa and $x_1 = 0.60$. Clearly, the given quality indicates that State 1 is in the *saturated mixture region*, which implies that $T_1 \equiv T_{sat} = 120.23°C$ from Table B.2 in App. B. Now, to find u, v, s and h, Eq. (1.25) has to be applied. Say for the specific internal energy we have $u_1 = u_f + x_1 u_{fg} = 504.5 + 0.6 \times 2025 = 1719.5$ kJ/kg. Finding values for v, h, and s proceeds similarly.

- *State 2:* $T_2 = 75°C$ and $p_2 = 500$ kPa. At such a low temperature, H_2O can only exist as compressed water, where $T_2 = 75°C$ (NOT the pressure) determines the water properties. Specifically, being *on the saturated liquid line*, $v_2 \equiv v_f = 0.001026$ m³kg, $u_2 = 314$ kJ/kg, etc.

- *State 3:* $T_3 = 1000°C$ and $h_3 = 4613.8$ kJ/kg. There is no doubt that State 3 is in the *superheated vapor region* (see Fig. 1.6). Interestingly enough, for a given T, the enthalpy of the steam, $h = u + pv$, decreases as the pressure goes up (see the Property Tables) because, the specific volume decreases more rapidly. Thus, a match for T_3 and h_3 can be found when $p_3 = 10$ MPa. That "pressure box" then also lists the v, u, and s values.

Example 1.7   Considering H_2O, provide the missing property values (i.e., either/or T, p, v, u, x) with phase descriptions for the following three states and indicate where they are located in diagrams. *Case (i):* $p = 400$ kPa, sat. vapor; *Case (ii):* $T = 50°C$, $v = 7.72$ m³/kg; and *Case (iii):* $T = 250°C$, $p = 500$ kPa.

Solution

- State (i) is on the saturated vapor line (i.e., $x = 1.0$) where $T = T_{sat} = 143.6°C$, $v = v_g = 0.4624$ m³/kg, and $u = u_g = 2553.1$ kJ/kg. In a p-v diagram State (i) is located at the intersection of the 400 kPa isobar with the saturated vapor line.

- State (ii) at that low temperature and relatively high specific volume is in the saturated mixture region with $p = p_{sat} = 12.35$ kPa and via Eq. (1.22) $x = (7.72 - 0.001012) / 12.025 = 0.64$. Hence, we have $u_1 = u_f + x_1 u_{fg} = 1638.7$ kJ/kg. In a T-v diagram, State (ii) is located in the saturated mixture region on the 50°C isotherm at $x = 0.64$.

- State (iii) is in the superheated vapor region because both the temperature and pressure are relatively high (see the Property Tables). Specifically, $v = 0.4744$ m³/kg and $u = 2723.3$ kJ/kg, while the specific enthalpy is, of course, higher, i.e., $h = u + pv = 2961$ kJ/kg. To locate State (iii) in a T-v diagram, the 250°C isotherm (horizontal line across all three phase regions) has to intersect the 500 kPa isobar, which runs *below* the $p_{sat} = 3976.3$ kPa line.

Example 1.8   The refrigerant R-134a undergoes phase change in cyclic devices such as refrigerators, air conditioners, and heat pumps. Briefly, R-134a vapor is compressed to a relatively high pressure and temperature and then isobarically liquefied in a condenser. Next, in a throttling device the temperature is radically reduced below the freezing point, which allows heat removal from a refrigerator space (or the cold ambient in the case of a heat pump) via an evaporator. Now in vapor form at low temperature and pressure, R-134a is then again compressed to a liquid to continue the cyclic process. So, find the missing information for R-134a when it is at the cyclic States 1 to 3.

1. 60°C and 600 kPa
2. 30°C and $v = 0.0065$ m³/kg
3. −12°C and 320 kPa

Solution

1. Superheated vapor region (in contrast to H_2O, for R-134a the temperature and pressure are quite low), where $v = 0.041389$ m³/kg, $u = 275.15$ kJ/kg, $h = 299.98$ kJ/kg, and $s = 1.0417$ kJ/kg · K.

2. Saturated mixture region because at that temperature we have $v_f < v < v_g$. Hence, $x = (0.0065 − 0.0008421) / 0.02578 = 0.22$ and so $= 126.63$ kJ/kg. Similarly, $h = 131.65$ kJ/kg and $s = 0.473$ kJ/kg · K.

3. Although $p > p_{sat} = 185.37$ kPa, State (3) is *compressed liquid where the temperature determines all other property values*. Thus, $y \equiv y_f$, i.e., $v = 0.0007499$ m³/kg, $u = 35.78$ kJ/kg, $h = 35.92$ kJ/kg, and $s = 0.14504$ kJ/kg · K.

Task   Indicate the cycle with the three states in a p-v and T-s diagram.

1.6   Homework Assignments

1.6.1   Concept Questions

Please answer the following questions by providing physical explanations as well as supportive math equations and sketches, if possible.

1 Looking at Sketches (ii) and (iii) to Answer 1.1a) of Example 1.1: why is $p_2 = p_1$; why does steam take up so much more volume than the same mass of water; and why does the boundary work expression for Case (ii) differ from Case (iii)?

2 Why is work performed equivalent to heat transferred, regardless of form or source?

3 Express spring work for $k = $ constant, $k = a\sqrt{z}$, and $k = bz$ and indicate it in a p-$\forall$ diagram; what are the units of constants a and b?

4   Explain $\Sigma E_{in/out}$ vs. ΔE_{total} inside a device/system/C∀ and justify that inside a C∀ we can approximate ΔE_{total} as ΔU.

5   Derive that $E_{kinetic} = m \cdot v^2/2$ from first principle; also, show that $1\,kJ = 1\,kPa \cdot m^3$ and $1\,kJ/kg = 1000\,m^2/s^2$.

6   Define and compare applications of internal energy vs. enthalpy on both the molecular and macro levels.

7   Focusing on *closed* systems: (i) construct a steady-state process; (ii) will enthalpy transfer ever appear as an energy form; and (iii) is the fluid pressure always constant in ideal (vertical) p-c devices?

8   Demonstrate graphically and mathematically that a *transient open uniform flow system* is the "mother of all thermodynamic systems/devices."

9   What is the *state postulate* for "simple substances"? Give three fluid examples with state functions or sources of property values.

10   How can a throttling device be isenthalpic?

11   What are the primary causes of irreversibilities, i.e., entropy increase?

12   Is it possible to construct a device/system that fulfills the 2nd Law of Thermodynamics but violates the 1st one?

13   Why should all transient systems be well mixed, e.g., using W_{stir} as an input?

14   Why is the customary assumption of *isobaric process* for tubular flow, mixing chambers, and heat exchangers basically flawed?

15   In Eq. (1.7), what happens physically when approximating dX/dt by $\Delta X/\Delta t$?

16   Why is heat, Q, path dependent, and are the Q of Eq. (1.10) and ΔH related?

17   For *polytropic gas processes*, derive the special cases of isothermal and isentropic compression and show that $W_{compr./isothermal} < W_{compr./isentropic}$. Explain!

18   When evaporating a liquid to a vapor isobarically, why does the temperature remain constant?

19   Why is the quality x confined to the saturated liquid/vapor lines and mixture region?

1.6.2   Fluid Property Values

1   Compare two natural gas furnaces (NGF) where NGF 1 has an efficiency of 82% and costs $1600, while NGF 2 features 95% efficiency at $2700. With an annual heating cost of $1200, how long will it take for NGF 2 to be more cost-efficient?

2   An insulated room, initially at 20°C, contains several heat sources, i.e., a resistance heater (1.2 kW), a fridge (300 W), a TV (110 W), and a bulb (40 W). Find the room's increase in energy rate.

3   Consider a container where one-third of the partitioned volume is filled with an IG at 1000°C, while the rest is evacuated. After removing the massless partition, heat is supplied until the gas has reached the initial pressure. What is T_{final}?

4   At one point 50 kg of saturated water in a tank is at 90°C. Find the fluid pressure and the tank volume.

5 Indicate in a p-v diagram, the evaporation of 0.5 kg of water at 100 kPa in a piston–cylinder device. Determine the volume change and the latent heat (i.e., enthalpy) of vaporization.

6 Consider a tank containing 8 kg of water and 2 kg of vapor at 90°C. Find the mixture pressure, volume, and quality. Show the state in a T-v diagram.

7 An 80-litre container has 4 kg of R-134a at 160 kPa. Find its temperature, quality, and enthalpy.

8 Find for the five given states of H$_2$O the remaining properties plus the state locations in a p-v or T-v diagram.

Case	T [°C]	p [kPa]	u [kJ/kg]	x [1]
1		200		0.6
2	125		1600	
3		1000	2950	
4	75	500		
5		850		0.0

9 Find for the four given states of H$_2$O the remaining properties plus points in p-v or T-v diagram.

Case	T [°C]	p [kPa]	v [m³/kg]
1	50		7.72
2		400	
3	250	500	
4	110	350	

10 Find for the four given states of R-134a the remaining properties plus points in a p-v or T-v diagram.

Case	T [°C]	p [kPa]	v [m³/kg]	Region
1	−4	320		
2	10		0.0065	
3		850		Saturated vapor
4	90	600		

11 Consider water (0.5 m³/kg and 300 kPa) in a p-c device with stops that are reached after all the water has turned into saturated vapor. Now the H$_2$O is further heated until the pressure is 600 kPa. Determine the property values for States 1, 2, and 3 and depict them in both the p-v and T-v diagrams.

12 An insulated vertical p-c device with a supply line contains at 200 kPa initially 4 kg of water and 6 kg of vapor. Now steam at 0.5 MPa and 350°C are supplied until all the water has vaporized. Find T_{final} and $m_{steam, in}$.

13 Consider a 0.3 m³ tank filled with saturated water at 200°C. Now, a valve at the tank bottom is opened and one-half of the total water mass is discharged. How much heat has to be supplied in order to keep the temperature constant during this process?

14 An insulated vertical p-c device with an outlet valve and spring-loaded piston contains initially 0.8 m³ of R-134a at 1.4 MPa and 120°C. Now the valve is opened and the piston descends until $p_{final} = 0.7$ MPa and $\forall_{final} = 0.5$ m³. Determine $m_{discharge}$ and T_{final}.

15 An adiabatic steam turbine (INLET: 12.5 MPa, 500°C, 25 kg/s; OUTLET: 10 kPa, $x = 0.92$) drives a generator plus an adiabatic air compressor (INLET: 98 kPa, 295 K, 10 kg/s; OUTLET: 1 MPa, 620 K). Find the net turbine power delivered to the generator.

CHAPTER 2

Macroscale Mass and Energy Balances with Applications

Chapter 1 reviewed some basic physics concepts and provided an overview of common energy transfer modes. Then the mass and energy conservation laws with some examples were provided to illustrate the different thermodynamic systems. Perhaps the only new material presented in Chapter 1 included:

1. A reminder that no device can function, or no process can proceed, without entropy generation (see Fig. 1.5).

2. The construction of phase diagrams for the two-phase fluids H_2O and *R-134a*, with ways of finding associated property values for all state variables.

Chapter 2 deepens the knowledge base and skill level dealing with the First Law of Thermodynamics, i.e., covering the derivation of a generalized balance equation for both mass and energy transfer, plus a series of supporting sample problem solutions.

2.1 Conservation of Energy and Mass

As indicated and mathematically stated with Eq. (1.1), the First Law of Thermodynamics postulates that energy forms can be converted but that the total energy is constant, i.e., conserved:

$$E_{\text{total}} = \underbrace{E_{\text{kinetic}} + E_{\text{potential}} + E_{\text{internal}}}_{=E_{\text{system}}}$$
$$\underbrace{+ \; E_{\text{mass flow}} + E_{\substack{\text{work} \\ \text{performed}}} + E_{\substack{\text{heat} \\ \text{transfered}}} \cdots}_{=E_{\text{surrounding}}} = \cancel{c} \tag{2.1}$$

Taking the total time-derivative of Eq. (2.1), we can write:

$$\frac{dE_{\text{total}}}{dt} \approx \frac{\Delta E_{\text{total}}}{\Delta t} = 0 = \frac{\Delta E}{\Delta t}\bigg|_{\text{system}} + \sum \dot{E}_{\text{net exchange}} \tag{2.2a}$$

31

Expressing the *net energy rate exchange as net efflux* $\sum \dot{E}_{out} - \sum \dot{E}_{in}$ (see also the Equation Sheets in App. A), we have:

$$\left.\frac{\Delta E}{\Delta t}\right|_{system} = \sum \dot{E}_{in} - \sum \dot{E}_{out} \tag{2.2b}$$

As discussed in Sec. 1.2, multiplying Eq. (2.2b) through by Δt and hence obtaining the energy balance for *observation (or process) time* $\Delta t = t_{final} - t_{initial}$, we can state:

$$\underbrace{\sum E_{in} - \sum E_{out}}_{\substack{\text{Net energy transfer by heat,}\\\text{work, and/or mass flow in/out}}} = \underbrace{\Delta E_{system} := [\Delta kE + \Delta pE + \Delta U]_{system}}_{\substack{\text{Change of total energy inside the system}\\\text{(e.g., kinetic, potential, and internal energies)}}} \tag{2.3}$$

Within a system or device ΔkE and ΔpE are typically negligible, i.e., $\Delta E|_{C\forall} \approx \Delta U|_{C\forall}$, using symbolically the *control volume* (C$\forall$) encompassing the given system or device.

Similarly, fluid-mass flow rates IN minus OUT of any system (or control volume) cause mass accumulation/depletion inside the system [see Eq. (1.2a)].

$$\sum \dot{m}_{IN} - \sum \dot{m}_{OUT} = \left(\frac{\Delta m}{\Delta t}\right)_{system} \tag{2.4}$$

Clearly, Eq. (2.4) is directly coupled to Eq. (2.2b), as it is required for the energy balance. Again, multiplying Eq. (2.4) through by Δt, yields:

$$\sum m_{IN} - \sum m_{OUT} = (\Delta m)_{system} = (m_{final} - m_{initial})_{system} \tag{2.5}$$

Following the deductive approach, applications are discussed first with *transient systems*, i.e., both open and closed; then, special cases of *steady open uniform flow devices* are illustrated.

Clearly, on the macroscale, energy transfer is closely coupled with mass transfer for any system. Thus conflating both mass and energy balances [see Eqs. (2.3) and (2.5)], the *generalized conservation equation* reads (see Equation Sheets):

$$\sum X_{in} - \sum X_{out} = \Delta X_{C\forall} = [X_{final} - X_{initial}]_{C\forall} \tag{2.6}$$

For *transient open system/device processes* over time-interval Δt, X can be the fluid mass m or any form of energy E, such as work and heat crossing the control surface (CS) and enthalpy and kinetic energy flowing into or out of the control volume (C$\forall$) via ports. Concerning the right-hand side of Eq. (2.6), $\Delta X_{C\forall} = [X_{final} - X_{initial}]_{C\forall}$ represents *temporal* changes in mass content, Δm, or internal energy ΔU, assuming $\Delta E \approx \Delta U$, inside the C$\forall$. (See Fig. 2.1a.)

Focusing on *mass transfer in a transient open system*, ΣX_{in} sums up over time all the mass-flow, i.e., uniform fluid streams entering the C$\forall$, while ΣX_{out} represents all the exiting streams. As a result, $\Delta X|_{C\forall} \equiv \Delta m|_{C\forall}$ is the mass-content change inside the control volume, i.e., system or device. Hence, by applying Eq. (2.6), the general mass balance (*M-B*) for process/observation time interval Δt is obtained as Eq. (2.5).

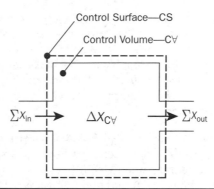

Figure 2.1a   Generalized sketch of the essence of Eq. (2.6).

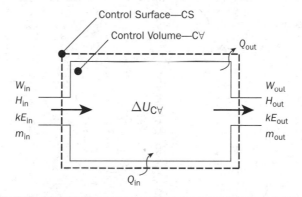

Figure 2.1b   Schematic illustrating Eq. (2.7).

Now, using Eq. (2.6) to describe *energy transfer for any transient open system,* we note that the uniform inlet/outlet streams are carrying enthalpy $H = m \cdot h$ and possibly kinetic energy kE, while heat Q and work W cross the control surface. Thus considering only changes of internal energy inside the C∀, Eq. (2.6) now reads in expanded form:

$$\sum \left[Q + W + H + \frac{mv^2}{2} \right]_{in} - \sum \left[Q + W + H + \frac{mv^2}{2} \right]_{out} = \Delta U_{C\forall}$$
$$= \left[(mu)_{final} - (mu)_{initial} \right]_{C\forall} \tag{2.7}$$

Notes

- The quantities in the mass balance (*M-B*) of Eq. (2.5) are needed in the energy balance (*E-B*) of Eq. (2.7), which is known as one-way coupling.
- Equation (2.7) can also be viewed as an application of Eq. (2.3).
- For *closed (typically transient) systems,* there are no inlet/outlet ports, implying that $m_{C\forall} =$ constant (as $m_{final} = m_{initial}$) and enthalpy $H_{in/out} = 0$ in Eq. (2.7).
- For *steady open flow systems,* the rate equations Eq. (2.2b) and Eq. (2.4) reduce in generalized form to $\sum \dot{X}_{in} = \sum \dot{X}_{out}$ as no mass/energy changes occur inside the system (see Sec. 1.3.2 and the Equation Sheets).

The chapter closes with a series of examples that have to be studied in depth and solved *independently*, following the problem-solving steps of Sec. 2.2. Specifically, first carefully read each example's problem statement and set up the three associated Prelims (Preliminaries), i.e., system Sketch, list of Assumptions, and Method/approach to be applied. *Correctly done Prelims should result in about 70% of the upcoming problem solution.* Next, go into the Solution phase on your own and then compare results, i.e., make corrections as needed.

Example 2.1: Filling of an Insulated Piston–Cylinder Device with Steam from a Supply Line

An idealized p-c device contains initially 4 kg of (liquid) water and 6 kg of vapor, to which superheated steam from a supply line (350°C and 500 kPa) is added until everything has turned into 100% vapor. The fluid pressure is constant at 200 kPa. What is the final vapor temperature and the amount (i.e., the mass) of steam that has to be supplied? Indicate the process in a phase diagram!

Sketches

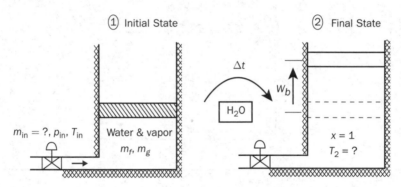

Assumptions

- Transient open uniform flow system.
- H_2O is a simple two-phase substance.
- Isobaric (frictionless piston), quasi-equilibrium expansion with boundary work, but zero heat transfer (adiabatic).
- Changes of potential and kinetic energies inside the cylinder are negligible, i.e., $\Delta E|_{CV} \approx \Delta U|_{CV}$.

Method

- Equation (2.6) applied to mass and energy transfer, i.e., $m_{in} - 0 = \Delta m_{system}$ and $(mh)_{in} - W_b = \Delta U_{system} = (mu)_{final} - (mu)_{initial}$, where with $p =$ constant so that $W_b = p\Delta V$ (see Table 1.1)
- $Q_{system} = 0$
- Steam-Property Tables for *initial State 1* and *final State 2* (which is clearly on the saturated vapor line as $x = 1$), as well as for *inlet State IN*.

Solution

- Properties for:

  State IN, i.e., $T_{in} = 350°C$ and $p_{in} = 0.5$ MPa → $h_{in} = 3168$ kJ/kg.

  State 1 with $p_1 = 200$ kPa and $x_1 = 6$ kg/10 kg $= 0.6$ → $h_1 = h_f + x_1 h_{fg} = 504.7 + 0.6 \times 2201.6 = 1825.6$ kJ/kg.

  State 2 with $p_2 = p_1 = 200$ kPa (because only the piston weight generates the fluid pressure) and $x_2 = 1.0$, as required → $h_2 = h_g = 2706.3$ kJ/kg at 200 kPa, which is associated with $T_{saturated} \equiv T_2 = 120.2°C$.

- Mass balance (M-B): $m_{in} = m_2 - m_1$, where only m_1 is known.

- Energy balance (E-B): Rewriting the E-B $(mh)_{in} - p\Delta\forall = \Delta U$ as $(m_2 - m_1)h_{in} = \Delta U + p\Delta\forall = \Delta H$.

Note

- The right-hand side is identical to ΔH!

- Hence, the E-B reads:

  $$\Delta H = (mh)_2 - (mh)_1 = (m_2 - m_1)h_{in}$$

- Solving for the final mass:

  $$m_2 = \frac{m_1(h_{in} - h_1)}{h_{in} - h_2} = 19 \text{ kg}$$

Comments

✓ Note that the contraction of $\Delta U + W_b = \Delta U + p\Delta\forall$ to ΔH for this *isobaric process* reduces the originally two unknowns to just one!

✓ There is no indication of any interest in how long the water-to-steam conversion process lasts; clearly, we took a macroscopic "black-box" approach for the problem solution.

✓ As the process is isobaric (constant piston weight during expansion), it is best to select a *p-v* diagram to depict the process:

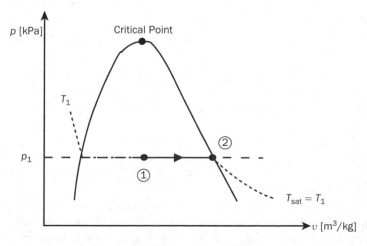

2.2   Problem-Solving Steps

Example 2.1 demonstrates the *necessary preliminaries*, i.e., *System Sketch, List of Assumptions,* and *Method to be used,* i.e., approach and reduced equations to be taken. Proper prelims may provide up to 70% of the problem solution! In general, the solution steps are as follows:

1. The *Sketch* should be a comprehensive depiction of the given problem statement, i.e., a full "translation" of the entire problem statement.
    Specifically, we need:

    •   A nice drawing of the device or system, indicating initial and final states for transient systems and the inlet and outlet states for steady flow devices.

    •   The symbols (no numbers, please!) for all *given* and *unknown* quantities/variables, i.e., $T, p,$ or x, as well as m and h.

    •   The symbols for all energy forms, including directions of $m, Q, W,$ and H (or their constant rate forms for steady open systems, i.e., $\dot{m}, \dot{W} \equiv P(power), \dot{H} = \dot{m}h$).

2. The *Assumptions* typically include:

    •   System identification, i.e., transient open or transient closed, or steady open uniform flow.

    •   Type of simple substance, e.g., two-phase fluids (such as H_2O or R-134a), gases (especially ideal gases), pure liquids or solids.

    •   Energy transfer mechanisms and their forms, e.g., heat $Q_{in/out} = 0$ if the system is well insulated (i.e., an adiabatic process); work forms, such as $W_{boundary}$ (if the C∀ changes with time), $W_{spring}, W_{electric}, W_{shaft},$ etc. (see Table 1.1); fluid properties that are constant, e.g., specific heats.

    •   Recognize the type of (quasi-equilibrium) process, e.g., isothermal, isobaric, isochoric, or isentropic (reversible + adiabatic), as well as adiabatic, compression, or expansion.

Notes

✓ It is always assumed that everything undergoes *quasi-equilibrium processes,* i.e., very slowly to preserve uniform property values in space for transient systems and at all times for steady-state processes.

✓ Typically, $\Delta E_{C\forall} \approx \Delta U_{C\forall}$ as changes in E_{kin} and E_{pot} inside the control volume are negligible.

✓ During fluid-discharging processes from tanks, $h_{exit} = h(t)$ and $p_{exit} = p(t)$, which can be approximated with their arithmetic means, respectively.

✓ In case of isobaric boundary work, $\Delta U + W_b = \Delta U + p\Delta\forall = \Delta H = (mh)_{final} - (mh)_{initial}$.

✓ Changes from State 1 to State 2 in *steady* flow devices are spatial, i.e., they occur over the device distance between Inlets and Outlets. For *transient* systems, changes in thermodynamic state occur over time, i.e., State 1 corresponds to INITIAL and State 2 corresponds to FINAL.

✓ For a two-component system, e.g., a turbine driving a compressor or two interacting tanks or two turbines, both components may be combined (forming the control volume) for simpler mass/energy/entropy balances.

3. The necessary *Method* to be implemented, or approach to be taken, results directly from the Sketch and Assumptions. Specifically, consider *problem-specific forms* as special cases of the generalized Eq. (2.6).

- Mass balance (*M-B*) according to the type of system identified and the physics indicated in the Sketch.

- Energy balance (*E-B*) tailored to the given system.

- Equations for unknowns in the *E-B*, e.g., for boundary-work, heat-exchange between streams, and property values using an equation of state. Some relations that prove to be especially useful for IGs include $p\forall = mRT$, $du = c_v dT$, and $dh = c_p dT$.

- Use of *Property Tables*, especially for two-phase fluids (i.e., H_2O and R-134a) and ideal gases.

4. *Solution Procedure:*

- Restate the problem-tailored *M-B* and *E-B* plus submodels, such as an equation of state, empirical correlations, etc.

- For two-phase fluids, provide parameter values from the Property Tables in App. B.

- Show the initial/final (or inlet/outlet) States of the fluid in a *T-v* or *p-v* or *T-s* diagram.

- For ideal gases, use the IG law plus IG relations, while for nonideal gases, employ designated relations (see the Equation Sheets in App. A).

Conclude your problem solution with a phase diagram for two-phase fluids and the required graphs, as well as comments providing physical insight.

Example 2.2   Consider a tank with a total volume of $\forall_{tank} = 0.12$ m³ that is initially filled with a saturated mixture of R-134a, of which 25% is liquid. The tank is heated so that the pressure remains constant throughout process time Δt. After the exit valve is opened, the pressure ($p = 800$ kPa) stays constant because heat is being added while all the liquid R-134a drains out of the tank. Find the heat transferred during this discharge process, and construct the associated phase diagram.

Sketches

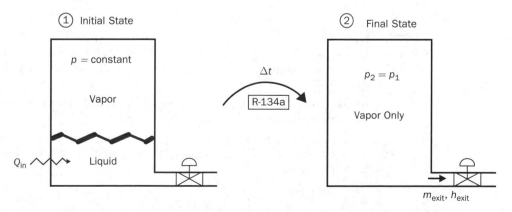

Assumptions

- Transient open uniform flow system.
- R-134a is a simple two-phase substance.
- Isobaric, quasi-equilibrium expansion with $h_{exit} = h_{liquid\ at\ 800\ kPa} = $ constant.
- No work; heat transfer $= Q_{in}$.
- Changes of potential and kinetic energies inside the tank are negligible, i.e., $\Delta E|_{CV} \approx \Delta U|_{CV}$.

Method

- Equation (2.6) applied to mass and energy transfer, i.e., $0 - m_{exit} = \Delta m_{system}$ and $-(mh)_{exit} + Q_{in} = \Delta U_{tank} = (mu)_{final} - (mu)_{initial}$.
- Initial mass $m_1 = m_{liquid} + m_{vapor} \equiv m_f + m_g$, where generally $m = \forall/v$ [see Eq. (1.18)].
- Note the liquid volume is $\forall_f = 0.25\forall_{tank}$ and the vapor volume is $\forall_g = 0.75\forall_{tank}$.
- Also, initial internal energy $U_1 = (mu)_1 = (mu)_f + (mu)_g$ in the tank.
- R-134a Property Tables for *initial State 1* (i.e., saturated mixture) and *final State 2* (i.e., saturated vapor), as well as *final exit State "ex"* (i.e., saturated liquid).

Solution

- R-134a properties for saturated liquid ($x = 0$) and saturated vapor ($x = 1$), both at $p = 800$ kPa (where temperature is $T_{sat} = 31.31°C$), are:

    Saturated liquid: $v_f = 0.0008458$ m³/kg, $u_f = 94.79$ kJ/kg; $h_f \equiv h_{exit} = 95.47$ kJ/kg.

    Saturated vapor: $v_g = 0.025621$ m³/kg, $u_g = 246.79$ kJ/kg.

- M-B balance: $m_{exit} = m_1 - m_2$, where $m_1 = 0.25\forall_{tank}/v_f + 0.75\forall_{tank}/v_g = 35.47 + 3.51 = 38.98$ kg, while $m_2 = \forall_{tank}/v_g = 4.68$ kg so that $m_{exit} = 34.30$ kg.

- E-B balance: $Q_{in} = (mh)_{exit} + (mu)_2 - (mu)_1$, where $(mh)_{exit} = 3274.62$ kJ, $(mu)_1 = (mu)_f + (mu)_g = 4229.2$ kJ, and $(mu)_2 = m_2 u_g = 1155$ kJ.

- Hence:

    $Q_{in} = 201.2$ kJ, needed during Δt to keep the tank-pressure constant.

Comments

✓ Finding the initial and final states is a little tricky as the saturated R-134a mixture changes to saturated vapor after all the liquid has been discharged.

✓ If we had assumed heat loss, the same numerical result would have been negative, indicating a change in heat-transfer direction.

✓ In a p-v diagram, the isobaric process at 800 kPa runs from $x_1 = 3.51/38.98 = 0.09$ (i.e., very close to the saturated liquid line) to $x_2 = 1.0$ (i.e., on the saturated vapor line).

Example 2.3: Superheated R-134a Discharges from a P-C Device with Stops Consider an ideal piston-cylinder device with stops, where 2 kg of R-134a is initially at 80°C and 800 kPa. While a valve at the bottom is opened, vapor escapes, and the piston starts descending when the vapor pressure drops to 500 kPa. The valve is closed when half the mass has exited and the final temperature is 20°C. Determine the necessary heat rejection, and show the process path in a phase diagram.

Sketches

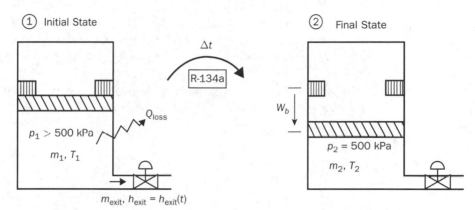

Assumptions

- Transient open uniform flow system.
- R-134a is a simple two-phase substance.
- Quasi-equilibrium expansion where h_{exit} is actually a function of time; thus, $h_{exit}(t)$ is approximated as $h_{exit} = 0.5(h_1 + h_2)$ over process time Δt.
- Boundary work at constant pressure when piston weight causes $p_2 = 500$ kPa plus heat transfer Q_{out}.
- Changes of potential and kinetic energies inside the p-c device are negligible, i.e., $\Delta E|_{CV} \approx \Delta U|_{CV}$.

Method

- Equation (2.6) applied to mass and energy transfer as $0 - m_{exit} = \Delta m_{system}$ and $W_b - (mh)_{exit} - Q_{out} = \Delta U_{tank} = (mu)_{final} - (mu)_{initial}$, where $h_{exit} = 0.5(h_1 + h_2)$.
- $W_b = p_2(\mathbb{V}_1 - \mathbb{V}_2)$ as $p = ¢$ during Δt, where $\mathbb{V}_i = (mv)_i; i = 1, 2$.
- R-134a Property Tables for *initial State 1* and *final State 2*, as well as *exit State*.

Solution

- Properties for State 1 at $p_1 = 800$ kPa and $T_1 = 80°C$ are: $v_1 = 0.03266$ m³/kg, $u_1 = 290.84$ kJ/kg, $h_1 = 316.97$ kJ/kg; for State 2 at $p_2 = 500$ kPa and $T_2 = 20°C$ are $v_2 = 0.04212$ m³/kg, $u_2 = 242.4$ kJ/kg, $h_2 = 263.46$ kJ/kg.
- M-B reads: $m_{exit} = m_1 - m_2 = 1$ kg.

- E-B reads: $W_b - m_{exit}[0.5(h_1 + h_2)] - Q_{out} = (mu)_2 - (mu)_1$ where $W_b = p_2(\forall_1 - \forall_2)$ with $\forall_1 = (mv)_1 = 0.06532$ m³ and $\forall_2 = (mv)_2 = 0.04212$ m³ so that $W_b = 11.6$ kg.

- Solving the E-B for the heat rejected yields $Q = 60.7$ kJ.

Comments

✓ The triple Prelims, i.e., Sketch, Assumption, and Method, take up more space than the actual Solution. However, *this professional approach provides a guide toward the correct problem solution and ultimately saves time.*

✓ The well mixed "black-box" with finite difference models for the mass and energy balances is simple, but it requires a major approximation for time-dependent variables; here, we assume $h_{exit} = 0.5(h_1 + h_2)$ being constant over Δt.

✓ Enlightening graphs are $p(\forall)$ showing the boundary work, as well as the *p-v* and *T-v* diagrams depicting the process paths.

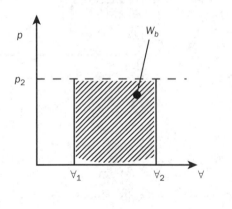
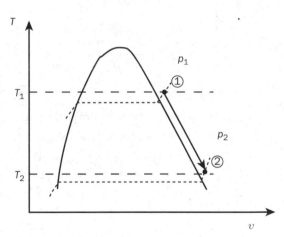

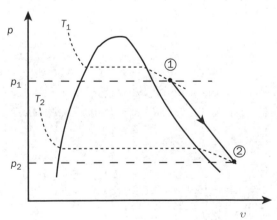

In Example 2.3, we ignored that h_{exit} is actually a function of time by assuming h_{exit} = const. Now we consider a problem solution where we consider time-dependence of a key variable, i.e., stirred fluid temperature $T = T(t)$.

Example 2.4: Development of an ODE for $T(t)$ Describing Watercooling in a Well Stirred Chiller [after Munro et al. (2018)]   Consider a tank with a capacity of $m_{tank} = 45$ kg of water ($c = 4.22$ kJ/kg·K), initially at $T_{initial} = 45°C$, and subject to a constant water stream, $\dot{m}_{in} = \dot{m}_{out} = \dot{m} = 0.075$ kg/s, also at $T_{in} = 45°C$. Mechanical mixing, $\dot{W}_{stir} = 0.6$ kW, and heat-withdrawal at a constant rate $\dot{Q}_{out} = 7.6$ kW via a cooling coil are, together with $\dot{H}_{exit}$, the energy transfer forms. It should be noted that the fluid pressure stays constant and that heat losses, kinetic energy, and potential energy changes can be neglected. Find $T(t)$, T_{final}, and graph the results.

Note   This is a pseudo-transient system as the fluid mass in the tank stays constant $\left(\dot{m}_{in} = \dot{m}_{out} \right)$ while $T = T(t)$ due to cooling.

Sketch

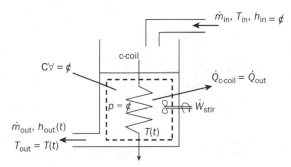

Assumptions

- Transient open system.
- Water (liquid) with constant specific heat, c, for both U and H.
- Uniform cooling of water, $T(t)$, where $\dot{Q}_{c\text{-}coil} \equiv \dot{Q}_{out}$.
- $p = ¢$ and ΔE_{kin} and $\Delta E_{pot} = 0$.

Method

- M-B: $\dot{m}_{in} = \dot{m}_{out} = \dot{m} = ¢$.

- E-B: $\Delta \dot{H} + \dot{W}_{in} - \dot{Q}_{out} = \left. \dfrac{dU}{dt} \right|_{C\forall}$; note we are going back to dU/dt because

  this problem asks us to develop an ODE [see Eq. (2.2b)].

- For liquids, as they are incompressible, $\dot{U} = \dot{m}\bar{c}T$ and $\dot{H} = \dot{m}\bar{c}T$.
- In general, $c = c(T)$.

Solution   Previously, we only computed end points, i.e., final (2) and initial (1). Now, the E-B reduces to:

$$\frac{dU}{dt}\bigg|_{CV} = \dot{W}_{stir} - \dot{Q}_{out} - \dot{m}\big[h(t)_{out} - h_{in}\big]$$

Thus, using the relations for U and H, we have:

$$mc\frac{dT}{dt} = \dot{W}_{in} - \dot{Q}_{out} - \dot{m}\big[cT(t) - cT_{in}\big]$$

This linear first-order inhomogeneous ODE can be directly integrated via separation of variables or recast as $g(x)y'_x = f_1(x)y + f_0(x)$, for which the general solution is:

$$y = Ae^F + e^F \int e^{-F}\frac{f_0(x)}{g(x)}dx; \quad F(x) = \int \frac{f_1(x)}{g(x)}dx$$

For the ODE, the following can be obtained:

$$g(x) = mc; \ f_1(x) = -\dot{m}c; \ f_0(x) = \big(\dot{W} - \dot{Q} + \dot{m}cT_{in}\big)$$

$$F(t) = \int \frac{-\dot{m}c}{mc}dt = -\frac{\dot{m}}{m}t$$

Thus:

$$T(t) = A\exp\left(-\frac{\dot{m}}{m}t\right) + \exp\left(-\frac{\dot{m}}{m}t\right)\int \frac{\dot{W} - \dot{Q} + \dot{m}cT_{in}}{mc}\exp\left(\frac{\dot{m}}{m}t\right)dt$$

$$\Rightarrow T(t) = A\exp\left(-\frac{\dot{m}}{m}t\right) + \frac{\dot{W} - \dot{Q} + \dot{m}cT_{in}}{\dot{m}c}$$

Substituting the initial condition, we get $A = \dfrac{(\dot{Q} - \dot{W})}{\dot{m}c}$. Thus:

$$T(t) = T_{in} + \frac{(\dot{W} - \dot{Q})}{\dot{m}c}\left(1 - \exp\left(-\frac{\dot{m}}{m}t\right)\right)$$

And $T(t \to \infty) = T_{\text{final}} = T_{\text{in}} + \dfrac{\left(\dot{W} - \dot{Q}\right)}{\dot{m}c} = 296\,\text{K}$, upon substituting the given data.

Graph

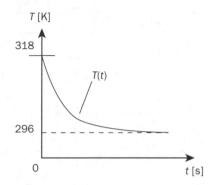

2.3 Transient Closed Systems

The moment an open system is closed, all mass-transfer terms are zero because only heat and work can cross boundaries (see Fig. 1.2b). Examples include closed tanks, piston-cylinder devices, and (partitioned) compartments where $m_{\text{system}}^{\text{total}} = \text{constant}$ at all times. Thus Eq. (2.7) reduces to:

$$\left(m_{\text{final}} - m_{\text{initial}}\right)_{\text{CV}} = 0, \quad \text{i.e.,} \quad m_{\text{final}} = m_{\text{initial}} = m_{\text{system}} = m = \text{\cent} \qquad \text{(2.8a–d)}$$

The energy balance Eq. (2.7) loses both the $H = m \cdot h$ and $E_{\text{kin}} = mv^2/2$ terms, as there are no fluid streams carrying energies into or out of the system over time.

Hence, the *E-B* reads for process time Δt:

$$\sum [Q + W]_{\text{in}} - \sum [Q + W]_{\text{out}} \approx \Delta U_{\text{CV}} = m \cdot \left[(u)_{\text{final}} - (u)_{\text{initial}}\right]_{\text{CV}} \qquad \text{(2.9)}$$

Example 2.5: Air Compression in a Spring-Loaded p-c Device with Heat Transfer (see Example 1.2) Consider an ideal piston-cylinder device with a linear spring, where air is initially at $p_1 = 100$ kPa (solely due to the piston weight) and $T_1 = 300$ K, occupying an initial volume of $\forall_1 = 0.15$ m³ (see Example 1.2). As 150 kJ is being supplied over process time Δt, the spring is compressed until the final state of $p_2 = 800$ kPa with $\forall_2 = 0.45$ m³ has been reached. Determine the work forms that are performed as well as the final temperature.

Sketches

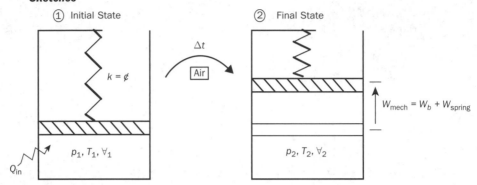

Assumptions

- Closed system.
- Air is an ideal gas.
- Linear spring (i.e., $k = \cancel{c}$).
- $W = W_{mech} = W_b + W_{spring}$ [with linear $p(\forall)$ for W_{spring}] and $Q = Q_{in}$.
- Quasi-equilibrium compression.
- Constant, averaged heat capacity $c_v(T = 300\,\text{K}) = 0.718\,\text{kJ/kg·K}$ from Property Tables.

Method

- M-B [Eq. (2.8)]: $m_{air} = m = $ constant (as for all closed systems).
- E-B [Eq. (2.9)]: $Q_{in} - W_{mech} = \Delta U_{system} = m(u_2 - u_1) = mc_v(T_2 - T_1)$, where

$$W_{mech} = \int_1^2 pd\forall = W_b + W_{spring}.$$

- From IG-law: $m_{air} \equiv m = (p\forall)/(R_{air}T)$.

Solution

- E-B is $Q_{in} - W_{mech} = mc_v(T_2 - T_1)$, where (see Table 1.1) $W_{mech} = \int_1^2 pd\forall$ is the trapezoidal area under the linear $p(\forall)$ line, i.e., $W_{mech} = 0.5(p_1 + p_2) \cdot (\forall_2 - \forall_1) = 135\,\text{kJ}$. Note that the W_b contribution is just $W_b = p_1 \Delta\forall = 30\,\text{kJ}$, equivalent to the rectangle under the $p_1(\forall)$ line. The air mass inside the cylinder is $m = (p\forall)_1/(R_{air}T_1) = 0.174\,\text{kg}$, so that with the E-B we have $T_2 = T_1 + (Q_{in} - W_{mech})/(mc_v) = 420\,\text{K}$.

Comments

- The value of the heat capacity, $c_v = 0.718\,\text{kJ/kg·K}$, is not quite correct as the reference temperature should be higher than 300 K. Thus one could recalculate with a new c_v value at $T_{ref} = 360\,\text{K}$.

- The $p(\forall)$ graph reveals the work forms:

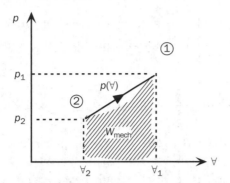

Example 2.6: Isothermal Helium Compression in an Ideal p-c Device Consider helium compression in an ideal p-c device from 100 kPa at 0.6 m³ with 60 kJ heat rejection to keep the (initial) gas temperature constant. Determine the final state, the required work input, and plot $p(\forall)$.

Sketch

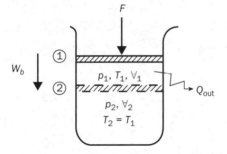

Assumptions

- Closed system.
- Helium (He) is an ideal gas.
- $W = W_b = \int_1^2 p d\forall$ and $Q = -Q_{out}$.
- Quasi-equilibrium compression at $T = \text{¢}$.

Method

- M-B: $m_{air} = m = \text{¢}$.
- E-B: $W_b - Q_{out} = \Delta U_{system} = m(u_2 - u_1) = mc_v(T_2 - T_1) = 0$, as $T_2 = T_1$; also $W_b = \int_1^2 p d\forall$.
- IG-law: $p\forall = mR_{He}T$; hence $p(\forall) = C/\forall$ where $C = mR_{He}T = \text{¢}$.

Solution

- From the *E-B* and with the IG-law:

- $Q_{out} = 60 \text{ kJ} = -W_b = -\int_1^2 pd\forall = mR_{He}T\int_1^2 d\forall/\forall = p_1\forall_1\ln(\forall_2/\forall_1)$, where $\forall_2$ is the unknown. Thus $\forall_2 = \forall_1 \cdot \exp[W_b/(p_1\forall_1)] = \forall_1 \cdot \exp[-60/60] = 0.3679\forall_1 = 0.221 \text{ m}^3$.

- Combining both states in the IG law, the final pressure equation reads for $T = \mathcal{C}$, $p_2 = p_1(\forall_1/\forall_2) = 272 \text{ kPa}$.

Comments

✓ It turns out that this analysis holds for *any* IG, as R_{He} was eliminated under the $T = \mathcal{C}$ condition!

✓ A sign switch from the E-B to the W_b computation was appropriate as compression work calculations are defined as being negative.

✓ In general, compression is associated with heat generation (recall your bicycle pump in action).

✓ The $p(\forall)$ graph looks as follows:

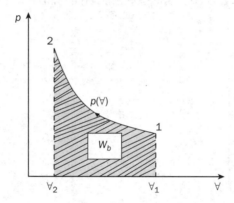

Example 2.7: Comparison of Constant Volume vs. Constant Pressure Heat Transfer Consider an isochoric vs. an isobaric heating increase of 10 kg of nitrogen (N_2) by 200°C either in a tank ($\forall = \mathcal{C}$) or in a p-c device ($p = \mathcal{C}$), using for $T_{ref} = 150$°C so that $c_v = 0.748 \text{ kJ/kg·K}$ and $c_p = 1.045 \text{ kJ/kg·K}$. Find the energy content (Q_{in}) in both cases for $\Delta T = 200$°C.

Sketches

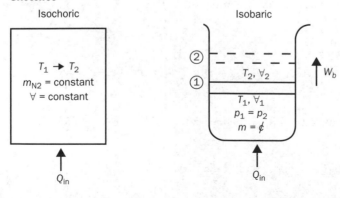

Assumptions

- Both are transient closed systems with N_2 being an IG.
- Quasi-equilibrium processes.
- For both, $Q = Q_{in}$.
- In the p-c device, $W = W_b = \int_1^2 p d\forall$.

Method

- M-B: $m_{tank} = m_{p.c.} = m = ¢$.
- E-B for the fixed control volume case (tank): $Q_{in} = mc_v \Delta T$.
- E-B for the moving control volume case (p-c device): $Q_{in} - W_b = \Delta U$, where $W_b = p\Delta\forall$ as $p = ¢$.

Solutions

- For the tank: $Q_{in} = mc_v \Delta T = 10 \cdot (0.748) \cdot 200 = 1496$ kJ.
- For the p-c device: $Q_{in} = \Delta U + p\Delta\forall \equiv \Delta H = mc_p \Delta T = 2090$ kJ.

Comments

✓ For the p-c device, when the internal energy change plus boundary work (at $p = ¢$) is contracted to a change in enthalpy, the number of unknowns is reduced by one.

✓ Heat transfer to an ideal p-c device is higher than to a tank because not only is the gas being heated up but also (boundary) work is being done by the system onto the surrounding.

Example 2.8: Thermal Equilibrium of Gases Initially at Different Temperatures   Consider two compartments (each $\forall = 1$ m³) filled with IGs (N_2 at 120°C and He at 40°C, both at 500 kPa). They are separated by a frictionless copper (Cu) piston ($m_{piston} = 8$ kg) in an insulated horizontal cylinder. Find the final temperature of both gases in the cylinder with and without the piston and plot $T_{final}(m_{piston})$.

Sketch

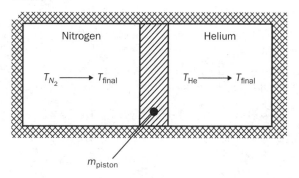

Notes

- $C\forall = \forall_{N_2} + \forall_{He} + \forall_{piston}$
- $W_{piston,net} = 0$
- $T_{mean} = 80°C = T_{ref}$

Assumptions   For this closed system, the $C\forall$ covers the two compartments plus the piston. Properties of the two IGs are evaluated at $T_{ref} = 0.5(120 + 40) = 80°C$, which is also assumed as the *initial Cu-piston* temperature. No heat is lost or work performed during the quasi-equilibrium process.

Method

- *M-B*: m_{N2} and m_{He} are constant and obtained from the IG law at State 1.
- *E-B*: $0 = \Delta U_{total} = \Delta U_{N2} + \Delta U_{Cu} + \Delta U_{He}$, where $\Delta U = mc\Delta T$ and
  $\Delta T \stackrel{\wedge}{=} T_{final} - T_{initial} = T_2 - T_1$.
- Property values at T_{ref} from tables.
- For the ideal gases, $c = c_v$ and $c = c$ for solids.

Solution

- $m_i = (p\forall/RT)_i$; $i = N_2$ and He. Hence $m_{N2} = 4.2871$ g and $m_{He} = 0.769$ g.
- $0 = (mc\Delta T)_{N2} + (mc\Delta T)_{Cu} + (mc\Delta T)_{He}$, from which for both gases and the Cu-piston $T_2 = T_{final} = 83.7°C$.
- Without the piston, i.e., with a mass-less membrane as the divider, the *E-B* reads:

  $$0 = (mc\Delta T)_{N2} + (mc\Delta T)_{He}, \text{ leading to } T_2 = T_{final} = 85.7°C.$$

- The *E-B* can be written as $T_{final}(m_p) = T_{Cu,initial} - \dfrac{\left((mc\Delta T)_{N2} + (mc\Delta T)_{He}\right)}{(mc)_{Cu}}$
  or $T_f(m_p) = T_{Cu,1} + \dfrac{K}{m_p}$.

Comments

✓ $T_{final} > T_{mean} = 80°C$ because the heat capacities of the IGs (and the Cu-piston) are quite different.

✓ Without the piston, the final temperature of the IGs is a little higher as no energy must be transferred to warm up the piston from 80°C.

✓ Plotting $T_f(m_p)$ results in a hyperbolic graph, as $T_f \sim \dfrac{1}{m_p}$.

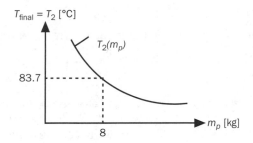

Example 2.9 An insulated two-compartment tank with a zero-mass membrane contains the same ideal gas (IG) but at different states, i.e., m_1, T_1, and m_2, T_2. Derive an expression for $T_3 = T_{final}$ after the massless membrane has been removed.

Sketch

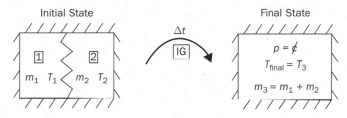

Assumptions

• Closed system.

• $p_1 = p_2 = \text{¢}$.

• $W = Q = 0$.

• Constant c_v.

Method

• M-B: $m_3 = m_1 + m_2$.

• E-B: $\Delta U\big|_{C\forall} = \Delta U_1 + \Delta U_2 = 0$ as $\Sigma E_{in/out} = 0$.

Solution From the *E-B*, we obtain:

$$\Delta U_1 = -\Delta U_2$$

$$m_1 c_v(T_3 - T_1) = m_2 c_v(T_3 - T_2)$$

$$=> (m_1 + m_2)T_3 = m_3 T_3 = m_1 T_1 + m_2 T_2$$

$$\therefore T_3 = T_{final} = \frac{m_1}{m_3}T_1 + \frac{m_2}{m_3}T_2$$

Comments

✓ The mass ratios greatly determine which compartment dominates.

✓ Check: If $m_2 = m_3$, $T_{final} = T_2$ as $m_1 = 0$.

✓ In reality, $c_v = c_v(T)$.

Example 2.10: Two Tanks Connected via a Valve   Tank A is insulated and contains initially 0.3 m³ of steam at 400 kPa with 60% quality. Tank B has initially 2 kg of steam at 200 kPa and 250°C. Now the valve is opened until the pressure is 200 kPa in both tanks, while 300 kJ of heat flows from tank B to the ambient, the pressure drop process in tank A is isentropic ($s = ¢$), i.e., adiabatic and reversible. Determine the total change in energy of tank B($\Delta E|_B$); the final temperature of tank A($T_{A,final}$); and the final masses of both tanks ($m_{A,final}$ and $m_{B,final}$). Show the results in a T-s diagram.

Sketch

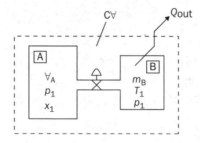

Notes

• $C\forall \triangleq \forall_A + \forall_B$.

• Initial State, 1.

• Final State, 2.

Assumptions

• Overall, this is a transient closed system, while the tanks alone are "transient system."

• Steam is a two-phase fluid.

• Tank A undergoes an isentropic process, i.e., $s_2 = s_1 = ¢$.

• $W = 0$; $Q = Q_B = Q_{out}$.

Method

• M-B: $m = \forall \dfrac{1}{v}$.

• E-B: $-Q = \Delta U|_{C\forall} = \Delta U_A + \Delta U_B$.

• $y = y_f + x y_{fg}$.

• Final state, 2, is at $p_2 = 200$ kPa, where $s_2 = s_1$ in tank A.

• Property Tables (App. B).

Solution Tank A initial with $p_1 = 400$ kPa and $x_1 = 0.6$:

- $v_1 = v_f + x_1 v_{fg} = 0.27788 \, \dfrac{\text{m}^3}{\text{kg}}$.

- $u_1 = u_f + x_1 u_{fg} = 1773.6$ kJ/kg.
- $s_1 = s_f + x_1 s_{fg} = 4.8479$ kJ/(kg·K).

Tank A final with $p_2 = 200$ kPa, $s_2 = s_1$, and x_2:

- $x_2 = \dfrac{s_2 - s_f}{s_{fg}} = 0.5928$·

- $v_2 = v_f + x_2 v_{fg} = 0.52552$ m³/kg.
- $u_2 = u_f + x_2 u_{fg} = 1704.7$ kJ/kg.
- State 1 is in the saturated mixture region. When the pressure is decreased and the process is isentropic, the final state, i.e., 2, will also lie in the saturated mixture region. Thus:

$$T_{A,\text{final}} \hat{=} T_{\text{sat}/p_2} = 120.2°\text{C}$$

- Mass in tank A:

$$m_{A,1} = \frac{\forall_A}{v_{A,1}} = 1.08 \text{ kg}; \quad m_{A,2} = \frac{\forall_A}{v_{A,2}} = 0.5709 \text{ kg}$$

Hence $m_{A,\text{out}} = (m_{A,1} - m_{A,2}) = 0.5091$ kg.

- Mass in tank B:

$$m_{B,2} = m_{B,1} + m_{A,\text{out}} = 2.509 \text{ kg}; \quad v_{B,2} = \frac{\forall_B}{m_{B,2}} = 0.9558 \text{ m}^3/\text{kg}$$

- *E-B* for the tank A and tank B system:

$$-Q = \Delta U|_{\text{CV}} = \Delta U_A + \Delta U_B.$$
$$\Delta U_B \approx \Delta E_B = (m_1 u_1 - m_2 u_2)|_A - Q = 642.275 \text{ kJ}$$

Comments

✓ As the tank A initial state is in the saturated mixture region and the pressure drop from p_1 to $p_2 = p_{\text{final}}$ occurs without any irreversibilities, i.e., $s_{\text{final}} = s_{\text{initial}}$, the final state of tank A has to be in the saturated mixture region.

✓ Although $Q = 300$ kJ is being lost from tank B, the energy increase ΔU_B is substantial, considering just "wet steam" supply from tank A.

T-s Diagram

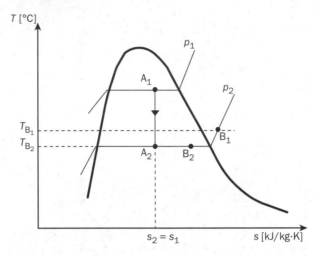

2.4 Steady Open Uniform Flow Devices

Because of the versatile applications driven by human needs, most thermodynamic devices fall into the category of *steady open uniform flow systems*. Examples include turbines and compressors, nozzles and diffusers, boiler tubes and throttles, heat exchangers and mixing chambers (see the Equations Sheets in App. A). Some of these devices are components of *steady cyclic operations* for heat engines (e.g., power plants), refrigerators, air conditioners, and heat pumps. Steady operation is desirable to achieve constant output, e.g., gas turbine power, flow of electricity from generators powered by steam turbines, hot or cold fluid streams of constant temperature, etc.

Under steady-state operational conditions, nothing accumulates or decreases inside the system (see Sec. 2.1 and the Equation Sheets). Hence, in Eq. (2.6) $\Delta X_{\text{control volume}} = (X_{\text{final}} - X_{\text{initial}})_{CV} = 0$ as a result. Thus, the *M-B* of Eq. (2.4) reduces to:

$$\sum \dot{m}_{\text{in}} = \sum \dot{m}_{\text{out}} \tag{2.10}$$

and the *E-B* of Eq. (2.7) reduces to:

$$\sum \left[\dot{Q} + \dot{W} + \dot{H} + \frac{\dot{m}v^2}{2} \right]_{\text{in}} = \sum \left[\dot{Q} + \dot{W} + \dot{H} + \frac{\dot{m}v^2}{2} \right]_{\text{out}} \tag{2.11}$$

It should be noted that *for steady open systems, all mass and energy quantities are in rate form*, i.e., the generalized $\dot{X}$ is constant at the inlets and outlets at all times. However, fluid properties, such as enthalpy rate $\dot{H}$ and mean velocity v, change *spatially* between device inlet and outlet. Hence it is a steady, uniform flow operation with all state variables being in quasi-equilibrium.

Example 2.11: Evaluation of a Steam Turbine   In an adiabatic turbine, steam (20 kg/s) is expanded from 6 MPa and 400°C with a mean velocity of 80 m/s to 40 kPa at 92% quality and 50 m/s. Determine (a) the ratio of change of kinetic energy to enthalpy change and (b) the turbine power. Plot the turbine power as a function of outlet pressure, i.e., for 10 kPa $< p_2 <$ 200 kPa, and comment.

Sketch

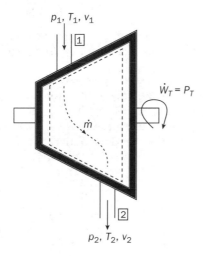

p_1, T_1, v_1

$\boxed{1}$

$\dot{W}_T = P_T$

$\dot{m}$

$\boxed{2}$

p_2, T_2, v_2

Assumptions

- Steady open system.
- H_2O is a two-phase fluid.
- Adiabatic expansion process i.e., $Q = 0$.

Method

- M-B: $\dot{m}_{in} = \dot{m}_{out} = \dot{m}$.

- E-B: $\dot{W}_T = -\dot{m}(h_2 - h_1) + \dot{E}_{kin,1} - \dot{E}_{kin,2}$.

Solution

- State 1 is in the superheated region. Therefore, $h_1 = 3177.2$ kJ/kg.
- State 2 is in the saturated region. At 40 kPa and $x = 0.92$, we have:

$$T_{sat} = 75.9°C; \; h_f = 317.6 \text{ kJ/kg}; \; h_{fg} = 2319.1 \text{ kJ/kg}$$

$$h_2 = h_f + xh_{fg} = 317.6 + 0.92 \times 2319.1 = 2451.2 \text{ kJ/kg}$$

- Change in $E_{kin} = \dfrac{\dot{m}}{2}\left(v_1^2 - v_2^2\right)$ and the change in enthalpy $= \dot{m}(h_1 - h_2)$:

$$\text{Ratio} = \frac{\left(v_1^2 - v_2^2\right)}{2\left(h_1 - h_2\right)} = \frac{\left(80^2 - 50^2\right)}{2\left(3177.2 - 2451.2\right) \times 10^3} = 0.00268$$

- Turbine power:

$$\dot{W}_T = -\dot{m}(h_2 - h_1) = -20(3177.2 - 2451.2) = -14520\,\text{kW} = -14.52\,\text{MW}$$

Comments

✓ The ratio of the change in kinetic energy (ΔkE) to change in enthalpy is very small, which is generally observed in steady-state devices like turbines.

Example 2.12: Ideal Gas in a Compressor   Air is compressed steadily in a compressor to 700 kPa and 500 K. At the inlet, the pressure and temperature of air are 100 kPa and 300 K, respectively. The mass flow rate of the air through the compressor is 0.1 kg/s, and a heat loss of 30 kJ/kg occurs during the process. Assuming that the velocity changes are negligible and that the inlet and the outlet are at the same level, estimate the power required to drive the compressor.

Sketch

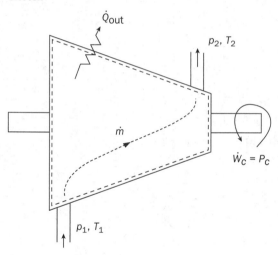

Assumptions

- Steady open system.
- Air is an ideal gas.
- Changes in kinetic and potential energies are negligible, i.e., $\Delta kE \approx 0$ and $\Delta pE \approx 0$.

Method

- M-B: $\dot{m}_{in} = \dot{m}_{out} = \dot{m}$
- E-B: $\Sigma \dot{E}_{in} = \Sigma \dot{E}_{out}$

Solution

- From E-B:

$$\dot{W}_{in} + \dot{m}h_1 = \dot{Q}_{out} + \dot{m}h_2$$

$$\dot{W}_{in} = \dot{m}q_{out} + \dot{m}(h_2 - h_1)$$

- The enthalpy of an ideal gas is a function of temperature only, and hence, from the Property Tables, the enthalpies of air are:

$$h_1 = h_{at\ 300\ K} = 300.19\ kJ/kg$$

$$h_2 = h_{at\ 500\ K} = 503.02\ kJ/kg$$

- Substituting the values in the energy balance equation:

$$\dot{W}_{in} = 0.1 \times 30 + 0.1 \times (503.02 - 300.19) = 23.283\ kW$$

Comments

✓ The work input, in this case, not only leads to an increase in the enthalpy but also heat loss. If the compressor was adiabatic, the same work input could be utilized to compress the ideal gas to a higher temperature.

Example 2.13: Flow through a Throttling Valve, i.e., a Flow Resistance   Refrigerant-134a enters a capillary tube in a refrigerator as a saturated liquid. It is throttled from an inlet pressure of 0.5 MPa to a final pressure of 0.1 MPa. Calculate the quality of the refrigerant at the exit of the throttle valve and the temperature drop during the process.

Sketch

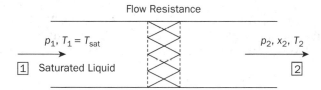

Assumptions

- The throttling process is adiabatic, i.e., $\dot{Q} = 0$.
- Steady open system.
- Changes in kinetic and potential energies are negligible, i.e., $\Delta kE \approx 0$ and $\Delta pE \approx 0$.

Method

- As there is no work or heat transfer and the changes in kinetic and potential energies are negligible, the E-B reduces to $h_2 = h_1$.
- M-B: Not needed for this problem because the results are per unit mass.
- $x_2 = \dfrac{h_2 - h_f}{h_g - h_f}$.

Solution

- At State 1, the refrigerant is a saturated liquid. Therefore:

$$P_1 = 0.5 \text{ MPa}; \, h_1 = h_f = 73.33 \text{ kJ/kg}; \, T_1 = T_{sat} = 15.71°C$$

- At State 2:

$$P_2 = 0.1 \text{ MPa}; \, h_f = 17.28 \text{ kJ/kg};$$

$$h_g = 234.44 \text{ kJ/kg}; \, T_{sat} = -26.37°C$$

- Now, as $h_2 = 73.33$ kJ/kg, $h_f < h_2 < h_g$. Therefore, the refrigerant is in the saturated state at State 2 and $T_2 = T_{sat} = -26.37°C$. Hence:

$$x_2 = \frac{h_2 - h_f}{h_g - h_f} = \frac{73.33 - 17.28}{234.44 - 17.28} = 0.258$$

- Temperature drop:

$$\Delta T = T_1 - T_2 = 15.71 - (-26.37) = 42.08°C$$

Comments

✓ A throttling device is commonly used in refrigerators to create a large temperature drop by restricting flow.

✓ The device is small in size, and hence there is neither enough time nor appreciable surface area for heat transfer to be significant. Hence the process can be considered to be adiabatic. A capillary tube is an example of a throttling device.

Example 2.14: Heat Exchanger   Refrigerant-134a enters a condenser with a mass flow rate of 10 kg/min at 1.4 MPa and 100°C and leaves at 40°C. It is cooled by water that enters the condenser at 500 kPa and 10°C and leaves at 20°C. Considering the system to be adiabatic and neglecting pressure drops determine (a) mass flow rate of the cooling water required and (b) heat transfer rate from refrigerant to water.

Sketch (Shell 1 → 2 and Tube 3 → 4 Heat Exchanger)

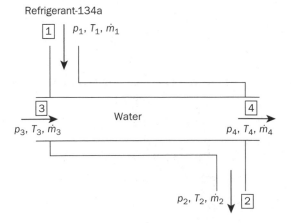

Assumptions

- Steady open system without any fluid mixing, where $\dot{Q}_{water} = \dot{Q}_{R\text{-}134a}$

- Changes in kinetic and potential energies are negligible, i.e., $\Delta kE \approx 0$ and $\Delta pE \approx 0$.

- No work performed.

- No heat transfer between the heat exchanger unit and the surroundings, i.e., $\dot{Q} = 0$.

Method

- M-B: $\dot{m}_1 = \dot{m}_2 = \dot{m}_w, \quad \dot{m}_3 = \dot{m}_4 = \dot{m}_r$.

- E-B: $\dot{m}_1 h_1 + \dot{m}_3 h_3 = \dot{m}_2 h_2 + \dot{m}_4 h_4$.

Solution

- For water, at $p = 500$ kPa, $T_{sat} = 151.83$. The temperatures of water at the inlet and the exit are below the saturation temperature, and hence it is in the compressed or subcooled state. For calculation purposes, the compressed liquid can be approximated to be in saturated state. Thus:

$$h_3 = h_{f \text{ at } 10°C} = 42.02 \text{ kJ/kg}$$

$$h_4 = h_{f \text{ at } 30°C} = 125.74 \text{ kJ/kg}$$

- For the refrigerant-134a, at $p = 1.4$ MPa, $T_{sat} = 52.40°C$. Thus, from the temperatures given, it can be inferred that the refrigerant enters in a superheated state and leaves in a compressed liquid state (approximated as saturation state). Thus, from the Property Tables, at $p = 1.4$ MPa:

$$T = 100°C, h_3 = 330.30 \text{ kJ/kg}$$

$$T = 40°C, h_4 = h_{f \text{ at } 40°C} = 108.26 \text{ kJ/kg}$$

- From E-B:

$$\dot{m}_w (h_4 - h_3) = \dot{m}_r (h_1 - h_2)$$

$$\Rightarrow \dot{m}_w = \dot{m}_r \frac{(h_2 - h_1)}{(h_3 - h_4)} = 10 \frac{(108.26 - 330.30)}{(42.02 - 125.74)} = 26.5 \text{ kg/min}$$

- Any heat loss from the entire heat exchanger unit is negligible. However, the purpose of a heat exchanger is to transfer heat between two fluids. Thus considering one of the fluids to be a system, one can estimate the heat transfer by using an energy balance equation. Let refrigerant-134a be a system. $\dot{Q}^r_{out} \equiv \dot{Q}^w_{in}$ because the problem statement requires the heat transferred from refrigerant to water. The E-Bs read:

$$\dot{Q}^r_{out} \equiv \dot{Q} = \dot{m}_r(h_2 - h_1) = \dot{Q}^w_{in} = \dot{m}_w(h_3 - h_4)$$

$$\Rightarrow \dot{Q} = \dot{m}_r (h_2 - h_1) = 10(330.30 - 108.26) = 2220.4 \text{ kJ/min} = 37 \text{ kW}$$

Comments

✓ Considering water as a system, the heat coming into the system has to be accounted for, and, when worked out, the E-B will still yield the same heat transfer value of 37 kW.

✓ A conversion rate was needed to turn kilojoules per minute into watts.

Example 2.15: Nozzles   The inlet area of an adiabatic nozzle is twice that of its exit. Air enters the nozzle with a velocity of 120 m/s and leaves with a velocity of 380 m/s. Considering the pressure and temperature of air to be 600 kPa and 500 K, respectively, at the inlet, calculate the temperature and the pressure at the exit of the nozzle.

Sketch

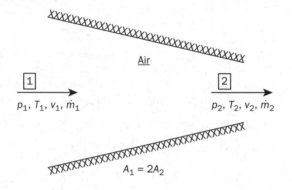

Assumptions

- Steady open system.
- Change in potential energy is negligible, i.e., $\Delta pE = 0$.
- Adiabatic system, i.e., $\dot{Q} = 0$.

Method

- M-B: $\dot{m}_1 = \dot{m}_2 = \dot{m}$, i.e., $\rho_1 v_1 A_1 = \rho_2 v_2 A_2$.

- E-B: $\dot{E}_{in} = \dot{E}_{out}$.

- IG law: $p = \rho RT$; $\rho = v^{-1}$

Solution

- At the inlet:

$$p_1 = \rho_1 RT_1$$

$$\Rightarrow \rho_1 = \frac{p_1}{RT_1} = \frac{600}{0.287 \times 500} = 4.181 \, \text{kg/m}^3$$

- From *M-B*:

$$\rho_1 v_1 A_1 = \rho_2 v_2 A_2$$

$$=> \rho_2 = \frac{\rho_1 v_1 A_1}{v_2 A_2} = \frac{4.181 \times 120 \times 2}{380} = 2.641 \, \text{kg/m}^3$$

- From the Property Tables, at $T_1 = 500$ K, $h_1 = 503.02$ kJ/kg.

- From *E-B*:

$$\dot{E}_{in} = \dot{E}_{out}$$

$$h_1 + \frac{v_1^2}{2} = h_2 + \frac{v_2^2}{2}$$

$$=> h_2 = \left(503.02 \times 10^3\right) + \frac{120^2 - 380^2}{2} = 438020 \, \frac{m^2}{s^2} = 438.02 \, \frac{kJ}{kg},$$

where 1 kJ/kg = 1000 m²/s²

- From the Property Tables, temperature T_2 at which $h_2 = 438.02$ kJ/kg is 437 K (upon interpolation).

- Using IG law:

$$p_2 = \rho_2 R T_2 = 2.641 \times 0.287 \times 437 = 331.23 \, \text{kPa}$$

Comments

✓ Due to the steady adiabatic nature of the device, an increase in the velocity results in a drop in the temperature; i.e., the internal energy is converted into kinetic energy.

✓ The same approach can be applied for a problem involving a diffuser, which operates in the opposite manner to that of the nozzle.

2.5   Homework Assignments

Part A probes the reader's knowledge base of the chapter via insight questions dealing with novel concepts, definitions, derivations of basic equations, and system visualizations.

Part B lists standard engineering thermodynamics problems, which probe the reader's understanding of the concepts, balance equations, correlations, and needed data sources. The challenge in solving such problems lies in *logical thinking*, after carefully reading the problem statement, rather than higher math skills. Additional, typically very similar problems may be found on the web or in introductory books, such as Cengel and Boles (the required text at NC State for the last 25 years), Moran and Shapiro, Borgnakke and Sonntag, and Turns and Pauley, which are (on and off) recommended texts at North Carolina State University.

Clearly, students are encouraged to consult the thermodynamics literature for alternative views and new material to further enrich their learning experience.

2.5.1   Part A: Insight

1   What is the advantage of casting Eq. (2.2b) in the form of Eq. (2.3)?

2   Sketch typical applications of Eqs. (2.3) and (2.5), and include well-placed control volumes (C∀s) with their control surfaces (CSs).

3   List and depict cases where the assumption $\Delta E_{C\forall} \approx \Delta U_{C\forall}$ would not be justifiable.

4   Provide a scenario for which a closed system/device is not transient.

5   Discuss the advantages of setting up the three Prelims, i.e., Sketch, Assumptions, and Method, before launching into the problem solution.

6   In Example 2.2 what would be a better assumption for $h_{exit} \approx \cancel{\mathbb{C}}$? What would an even more realistic solution approach look like?

7   Provide T-v and p-v diagrams for the results in Example 2.2.

8   A simpler transient process than the one in Example 2.4 is the problem of initial temperature rise in a one-dimensional heated wire (m, c) subject to $Q_{net,in}(t)$. Thus, find $dT/dt @ t = 0$, and plot $T(t)$. Comment!

9   Complete Example 2.9, i.e., check if $T_{final} = T_2$ when $m_2 = m_3$.

10   With respect to Example 2.10, discuss the usefulness of the C∀ concept.

11   Why are the conservation laws for steady flow devices written in rate form [see Eqs. (2.10) and (2.11)]? Why is *uniform flow* an important underlying assumption for such systems?

12   Looking at Example 2.11, why does ΔkE provide so little to turbine power generation?

13   Explain the inner working and use of throttling devices, say, in refrigerators (see Example 2.13).

14   Consider a nozzle with heat transfer: (a) Is heat transfer desirable to or from the fluid? (b) How will it affect the exit velocity?

2.5.2   Part B: Problems

15   A diffuser is to be designed to decrease the kinetic energy of air entering a compressor. At the inlet, the pressure and temperature are 100 kPa and 30°C, respectively, and the air enters the diffuser with a velocity of 350 m/s. If the pressure and temperature at the exit are 200 kPa and 90°C, respectively, determine the velocity of air at the exit.

16   Steam enters a nozzle with a velocity of 10 m/s and loses heat at the rate of 25 kW as it flows through it. The temperature and pressure at the inlet are 400°C and 800 kPa, respectively, while at the exit, the corresponding values are 300°C and 200 kPa. If the inlet area is 800 cm², determine the volumetric flow rate and the velocity of the steam at the exit.

17   Refrigerant-134a flows through an adiabatic compressor at the rate of 1.2 kg/s. It enters the device as a saturated vapor at −24°C and leaves at 0.8 MPa and 60°C. Determine the volumetric flow rate of the refrigerant at the compressor inlet and the power input to the compressor.

18 Steam enters a turbine at a pressure and temperature of 10 MPa and 500°C. It leaves the device at a pressure of 10 kPa with a quality of 90%. Neglecting heat loss and any changes in kinetic and potential energies, determine the mass flow rate required to produce a power output of 5 MW.

19 Oil ($c_p = 2.20$ kJ/kg°C), flowing at the rate of 2 kg/s, is cooled from 150°C to 40°C in a counterflow heat exchanger by water ($c_p = 4.18$ kJ/kg°C). If the inlet temperature and mass flow rate of water are 22°C and 1.5 kg/s, determine the rate of heat transfer between the two fluids and the exit temperature of water. Assume negligible heat loss between the heat exchanger and the surroundings.

20 Air enters a condenser at 100 kPa and 27°C with a volume flow rate of 600 m³/min and leaves at 95 kPa and 60°C. It is used to cool R-134a, which enters the condenser at 1 MPa and 90°C. Given the final state of the refrigerant as 1 MPa and 30°C, determine the mass flow rate of the refrigerant.

21 A balloon is connected by a valve to a large reservoir that supplies steam at 150 kPa and 200°C. The balloon initially contains 50 m³ of steam at 100 kPa and 150°C. When the valve is opened, steam is allowed to enter the balloon until the pressure equilibrium with the steam at the supply line is attained. The volume of the balloon increases linearly with pressure while the mass doubles at the end of the process. Heat transfer also takes place between the balloon and the surroundings. Determine the final temperature and the boundary work during this process.

22 A vertical piston-cylinder device initially contains 0.25 m³ of air at 600 kPa and 300°C. When the valve, which is connected to the cylinder, is opened, air escapes until three-quarters of the mass leaves the cylinder and the volume has been reduced to 0.05 m³. Determine the final temperature in the cylinder and the boundary work during the process.

23 A vertical piston-cylinder device initially contains 0.8 m³ of refrigerant-134a at 1.4 MPa and 120°C. The device is insulated, and a linear spring applies full force to the piston. At this point, a valve connected to the cylinder is opened and the refrigerant is allowed to escape. Due to this, the piston moves down, and the spring unwinds. At the end of the process, the pressure and the volume are reduced to 0.7 MPa and 0.5 m³. Determine the amount of refrigerant that has escaped and the final temperature of the refrigerant.

An adiabatic steam turbine is directly coupled to an adiabatic air compressor. The turbine is used to drive the compressor and a generator. For the case of the turbine, steam enters at 12.5 MPa and 500°C at a rate of 25 kg/s and exits at 10 kPa and a quality of 0.92%. Air enters the compressor at 98 kPa and 295 K at a rate of 10 kg/s and exits at 1 MPa and 620 K. Determine the net power delivered to the generator by the turbine.

24 Refrigerant-134a enters a compressor as saturated vapor at 10°C and leaves at 1400 kPa with an enthalpy of 281.39 kJ/kg and a velocity of 50 m/s. The mass flow rate of the refrigerant is 5 kg/s, and the work done on the fluid is found to be 132.4 kW. Assuming that the refrigerant enters and leaves with negligible velocities and that the inlet and exit are at the same level, determine the rate of heat transfer during this process.

25 A piston-cylinder device, containing 2 kg of refrigerant-134a at 800 kPa and 80°C, has a piston touching a pair of stops at the top. The piston is designed in such a way that a 500-kPa pressure is required to move it. The device also contains a valve at the bottom, which is opened to let the refrigerant out of the cylinder. After some time, the piston is observed to move, and the valve is closed when half the refrigerant oozes out of the tank. At this instant, the temperature recorded inside the cylinder is 20°C. Determine the work done and the rate of heat transfer.

26 Geothermal water enters the flash chamber (a throttling valve) in a single-flash geothermal power plant at 230°C as a saturated liquid at a rate of 50 kg/s. The steam, which comes out from the chamber, enters a turbine and leaves at 20 kPa with a moisture content of 5%. Calculate the temperature of the steam after the flashing process and the power output from the turbine if the pressure of the steam at the exit of the flash chamber is (a) 1 MPa, (b) 500 kPa, (c) 100 kPa, and (d) 50 kPa.

27 A turbocharger in an internal combustion engine is a device that contains a coupled system of a turbine and a compressor. Exhaust gases enter the turbine at 400°C and 120 kPa at a rate of 0.02 kg/s and leave at 350°C. Air enters the compressor at 50°C and 100 kPa and leaves at 130 kPa at a rate of 0.018 kg/s. While the compressor is primarily used to increase the pressure of incoming air, it also increases the temperature, which could resulting in "knocking" in the engine. Thus air is made to flow through an aftercooler to reduce the temperature below 80°C to avoid a knock. The cold ambient air enters the aftercooler at 30°C and leaves at 40°C. Neglecting any frictional losses inside the turbocharger and assuming that the exhaust gases have the same properties as that of air, determine the temperature of air at the outlet of the compressor and the minimum volume flow rate of ambient air required to avoid knock.

CHAPTER 3

The Second Law of Thermodynamics

The concept of entropy was casually referred to throughout Chaps. 1 and 2. Indeed, no device works, no process proceeds, no system is real without exhibiting irreversibilities. Thus, entropy generation has two main sources: (1) entropy change of the system/device and (2) the entropy changes of the coupled surrounding, i.e., $S_{gen} = \Delta S_{system} + \Delta S_{surroundings} > 0$ (see Fig. 1.5).

In a way, Chap. 3 is just Chap. 2 with a very important addition. Specifically, we are adding to the mass and energy balances for open and closed systems an entropy balance (*S-B*) to find ΔS_{system} or S_{gen} [see Eq. (3.6) and recall Eq. (1.10)].

3.1 An Entropy Concept and the 2nd Law in Balance Form

Entropy, S [kJ/K] or specific entropy *s* [kJ/K·kg], is a measure of the degree of molecular disorder, i.e., random activities of the molecules, as a result of heat transfer, frictional effects, decay, turbulence, etc.; in short, due to *irreversibilities*. Like energy, it is a very important property of matter, i.e., our simple substances. It should be noted that Boltzmann and Gibbs, among others, have postulated more stringent, alternative definitions (Jaynes, 1965).

So it is not surprising that a pure crystal at 0 K (see Sec. 1.2) has $S = 0$ and that 1 kg of ice when heated and first turning into water and then into steam experiences significant entropy increases, i.e., $S_{ice} << S_{water} << S_{steam}$. In 1865, Clausius expressed this observation mathematically:

$$dS = \left(\frac{\delta Q}{T}\right)_{\text{internally reversible}} \tag{3.1}$$

In words, a path-dependent, incremental change in heat at the associated temperature *T* supplied to a substance in a *reversible* fashion causes a differential entropy increase. Clearly, for real processes:

$$\Delta S = S_2 - S_1 > \int_1^2 \left(\frac{\delta Q}{T}\right)_{\text{real}} \tag{3.2a}$$

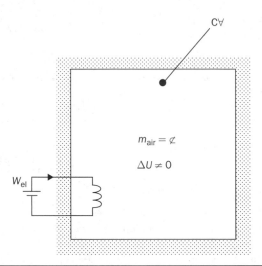

Figure 3.1   Heat transfer via an electric resistance heater in an adiabatic tank during Δt.

because of irreversibilities—implying energy/work losses. Thus, Eq. (3.2a) can be expressed as:

$$\Delta S > \int_1^2 \left(\frac{\delta Q}{T}\right)_{real} = \int_1^2 \left(\frac{\delta Q}{T}\right)_{internally\ reversible} + \int_1^2 \left(\frac{\delta Q}{T}\right)_{waste} \tag{3.2b}$$

As enthalpy and internal energy, *entropy is a state function*, i.e., independent of any particular process path, including reversible ones. Nevertheless, *all substances/systems/ devices undergoing real processes interact with the surrounding*. Hence, for calculating the *total entropy produced*, changes in both the substance *and* the connected surrounding have to be accounted for, i.e., $\Delta S_{total} \equiv S_{gen} = \Delta S_{system} + \Delta S_{surrounding} > 0$. As an aside, this simple fact and the definition of Eq. (3.1) allowed Gibbs to derive the *T-ds* equations for computing entropy changes of gases as well as pure liquids and solids (see Sec. 3.2).

Before getting to that point, an even more profound observation is the fact that the 1st law is insufficient to fully describe device processes and the direction they naturally proceed. For example, consider heating of a fluid in an insulated tank (Fig. 3.1).

The *E-B* just assures that energy is conserved; i.e., electric work done causes a change in internal energy of the fluid. Specifically, during process time Δt:

$$W_{el} = \Delta U = m(u_2 - u_1) \tag{3.3a}$$

Now, Eq. (3.3a) is also correct when viewed the other way around; i.e., lowering the air's internal energy from U_2 to U_1 will recover W_{el} again:

$$\Delta U = U_2 - U_1 = W_{el} \tag{3.3b}$$

3.1.1 The Second Law

It is axiomatic that such a process [i.e., its reverse direction indicated with Eq. (3.3b)] is impossible because of losses when heating up the gas via W_{el}. As all processes are associated with irreversibilities, another law of thermodynamics is required to indicate realistic process directions (see Fig. 1.5). Such energy losses (actually wasted work potential) are quantitatively captured in terms of an inequality, known as the *increase-in-entropy principle*. Mathematically, the Second Law of Thermodynamics states:

$$\Delta S_{total} \equiv S_{gen} = \Delta S_{system} + \Delta S_{surrounding} > 0 \qquad (3.4)$$

It is emphasized again that S_{gen} *encompasses entropy changes of both the substance/device/fluid under consideration and those of the surrounding interacting with the system.* For example, freezing water (being the system) will generate $\Delta S_{system} < 0$, i.e., a drop in entropy level; however, $\Delta S_{surrounding} >> 0$ due to the necessary heat transfer; hence, $S_{gen} = \Delta S_{total} > 0!$

Similarly, efficiencies of real processes/systems/cycles are always less than 100%. Thus:

$$\eta \equiv \frac{\text{desired OUTPUT}}{\text{required INPUT}} < 1.0 \qquad (3.5)$$

The implications of Eqs. (3.4) and (3.5) include:

- All real processes are afflicted with (internal) irreversibilities that are encapsulated macroscopically by S_{gen}; thus, for the process to proceed, S_{gen} has to increase and 100% efficiency can never be achieved.

- Only for *reversible* processes, $S_{gen} = 0$, i.e., both $\Delta S_{system} = 0$ (internaly reversible) and $\Delta S_{surrounding} = 0$ (externally reversible).

- It is possible that $\Delta S_{system} < 0$, as in ice making, gas compression, house construction, etc.; however, we always have $\Delta S_{total} = S_{gen} > 0$ for all real processes.

- As S_{gen} is a measure of energy loss or derived work-potential wasted, *minimizing* ΔS_{total} *is a modern design tool for performance optimization (see Chap. 4).*

3.1.2 Entropy Balance

The inequality (3.4) can be recast as an entropy "balance" by summing up all entropy contributions causing a change in entropy level inside the system or control volume. Specifically, we account for entropy inflow and outflow via fluid stream (the mass transfer mechanism), heat transfer to or from the system with associated temperatures, and, of course, the entropy generated inside the CV due to irreversibilities, i.e., S_{gen}. Hence, for a transient open system operating during time interval Δt(see Fig. 3.2):

$$\underbrace{\sum_{in} m \cdot s - \sum_{out} m \cdot s + \sum \frac{Q}{T_{ambient}}}_{\Delta S_{surrounding}} + \underbrace{S_{gen}}_{\substack{\text{system} \\ \text{irreversibilities}}} = \Delta S \big|_{CV} \qquad (3.6a)$$

Here, terms in $\Sigma Q/T_{ambient}$ are either "positive" for Q_{in} or "negative" when $Q = Q_{out}$. Also:

$$\Delta S \big|_{CV} = \underbrace{(ms)_{final} - (ms)_{initial}}_{\Delta S_{system}} \qquad (3.6b)$$

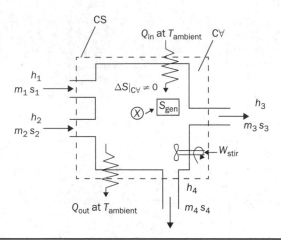

FIGURE 3.2 Schematic of a transient open system.

It should be noted that Eq. (3.6a) is a special form of Eq. (1.7b), where S (or m·s or Q/T) is X:

$$\sum X_{\text{in}} - \sum X_{\text{out}} + X_{\text{source}} = \Delta X_{\text{control volume}} = \left(X_{\text{final}} - X_{\text{initial}}\right)_{\text{CV}} \tag{3.6c}$$

where the most important (new) term is $X_{\text{source}} \equiv S_{\text{gen}}$. From Eq. (3.6a or 3.6c), the key unknown can be isolated as:

$$S_{\text{gen}} = (ms)_{\text{final}} - (ms)_{\text{initial}} + \sum(ms)_{\text{out}} - \sum(ms)_{\text{in}} - \sum\left(\frac{Q}{T}\right)_{\text{ambient}} > 0 \tag{3.7}$$

in which ΔS_{system} and $\Delta S_{\text{surrounding}}$ can be readily identified. *It should be noted again that* $\left(\frac{Q}{T}\right)_{\text{ambient}}$ *can be positive or negative as physically best determined with Eq. (3.6), rather than using Eq. (3.7).* Clearly, work performed onto or by a system/device does not appear in the entropy balance. *Work is an entropy-free form of energy transfer.* The reason is that *work performed is microscopically organized—quite in contrast to heat transfer.*

The special cases of Eq. (3.7) are:

1. For *steady open* systems, S_{gen} is not equal to 0 due to irreversibilities, but $\Delta S_{\text{system}} = 0$ because nothing changes inside the system or CV. As for all steady open systems, the reduced Eq. (3.7) is *written in rate form.*

2. For *closed* systems, $\Sigma(ms)_{\text{OUT}} = \Sigma(ms)_{\text{IN}} = 0$ because there is no mass flow carrying specific entropy.

3. For *adiabatic* (i.e., insulated) devices $\Sigma\left(\frac{Q}{T}\right)_{\text{ambient}} = 0$ as no heat crosses the CS.

4. For *isentropic* processes, which are adiabatic *and* reversible, $S_{\text{system}} = 0$.

As a little illustrative example, consider the cooling of a cup of coffee. We have $S_{gen} = \Delta S_{system} + \Delta S_{surroundings} - Q_{cooling}/T_{coffee} + Q_{cooling}/T_{ambient} = Q(1/T_a - 1/T_c) > 0$, as $T_c > T_a$. An alternative solution approach would be to use Eq. (3.6) after system identification.

As mentioned, the larger S_{gen}, the more inefficient a process/device/system is; i.e., S_{gen} is directly related to the "amount of wasted energy generated." In convection heat transfer, this "energy destruction" appears as viscous dissipation (i.e., friction being a mechanical source) and random disorder (due to heat input being a thermal source):

$$S_{gen}^{total} = S_{gen}^{friction} + S_{gen}^{thermal} \tag{3.8}$$

For optimal system/device design, it is important to find typically the *best possible system geometry and operational conditions* so that, S_{gen}^{total} is a minimum.

3.2   The *T-s* Diagram and Entropy Calculations

It should be evident by now that a proper evaluation of a system or device requires the solutions of the mass and energy balances as well as the entropy balance, i.e., in general, Eqs. (2.7), (2.9), and (3.6). If Eq. (3.6) results in a zero or a negative value, the process/device is impossible. There are three basic sources for obtaining entropy values for different simple substances, ranging from ideal gases to two-phase fluids, as needed in a specific entropy balance:

- *Property Tables* (see App. B), which contain specific entropy values for H_2O and R-134a as a function of pressure and temperature, as well as "ideal gases."

- The graphical representation of some of the tabulated property values in *T-s diagrams* (see Fig. 3.3c for H_2O).

- The *T-ds relationships* (after Gibbs), typically applied to ideal gases as well as pure liquids and solids [see Eqs. (3.10) and (3.11)].

A *T-s* diagram for H_2O or R-134a looks somewhat similar to a *T-v* diagram (or a *p-v* diagram) with its three regions—i.e., compressed liquid, saturated mixture, and superheated vapor—separated by the saturated liquid line and the saturated vapor line (Figs. 3.3a to 3.3c). Again, the liquid–vapor mixture properties under the various "domes" are determined via Eq. (1.11).

Clearly, real vs. ideal processes can be well depicted in *T-s* diagrams (see Fig. 3.4). For example, in expansion processes (see Fig. 3.4a for a turbine), $\Delta \dot{H}_{ideal} > \Delta \dot{H}_{real}$ and hence real turbine power production is always less than for (nonexisting) isentropic turbines.

In contrast, Fig. 3.3b depicts air being compressed from p_1 to p_2 under isentropic (adiabatic and reversible) conditions with $\Delta S = 0$. More realistically, $\Delta S = S_2 - S_1 < 0$ for most compressors because compressed gas molecules exhibit a lower degree of disorder when $T_2 < T_{2s}$. However, depending on the interplay of compressor inlet vs. outlet temperatures and pressures, the gas entropy change can be positive [see Eq. (3.14)].

3.2.1   The *T-s* Equations after Gibbs

Recalling that entropy is a state function, i.e., path independent, we can select the simplest (idealized) system to evaluate entropy changes. For example, an internally

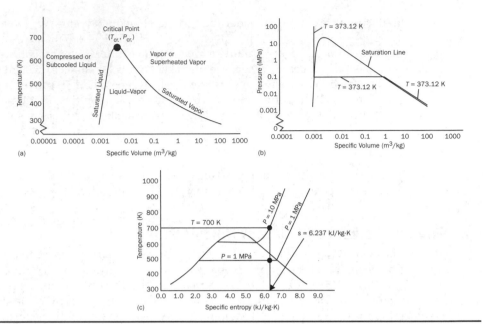

Figure 3.3 H$_2$O phase diagrams: (a) *T-v*, (b) *p-v*, and (c) *T-s*.

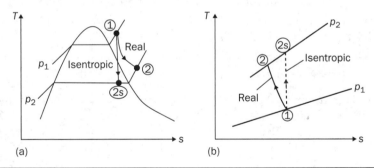

Figure 3.4 Two *T-s* diagrams for two-phase H$_2$O: (a) ideal vs. real turbine processes; (b) ideal vs. real compressor processes.

reversible piston-cylinder device is subjected to (reversible) heat transfer causing (reversible) boundary work and a change in internal energy of the fluid (say, an *IG*) inside the cylinder (see Fig. 3.5). The simple *E-B* on a differential basis then reads:

$$\delta Q - \delta W = dU \tag{3.9a}$$

Recalling that for reversible heat transfer [see Clausius, Eq. (3.1)] $\delta Q = TdS$ and for reversible boundary work $\delta W = pd\forall$, we have:

$$TdS - pd\forall = dU \tag{3.9b}$$

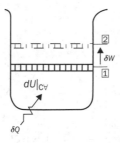

FIGURE 3.5   Reversible processes in an ideal piston-cylinder device.

Equation (3.9b), written per unit mass, yields Gibbs I:

$$ds = \frac{du}{T} + \frac{p}{T}dv \qquad (3.10)$$

Similarly, with $dh = du + d(pv)$, we obtain Gibbs II:

$$ds = \frac{dh}{T} - \frac{v}{T}dp \qquad (3.11)$$

3.2.2   Special Applications

Although the Gibbs relations hold for any simple substance undergoing a quasi-equilibrium process, applications can be most frequently found for liquids/solids as well as for ideal gases.

(1) For *liquids and solids* being incompressible ($p = \cancel{c}$) and assuming constant specific heats:

$$ds = \frac{c(T)\,dT}{T} \text{ or } \Delta s = \int_{1}^{2}\frac{c(T)\,dT}{T} = c_{\text{avg}}\ln\left(\frac{T_2}{T_1}\right) \qquad (3.12\text{a–c})$$

It should be noted that for isentropic liquid/solid processes (i.e., $s_2 = s_1$) $T_2 = T_1 = T = \cancel{c}$.

(2) For *ideal gases*, employing $pv = RT$ as well as $du = c_v\,dT$ and $dh = c_p\,dT$, Gibbs I and Gibbs II yield:

$$ds = \frac{c(T)\,dT}{T} + \frac{R\,dv}{v} \qquad (3.13\text{a})$$

which, integrated with an averaged heat capacity c_v, results in:

$$s_2 - s_1 = \overline{c_v}\ln\left(\frac{T_2}{T_1}\right) + R\ln\left(\frac{v_2}{v_1}\right) \qquad (3.13\text{b})$$

In case temperatures and pressures are given:

$$ds = \frac{c(T)\,dT}{T} - \frac{R\,dp}{p} \qquad (3.14\text{a})$$

and hence with an averaged c_p:

$$s_2 - s_1 = \overline{c}_p \ln\left(\frac{T_2}{T_1}\right) - R \ln\left(\frac{p_2}{p_1}\right) \tag{3.14b}$$

(3) For *isentropic processes* $\Delta s = s_2 - s_1 = 0$, i.e., $s = \mathcal{c}$, and using an *ideal gas* we obtain:

$$\overline{c}_v \ln\left(\frac{T_2}{T_1}\right) = -R \ln\left(\frac{v_2}{v_1}\right) \text{ or } \ln\left(\frac{T_2}{T_1}\right) = \ln\left(\frac{v_1}{v_2}\right)^{\frac{R}{\overline{c}_v}}$$

so that:

$$\left(\frac{T_2}{T_1}\right) = \left(\frac{v_1}{v_2}\right)^{k-1} \tag{3.15}$$

Here $k = c_p/c_v$ and from $dh = du + d(pv)$, i.e., for an ideal gas $c_p\, dT = c_v\, dT + R\, dT$, we have $R = c_p - c_v$. Similarly:

$$\left(\frac{T_2}{T_1}\right) = \left(\frac{p_2}{p_1}\right)^{\frac{k-1}{k}} \tag{3.16}$$

(4) For *polytropic processes of gases*, i.e., $pv^n = C$ [see also Eq. (1.12)], k in the previous equations is replaced by n in order to cover a wider range of compression or expansion pathways. For all practical purposes, the polytropic exponent ranges from $n = 1$ for isothermal gas processes (just check $pv = RT$) to $n = k = c_p/c_v$ for isentropic processes.

Note that Eqs. (3.15) and (3.16) can be combined to eliminate the temperatures. In any case, after the property values have been obtained and employing Gibbs I and II, *ideal-gas-entropy changes* are computed as:

$$\Delta S = m\left[\overline{c}_v \ln\left(\frac{T_2}{T_1}\right) + R \ln\left(\frac{v_2}{v_1}\right)\right] \tag{3.17}$$

or

$$\Delta S = m\left[\overline{c}_p \ln\left(\frac{T_2}{T_1}\right) - R \ln\left(\frac{p_2}{p_1}\right)\right] \tag{3.18}$$

where for isentropic processes Eqs. (3.15) and (3.16) are recovered.

In summary:

- The *T-ds* equations can be applied for gases, especially for IGs, as well as for pure liquids and solids.

- For air, the necessary $c_v(T)$ and $c_p(T)$ values can be obtained from the Property Tables in App. B.

- Once integrated, they provide ΔS_{system} and thus only part of the entropy balance [see Eq. (3.6)].

Note that for two-phase fluids, i.e., H_2O and R-134a, the Property Tables have to be used to find s-values.

3.3 Entropy Generation in Transient and Steady-State Systems

The 2nd Law has already been applied to entropy changes in liquids and solids [see Eqs. (3.12a–c)] and will be further illustrated in Sec. 3.4. In this section, somewhat ideal-ized thermal devices are discussed mainly via solved examples. We consider *transient systems*, such as closed and open tanks (or compartments) and piston-cylinder devices (see Figs. 1.4 a, b), as well as *steady open systems*, such as uniform flow turbines, com-pressors, nozzles, diffusors, throttles, boiler tubes, mixing chambers, and heat exchang-ers (see Figs. 1.5 a–f). *Mass, energy, and entropy balances in light of stated simplifications go typically hand in hand in order to solve a problem.*

In Examples 3.1 to 3.5, proper identification of the type of system, fluid, energy forms, transfer directions, and process is most important. As always, these tasks are readily accomplished via the triple Prelims: Sketch, Assumptions, and Method (see Sec. 2.1).

Closed Systems   We start with basic applications of the T-ds relations for ideal gases and then expand to typical p-c devices, focusing on S_{gen} calculations.

Example 3.1   Consider two isothermal, isolated heat reservoirs at temperatures T_1 and T_2. Heat Q is transferred from the reservoir at T_1 to the one at T_2. Derive an expression for the total entropy change of the system.

Sketch

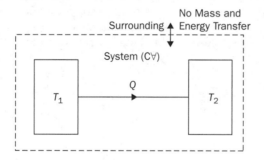

Assumptions   The system is isolated, i.e., no interaction with the surrounding.

Method $\Delta S_1 = -\dfrac{Q}{T_1}$ and $\Delta S_2 = \dfrac{Q}{T_2}$ during process time Δt.

Solution $\Delta S_{total} = \Delta S_1 + \Delta S_2 = -\dfrac{Q}{T_1} + \dfrac{Q}{T_2} = Q\left(\dfrac{1}{T_2} - \dfrac{1}{T_1}\right).$

Comments
- If $T_1 = T_2$, $\Delta S = 0$ as no heat is being transferred.
- If $T_2 > T_1$, $\Delta S < 0$, implying that this process violates the Second Law of Thermodynamics and hence is impossible.

Example 3.2 In a piston-cylinder device, air is compressed from 1 bar at 30°C to 3 bar at 120°C. Calculate the entropy change per unit mass for this process.

Sketch

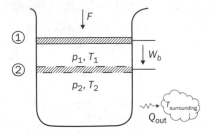

Assumptions Air behaves as an ideal gas. C_p and C_V are constant over the given temperature range.

Method With Gibbs II, $\Delta s = C_p \ln\left(\dfrac{T_2}{T_1}\right) - R \ln\left(\dfrac{p_2}{p_1}\right)$; see Property Tables.

Solution $\Delta s_{air} = 1.008 \times \ln(393/303) - 0.287 \times \ln(3/1) = -0.0531$ kJ/kg·K.

Comment As expected, gas compression reorders the molecules and hence $\Delta s < 0$; however, S_{gen} will be greater than zero.

Example 3.3 An insulated tank has two compartments of equal volume. An ideal gas is contained in one compartment while the other compartment is a vacuum. The membrane between the two compartments is punctured (or magically removed), and the ideal gas now occupies both the compartments. Selecting Argon (Ar) as the ideal gas, calculate the entropy change per unit mass for this process.

Sketch

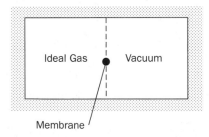

Assumptions No heat is transferred or work performed. Also, there is no change in the internal energy of the gas after free expansion in the insulated tank.

Method For an ideal gas, $\Delta U = mc_V \, \Delta T = 0$. Clearly, $u_2 = u_1$ as there is no change in the internal energy of the gas after free expansion in the insulated tank. It implies that this is an isothermal process where $T_1 = T_2$.

Solution So $\Delta s = R \ln\left(\dfrac{V_2}{V_1}\right)$ and $V_2 = 2V_1$. Hence:

$$\Delta s = (8.3144/39.948) \times \ln(2) = 0.144 \text{ kJ/kg·K}$$

Comment   For the given process conditions, only the tank volume changes during the time interval, so that $\Delta s \sim \ln(V_2/V_1)$ only.

Example 3.4   A fixed mass of an IG undergoes a process from one specific state (say, T_1 and v_1) to another specific state (T_2 and v_2). Derive ΔS_{system} under (a) irreversible and (b) reversible conditions.

Note   The quick answer is that ΔS_{IG} is the same for both types of processes as entropy is path independent.

Solution   Using Gibbs I for IGs, i.e., $ds = c_v \, dT/T + R \, dv/v$, and assuming c_v to stay constant, we have:

$$\Delta S_{IG} = m(c_v \ln T_2/T_1 + R \ln v_2/v_1)$$

Example 3.5   Consider a gas turbine (3.5 MPa, 500°C, $k = 1.381$) expanding adiabatically to 0.2 MPa. Find the (theoretically) maximum specific work [kJ/kg] that this device can produce.

Sketch

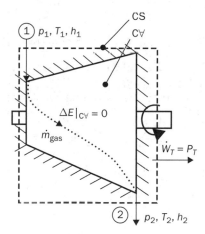

Assumptions
- Steady open system.
- I.G.
- Quasi-equilibrium expansion process.
- $\Delta kE \approx 0$.
- $\Delta pE = 0$.
- Adiabatic, isentropic expansion: $\dot{Q} = 0, ds = 0$.

  Note: For P^T_{max} to be achieved, the expansion process must be isentropic, i.e., $ds = 0$!

Method

- $(\Delta \dot{m})|_{C\forall} = 0$.

- $\sum \dot{E}_{in} - \sum \dot{E}_{out} = 0$.

- Gibbs equations.

Solution

The *E-B* (from Chap. 2) reads: $P_T = \dot{m}(h_2 - h_1)$.

Thus, $w_T = \Delta h = c_p(T_2 - T_1)$, where T_2 is unknown.

From Gibbs II with $ds = 0$, we obtain:

$$\ln\left(\frac{T_2}{T_1}\right) = \frac{R}{c_p} \ln\left(\frac{p_2}{p_1}\right)$$

from which, recalling that $R = c_p - c_v$, we find:

$$\frac{T_2}{T_1} = \left(\frac{p_2}{p_1}\right)^{\frac{k-1}{k}} , \text{ where } k = \frac{c_p}{c_v}$$

With the given numbers, $T_2 = 351$ K, and hence $w_T = 439$ kJ/kg.

Example 3.6   Via a supply line, air at 500 kPa and 70°C is pressed into an insulated p-c device, initially with 0.40 m³ containing 1.3 kg of air at 30°C. The process stops when the cylinder volume has increased by 50%. Assuming constant specific heats [i.e., $c_v = 0.718$ kJ/(kg·K) and $c_p = 1.005$ kJ/(kg·K)], develop an equation for T_{final}, solve for T_{final}, and find the entropy generated.

Sketch

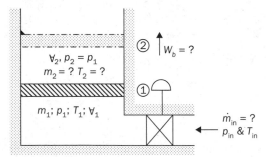

Assumptions

- Transient Open system.
- Air $\cong$ I.G.
- $Q = 0$; $W = W_b$.
- Quasi-equilibrium expansion at $p|_{C\forall} = \mathcal{C}$.
- Constant c_v and c_p.

Method

- Reduced M-B, E-B, and S-B.
- $p\forall = mRT$.
- Gibbs II for Δs.
- $W_b = p\Delta\forall$.
- Expect $T_1 < T_2 < T_{in}$.
- $h = c_p T$ and $u = c_v T$.

Solution

- M-B → $m_{in} = m_2 - m_1$; $m_2 = \dfrac{p_2 \forall_2}{RT_2}$; T_2 in K! $p_2 = p_1$.

$$p_2 = p_1 = \frac{m_1 RT_1}{\forall_1} := 283.6\,\text{kPa}$$

$$\text{w.}\mid \forall_2 = 1.5\forall_1: m_{in} = m_2 - m_1 = \frac{598.8}{T_2} - 1.3\,[\text{kg}]$$

- E-B → $(mh)_{in} - W_b = \Delta U\big|_{\text{system}} = m_2 u_2 - m_1 u_1$, where $W_b = p_1 \Delta\forall = 56.52$ kJ.
  Hence we balance:

$$m_i(T_2)c_p T_{in} - W_b = m_2(T_2)c_v T_2 - m_1 c_v T_1$$

- Numerically:

$$\left(\frac{590.8}{T_2} - 1.3\right) * 1.005(70 + 273) - 56.52 = \left(\frac{590.8}{T_2}\right) * 0.718 T_2 - 1.3 * 0.718(30 + 273)$$

Solving for T_2 yields $T_{final} = T_2 = 315$ K or $42°C = \begin{cases} > T_1 \\ < T_{in} \end{cases}$.

Now:

$$m_2 = \frac{590.8}{315} = 1.87\,\text{kg} \rightarrow m_{in} = 0.574\,\text{kg}$$

- S-B → $S_{gen} = \Delta S_{system} + \sum S_{out} - \sum S_{in}$

$$= (ms)_2 - (ms)_1 + 0 - (ms)_{in}$$
$$\text{or } S_{gen} = m_2(s_2 - s_{in}) - m_1(s_1 - s_{in})$$

Note With Gibbs II for an I.G.:

$$S_{gen} = m_2\left[c_p \ln\left(\frac{T_2}{T_{in}}\right) - R\ln\left(\frac{p_2}{p_{in}}\right)\right] - m_1\left[c_p \ln\left(\frac{T_1}{T_{in}}\right) - R\ln\left(\frac{p_1}{p_{in}}\right)\right]$$

$$S_{gen} = 0.0971\,\text{kg/K}$$

Comments

✓ For the supply line, $h_{in} = ¢$ as p_{in} and T_{in} are constant.

✓ The *E-B* delivers T_2 implicitly in conjunction with the *M-B* using an iterative program, or solve for T_2 explicitly.

✓ The S_{gen} equation is revealing which action/terms contribute to and which decrease S_{gen}.

✓ Again, physically more intuitive is the $\Delta S_{system} = (S_{final} - S_{initial})_{cv} = \Sigma S_{in} - \Sigma S_{out} + S_{gen}$ to track entropy increase vs. decrease.

Example 3.7   An adiabatic air compressor (IN: 98 kPa, 295 K, and 10 kg/s of air; OUT: 1 MPa and 620 K) is powered by an adiabatic turbine (IN: 12.5 MPa, 500°C, and 25 kg/s of steam; OUT: 10 kPa and a quality of 92%). Determine the net power delivered to the generator as well as the entropy generated within the compressor (employing averaged c values) and the turbine. Provide a *T-s* diagram for the turbine and a $p(\forall)$ plot for the compressor work.

Sketch

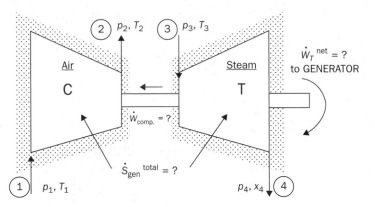

Assumptions

- Steady open systems.
- $Air|_G \cong$ I.G.; $H_2O|_T =$ 2-phase.
- $\dot{Q} = 0$; $\dot{W}_T^{net} = P_T - \dot{W}_c$.
- Quasi-equilibrium expansion/compression processes.
- Averaged specific heats.

Method

- E-Bs: $\Sigma \dot{E}_{in} = \Sigma \dot{E}_{out}$ taken separately for C&T; M-B: $m_{in} = ¢$ and $m_{steam} = ¢$; $\dot{m}_{air} = 10$ kg/s, $\dot{m}_{steam} = 25$ kg/s.

- S-Bs: $\dot{S}_{gen}^{total} = \dot{S}_{gen}^{T} + \dot{S}_{gen}^{C}$; using Gibbs II for $\dot{S}_{gen}^{C}$;

- Take $\bar{c} = c\left(\frac{1}{2}(T_{high} + T_{low})\right)$; Property Tables for both air and steam.

Solution

- Balances → $\dot{m}_{air}$ and $\dot{m}_{steam}$ are constant; $\Sigma\dot{E}_{in} = \Sigma\dot{E}_{out}$ for C and T.

  Thus $\dot{W}_C + \dot{m}_{air}h_1 = \dot{m}_{air}h_2$ and $\dot{W}_C = \dot{m}_{air}(h_2 - h_1)$

  $\dot{m}_{steam}h_3 = \dot{P}_T + \dot{m}_{steam}h_4$ and $\dot{W}_T = \dot{m}_{steam}(h_3 - h_4)$.

  S-B: $\dot{S}_{gen}^{total} = \dot{S}_{gen}^{T} + \dot{S}_{gen}^{C} = \dot{m}_{air}(s_2 - s_1) + \dot{m}_{steam}(s_4 - s_3)$.

- Properties:

  - C-props for air at T_1 and T_2 $\begin{cases} \text{State } 1 \to h_1 = 295.17 \dfrac{kJ}{kg} \text{ at } 295 \text{ K} \\[2mm] \text{State } 2 \to h_2 = 628.07 \dfrac{kJ}{kg} \text{ at } 620 \text{ K} \end{cases}$

  - T-props for H_2O at p_3 and T_3 → $h_3 = 3343.6 \dfrac{kJ}{kg}; s_3 = 6.4651 \dfrac{kJ}{kg\cdot K}$.

    p_4 and T_4 → $h_4 = h_f + x_4 h_{fg} = 2392.5 \dfrac{kJ}{kg}; s_4 = 7.5489 \dfrac{kJ}{kg\cdot K}$.

Now, $W_c = 10 * (628.07 - 295.17) = 3329 \text{ kW}$ and $P_T = \dot{W}_T = 23{,}777 \text{ kW}$

$$\therefore \dot{W}_T^{net} = 20{,}448 \text{ kW}$$

$$\dot{S}_{gen}^{total} = \dot{m}_{air}(s_2 - s_1) + \dot{m}_{steam}(s_4 - s_3)$$

W. | Gibbs II:

$$\dot{S}_{gen}^{C} = \dot{m}_{air}(s_2 - s_1) = \dot{m}_{air}\left[\int_1^2 c_p(T)\frac{dT}{T} - R_{air}\ln\frac{p_2}{p_1}\right]$$

where $\int_1^2 c_p(T)\dfrac{dT}{T}$ was $c_p(T) \cong \bar{c}_p = c_p\left(\dfrac{T_1 + T_2}{2}\right) = c_p(467.5 \text{ K}) \approx 1.012\dfrac{kJ}{kg\cdot K}$.

Hence:

$\dot{S}_{gen}^{C} = 0.92 \text{ kW/K}$

$\dot{S}_{gen}^{T} = 25(7.5489 - 6.4651) = 27.1 \text{ kW/K}$

Thus $\dot{S}_{gen}^{total} = 28 \text{ kW/K}$

Graphs

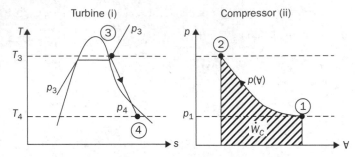

Example 3.8 Consider the turbocharger of an internal combustion engine. The exhaust gases enter the turbine at 450°C at a rate of 0.02 kg/s and leave at 400°C. Air enters the compressor at 70°C and 95 kPa at a rate of 0.018 kg/s and leaves at 135 kPa. The mechanical efficiency between the turbine and the compressor is 95% (5% of the turbine work is lost during its transmission to the compressor). Using air properties for the exhaust gases, determine (a) the air temperature at the compressor exit and (b) the isentropic efficiency of the compressor.

Sketch

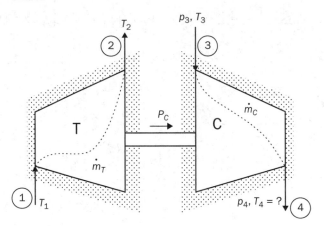

Assumptions

- Steady open coupled system.
- Exhaust gases $\cong$ Air $\cong$ I.G.
- Adiabatic devices w. | constant c-values.
- Compressor runs isentropically; thus, determine $\eta_{\text{isen}} = \Delta h_{\text{ideal}}/\Delta h_{\text{real}}$.

Method

- E-Bs for turbine and compressor.
- $\dot{W} = \dot{m}c_p\Delta T; \quad \dot{W}_C = \eta_{\text{TC}}\dot{W}_T.$

- $\eta_{\text{isen.}} = \dfrac{T_{2s} - T_1}{T_2 - T_1}$.

Solution

- E-B → $\dot{W}_T = m_{\text{exit}} c_p \left(T_2 - T_1 \right)_T$; $\dot{W}_C = \dot{m}_{\text{air}} c_p \left(T_4 - T_3 \right)_C$.

- Properties: $c_p \Big|_{\text{exh.}} \left(T_{\text{avg}} = 425°C \right) = 1.075 \text{ kJ/kg·K}$ and $c_{p,\text{air}} \left(T_{\text{avg}} = 100°C \right) = 1.011 \text{ kJ/kg·K}$ ∴ $\dot{W}_T = 1.075 \text{ kW} > \dot{W}_C = \eta_{TC} \dot{W}_T = 1.021 \text{ kW}$.
  Now:

$$T_4 \big|_C = T_3 + \frac{\dot{W}_c}{m_{\text{air}} c_p} := 126°C$$

- Isentropic compression in C: $T_{4s} = T_3 \left(\dfrac{p_4}{p_3} \right)^{\frac{k-1}{k}}$; $k = 1.397$ ∴ $T_{4s} = 379 \text{ K} \cong 106°C < T_{2,\text{real}}$.

  Now:

$$\eta_{\text{isen.}} = \frac{\Delta h_{\text{isen.}}}{\Delta h_{\text{real}}} = \frac{T_{4s} - T_3}{T_{4,\text{real}} - T_3} = 0.642$$

Graphs

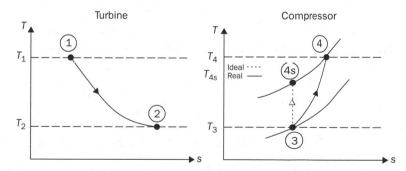

Comments

✓ Turbine pressures are not needed because for IGs $\Delta h \sim \Delta T$ only.

✓ 35.8% more power is required to derive a real compressor compared to the ideal.

Example 3.9 Consider the compression strokes of an ICE (internal combustion engine).

In this ICE, at State 1 the gas has a pressure of 100 kPa, a temperature of 300 K, and a volume of 6.543×10^{-4} m³. At State 2, the gas has a volume of 0.8179×10^{-4} m³. Find the boundary work, $W_b = W_{1-2}\big|_{\text{compression}}$.

Sketch

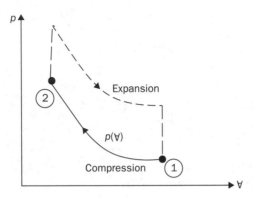

Assumptions

- Closed reversible adiabatic system.
- Air-gasoline mixture $\cong$ I.G.
- Polytropic compression: $p(\forall) = C\forall^{-n}$.

Method $W_{1-2} = \int_1^2 p \, d\forall; \; p(\forall) = C\forall^{-n}; \; n = k = \gamma = \dfrac{c_p}{c_v}$ <isentropic>.

Solution

- Constant: $C = p_1\forall_1^{\gamma}$; mass $m = \dfrac{p_1\forall_1}{RT_1}$.

- $W_{1-2} = C\int_1^2 \dfrac{d\forall}{\forall^{-\gamma}} = p_1\forall_1 \left[\dfrac{\forall^{1-\gamma}}{1-\gamma}\right]_{\forall_1}^{\forall_2} = \dfrac{p_1\forall_1}{1-\gamma}\left(\forall_2^{1-\gamma} - \forall_1^{1-\gamma}\right)$.

- Numbers: $W_{1-2} = -212.2$ J; negative, as work is done onto the surrounding air.

Example 3.10 A 50-kg copper block initially at 140°C is dropped into an insulated tank that contains 90 L of water at 10°C with a density of 997 kg/m³. The specific heat of copper is 0.386 kJ/(kg·°C) and for water, 4.18 kJ/(kg·°C).

Determine the final equilibrium temperature and the total entropy change for the process.

Sketch

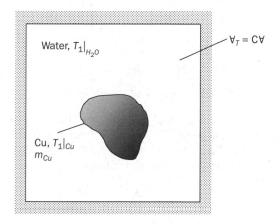

Assumptions

- Transient closed system.
- Incompressible liquid and solid.
- $Q = 0; W = 0; p = ¢.$
- Constant c-values.

Method

- Reduced E-B to find that $T_2 = T_{equ.}$; thus, $0 = \Delta U |_{CV}$.
- $m_{water} = \rho_{water} \forall_T$.
- Gibbs I for $\Delta S_{total} = \Delta S_{Cu} + \Delta S_{water}$.

Solution

- E-B: $\Delta U_{Cu} + \Delta U_W = \left[mc(T_2 - T_1) \right]_{Cu} + \left[mc(T_2 - T_1) \right]_w = 0,$ where $m_{water} =$

  $\rho_{water} \forall_T := 89.73 \text{ kg}.$

  Thus, solving for T_2 yields:

$$T_2 = T_{equ.} = T_{final} = 16.4°C = 289.4 \text{ K}$$

- ΔS_{system}: As $p = ¢$, Gibbs I reduces to:

$$\Delta S \Big|_{\substack{Cu \\ water}} = mc \ln \frac{T_2}{T_1} := \begin{cases} -6.864 \text{ kJ/K} \\ 8.399 \text{ kJ/K} \end{cases}$$

Hence:

$$\Delta S_{total} = \Delta S_{Cu} + \Delta S_{water} = 1.52 \text{ kJ/K}$$

Note

The original S-B:

$$\sum S_{in} - \sum S_{out} + \sum \frac{Q}{T_{surr}} + S_{gen} = \Delta S \big|_{C\forall}$$

Reduces to:

$$0 - 0 + 0 + S_{gen} = \Delta S_{total}$$

Hence, $\Delta S_{total} = S_{gen} > 0$, i.e., a realistic system/process!

Example 3.11   Air is discharged from an insulated tank with a volume of 5 m³, an initial pressure of 500 kPa, and a temperature of 57°C until the pressure reaches 220 kPa. During the process, an electric resistance heater maintains a constant temperature. Find the electric energy supplied as well as the total entropy change during this process.

Sketch

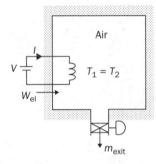

Notes

- $h_{exit} = ¢$, as for I.G.s $h = h(T)$ only but s_{exit} at $p_{exit} = \frac{1}{2}(p_1 + p_2)$.

Assumptions

- Transient open system.
- Air $\cong$ I.G.; hence, $h = h(T)$ only.
- $Q \approx 0$; $W = W_{el}$; $\Delta E_{kin} \approx 0$
- Isothermal quasi-equilibrium process.
- $\Delta E \big|_{system \atop C\forall} \approx \Delta U \big|_{system \atop C\forall}$.

Method

- $m_{exit} = m_1 - m_2$.

- $m = \dfrac{p\forall}{RT}$.

- $W_{el} - (mh)_{exit} = \Delta U \big|_{C\forall}$.

- $(-ms)_{exit} + S_{gen} = \Delta S \big|_{CV}.$
- Property Tables.

- $\Delta s = c_p \ln \dfrac{T_1}{T_{exit}} - R \ln \dfrac{p_i}{p_{exit}}.$

Solution

- M-B: $m_{exit} = m_1 - m_2; \ m_i \dfrac{p_i \forall}{RT}; i = 1, 2.$

$$\therefore m_1 = \frac{500 * 5}{0.287 * 330} = 26.4 \text{ kg and } m_2 = 10.56 \text{ kg} \rightarrow m_{exit} = 15.84 \text{ kg.}$$

- To keep $T = \cancel{c}$, $W_{el} = \Delta U_{tank} + (mh)_{exit}$, where $\Delta U_{tank} = (mu)_2 - (mu)_1.$

$$W_{el} = 10.56 * 235.62 - 26.4 * 235.61 + 15.84 * 330.3 \rightarrow W_{el} = 1501 \text{ kJ}$$

Note This could also be done using $u = c_v(T)$ and $h = c_p(T)$.
S-B: $S_{gen} = \Delta S_{system} + (ms)_{exit}$ or $S_{gen} = m_2(s_2 - s_{exit}) - m_1(s_1 - s_{exit}).$

- Numerical results:
  - Use Property Table values for s and/or Gibbs II.
  - For s_{exit}, we need $p_{exit} \approx \dfrac{1}{2}(p_1 + p_2) = 360 \text{ kPa.}$

    Thus using Gibbs II:

$$\Delta s = s_i - s_{exit} = \underbrace{c_p \ln \frac{T_1}{T_{exit}}}_{\equiv 0} - R \ln \frac{p_i}{p_{exit}}; i = 1, 2$$

$$S_{gen} = 10.56 * 0.1606 - 26.4(-0.1024) = 4.40 \text{ kJ/K}$$

Example 3.12 A piston-cylinder device contains 12 kg of R-134a saturated vapor at 240 kPa. Both 300 kJ of heat and electric work (110-voltage source) are supplied for 6 min until the fluid is at 70°C. Determine the necessary current and sketch the process in a T-v diagram.

Sketch

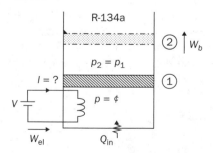

Assumptions

- Closed (transient) system with deforming C∀.
- R-134a $\cong$ 2-phase.
- Q_{in}; $W = W_{el} - W_b$.
- $\Delta E|_{CV} \approx \Delta U|_{CV}$.
- Quasi-equilibrium isobaric expansion.

Method

- $m_{R\text{-}134a} = m = ¢$.
- $Q_{in} + W_{el} - W_b = \Delta U|_{CV}$.
- $W_{boundary} = p\Delta\forall$.
- Property Tables.

Solution

- Properties: $\begin{cases} \text{State 1} := p_1 \text{ and } x_1 > h_1 = h_g|_{p1} = 247.28 \\ \text{State 2} := p_2 = p_1 \text{ and } T_2 > h_2 = 314.51 \end{cases}$.

- E-Bs: $Q_{in} + W_{el} - p\Delta\forall = \Delta U$ or $W_{el} = \Delta U + p\Delta\forall - Q = \Delta H - Q$. Hence:

$$W_{el} = VI\Delta t = m(h_2 - h_1) - Q$$

or:

$$I = \frac{[m(h_2 - h_1) - Q]}{V\Delta t}, \text{ with } 10^3 \text{ VA} = 1 \text{ kJ/s} = 1 \text{ kW}$$

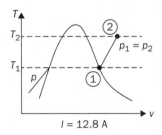

$I = 12.8$ A

Example 3.13 A horizontal cylinder is separated into two compartments by an adiabatic, frictionless piston. One side contains 0.2 m³ of nitrogen, and the other side has 0.1 kg of helium, both initially at 20°C and 95 kPa. Only the helium end of the cylinder is insulated. Now heat is added to the nitrogen from a reservoir at 500°C until the helium pressure reaches 120 kPa.

Determine:

   a. The final helium temperature.

   b. The final nitrogen volume.

c. The amount of heat transferred to the nitrogen.

d. The entropy generation during the process.

Sketch

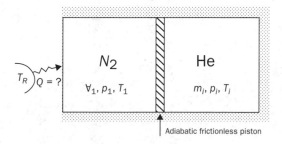

Adiabatic frictionless piston

Assumptions

- Closed system.
- N_2 and He are I.G.s.
- $Q = Q_{in}$; $W = 0$.
- He undergoes isentropic compression, i.e., reversible and adiabatic

Method

- Reduced E-B and S-B.

- $T_2^{He} = T_1 \left(\dfrac{p_2}{p_1} \right)^{\frac{\gamma-1}{\gamma}}$.

- $p\forall = mRT$ for He and N_2.
- Property Tables.

Solutions

- Table B.2 in App. B:
$$\begin{cases} N_2 \rightarrow R = 0.2968\dfrac{kPa \cdot m^3}{kg \cdot K}; \gamma = \dfrac{c_p}{c_v} = 1.4 \\[2ex] He \rightarrow R = 2.0769\dfrac{kPa \cdot m^3}{kg \cdot K}; \gamma = 1.667 \end{cases}.$$

- Final temperature of He $\rightarrow T_2^{He} = T_1 \left(\dfrac{p_2}{p_1} \right)^{\frac{\gamma-1}{\gamma}} = 293 \left(\dfrac{120}{95} \right)^{\frac{0.667}{1.667}}$.

$$T_2^{He} = 321.7 \text{ K}$$

- Final volume of $N_2 \rightarrow \forall_2^{N_2} = \forall_1^{N_2} + \left(\forall_1^{He} - \forall_2^{He} \right)$, where:

$$\forall_{1,2}^{He} = \left(\dfrac{mRT}{p} \right)_{1,2} = \begin{cases} 0.6406 \text{ m}^3 \text{ for } \forall_1^{He} \\ 0.5568 \text{ m}^3 \text{ for } \forall_2^{He} \end{cases}$$

Hence, $V_2^{N_2} = 0.2 + (0.6406 - 0.5568) = 0.2838 \text{ m}^3$.

- *E-B* for cylinder: $Q_{in} = \Delta U_{N_2} + \Delta U_{He} = [mc_v \Delta T]_{N_2} + [mc_v \Delta T]_{He}$.

With $m_{N_2} = \left(\dfrac{p_1 V_1}{RT_1}\right)_{N_2} = 0.2185 \text{ kg}$ and $T_2^{N_2} = \left(\dfrac{p_2 V_2}{mR}\right)_{N_2} = 525.1 \text{ K}$, we'll have

$Q_{in} = 46.6 \text{ kJ}$.

- *S-B* for N_2 compartment as the He gas process is isentropic ($\Delta S|_{He} = 0$).

$$\frac{Q_{in}}{T_R} + S_{gen} = \Delta S|_{N_2} \rightarrow S_{gen} = \Delta S_{N_2} + \Delta S_{surr} = m_{N_2}\left(c_p \ln\frac{T_2}{T_1} - R \ln\frac{p_2}{p_1}\right) - \frac{Q_{in}}{T_R}$$

$$S_{gen} = 0.057 \text{ kJ/K}$$

3.4   Introduction to Cycles and Their Components

A wide variety of cyclic processes is carried out with steady open uniform-flow devices. Naturally, Secs. 3.4 and 3.5 provide the theory and applications of how the 2nd Law relates to such components of simple closed and open cycles. Furthermore, solving simple cycle problems strengthens the understanding of the material discussed so far. If there is extra time for lectures on cycles and/or interest in course projects dealing with more complex cycles, Chap. 5 discusses the steady cyclic conversion of energy in more depth.

Thermodynamic cycles are steady operations where the enclosed fluid, typically subject to phase changes, circles around and around through various components, forming a loop. Cyclic processes include steam-powered plants or appliances such as air conditioners, refrigerators, and heat pumps. The specific components are all steady uniform flow devices, ignoring any (very short) device turn-on and turn-off periods. Examples include:

- *Heat exchangers*, such as boilers, evaporators, condensers, and intercoolers.
- *Expansion devices*, such as turbines and throttling tubes.
- *Compression devices*, such as compressors and pumps.

There are several distinct types of cycles (or cyclic machinery): the heat engine (HE), the refrigerator (R) or air conditioner (AC) or heat pump (HP), and their idealized versions, i.e., the Carnot cycles, abbreviated as C-HE, C-R, C-AC, and C-HP.

(i) Heat Engine (HE)

A well-known form of the heat engine is the steam power plant (see Fig. 3.5) (https://www.youtube.com/watch?v=IdPTuwKEfmA). From the high-temperature reservoir, i.e., the heat source at T_H = constant, heat (Q_H) is supplied to the boiler tubes, which produce *high-enthalpy* steam, which expands in the turbine to generate net work $\left(\dot{W}_{net} \equiv \dot{W}_{HE} = \dot{W}_{turbine} - \dot{W}_{pump}\right)$ to drive a generator (not shown). The exhaust (wet) steam is recycled via the condenser, which, in order to produce water for the pump, rejects heat $\left(\dot{Q}_L\right)$ via cooling into the ambient low-temperature reservoir at T_L.

The pump then closes the clockwise cycle by delivering the water to the boiler at high pressure. The H-reservoir can be due to fossil fuel combustion, solar energy, or nuclear chain reaction, generating high-temperature vapor, while the L-reservoir could be a large water body or the atmosphere turning the vapor (or mixture) back into pure liquid. So the *global energy balance* for a heat engine is simply based on Eq. (2.3):

$$\sum E_{in} - \sum E_{out} = 0 \tag{3.19}$$

Hence, on a rate basis (see Fig. 3.6):

$$\dot{Q}_H - \dot{W}_{HE} - \dot{Q}_L = 0 \text{ or } \dot{W}_{HP} = \dot{Q}_H - \dot{Q}_L \tag{3.20a, b}$$

Thus, based on the definition for a system's thermal efficiency [see Eq. (3.5)], we have:

$$\eta = \frac{\dot{W}_{HE}}{\dot{Q}_H} = 1 - \frac{\dot{Q}_L}{\dot{Q}_H} < 1.0 \tag{3.21a, b}$$

Notes

- $\dot{m}_{steam} = \dot{m} = ¢.$
- $\dot{W}_{HE} = \dot{W}_{turbine} - \dot{W}_{pump}.$
- $\dot{W}_{HE} = \dot{Q}_H - \dot{Q}_L.$

Clearly the lower Q_L and/or the higher Q_H, the better the system's efficiency is. However, there are succinct limits:

- For a cyclic HE operation, heat must be rejected to a cold reservoir, i.e., $\dot{Q}_L = 0$ is impossible; this is known as the Kelvin-Planck statement. Specifically, the

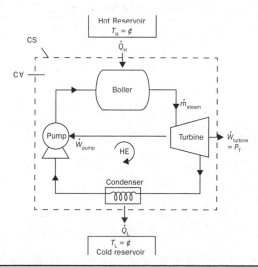

FIGURE 3.6 Schematic diagram of a steam power plant.

exhausted steam exiting the turbine must be fully liquefied in the condenser (i.e., heat rejected to the ambient) for the pump to work and to generate a cyclic process. If the exhaust steam were directly compressed again back to the turbine inlet pressure, more (compressor) work would be needed than ever produced by the turbine.

- Increasing the rate of Q_H is desirable but limited by the type of material the boiler tubes are made of; things may melt.

It should be noted that *mechanical engines* for propulsion (i.e., ICEs, jet engines, etc.) *are not cyclic* as they reject the working fluid (e.g., a gasoline–air mixture) through the exhaust pipe into the environment.

(ii) Refrigerator (R) and Heat Pump (HP)

Refrigerators and heat pumps are lumped together because in principle they are the same counterclockwise cycles (see Fig. 3.7). They differ in terms of the emphasis of *desired heat transfer*. That is, for refrigerators we want $\dot{Q}_L$ to be removed from the R-space to maintain a low temperature, while for heat pumps we want a sufficiently high $\dot{Q}_H$ to warm our living room. In both cases, the energy balance reads:

$$\dot{W}_{in} + \dot{Q}_L - \dot{Q}_H = 0 \tag{3.22}$$

Notes

- $\dot{m}_{refr.} = \dot{m} = ¢.$

- $\dot{W}_{HE} = \dot{W}_{turbine} - \dot{W}_{pump}.$

- $\dot{W}_{compr.} = \dot{Q}_H - \dot{Q}_L.$

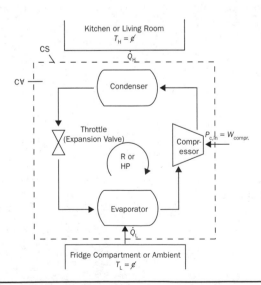

FIGURE 3.7   Refrigeration cycle/heat pump cycle.

Focusing on *Rs*, the working fluid is typically R-134a, which in vapor form is compressed $(\dot{W}_{in})$, resulting in high vapor pressure and temperature. Vapor condensation in the condenser implies heat release, i.e., $\dot{Q}_H$, is rejected into the ambient (notice at home the hot coils behind the fridge). The liquid R-134a expands in a throttling device (a tube with a local hydraulic resistance, e.g., a partially open valve or a porous plug) to a lower pressure and very low temperature. Hence, in the evaporator, a phase change occurs again, where the heat needed for the vaporization (desired $\dot{Q}_L$) is being absorbed from the R-space or freezer.

Focusing on *HPs*, the cycle is the same, except that now $(\dot{Q}_H)$ is typically desired.

Clearly, for both cyclic operations $\dot{W}_{in} = 0$ is impossible, which is known as the *Clausius statement*.

Following the performance definition of *desired Output/required Input*, coefficients of performance (COP) are introduced:

$$COP_R = \frac{\dot{Q}_L}{\dot{W}_{in}} \text{ and } COP_{HP} = \frac{\dot{Q}_H}{\dot{W}_{in}} \text{ where } \dot{Q}_{in} = \dot{Q}_H - \dot{Q}_L \qquad (3.23a\text{–}c)$$

Typically, the COPs are greater than unity, implying that any "thermal efficiency" would not work as a performance description of cycles requiring work-input.

(iii) Carnot HEs, Rs, and HPs

A Carnot cycle is reversible, i.e., no losses, no waste, and hence no irreversibilities. Their purpose is to have an idealized/unattainable goal to compare and judge the performance of *real cycles*. Considering first a *Carnot heat engine (C-HE)*, the ideal cycle runs on reversible isothermal/adiabatic expansions and compressions (see Figs. 3.8 to 3.10). Under reversible conditions, the heat/flow ratio is the same as its corresponding temperature ratio (see Sec. 3.5.1), i.e.:

$$\left(\frac{Q_L}{Q_H}\right)_{rev} = \frac{T_L}{T_H} \qquad (3.24)$$

Hence the *Carnot thermal efficiency* [see Eq. (3.20)], *which is the maximum possible*, reads

$$\eta = 1 - \frac{T_L}{T_H} \qquad (3.25)$$

A counterclockwise C-HE (which works on a reversed Carnot cycle) is a Carnot refrigerator or heat pump, i.e., C-R or C-HP. As mentioned, their performance is measured in terms of the coefficient of performance (COP). The thermal efficiency cannot be employed any longer as the generalized performance definition (*desired Output*)/(*required Input*) for Rs and HPs is greater than one. Following Eq. (3.22a–c) and Eq. (3.23), we have for work-requiring Carnot cycles:

$$COP_{R,rev} \equiv COP_{C-R} = \frac{1}{\dfrac{T_H}{T_L} - 1} \text{ and } COP_{HP,rev} \equiv COP_{C-HP} = \frac{1}{1 - \dfrac{T_L}{T_H}} \qquad (3.26a, b)$$

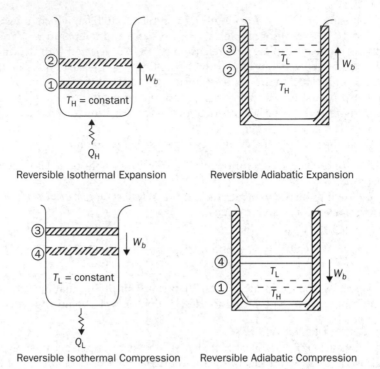

Reversible Isothermal Expansion   Reversible Adiabatic Expansion

Reversible Isothermal Compression   Reversible Adiabatic Compression

Figure 3.8   Stages in a closed Carnot cycle.

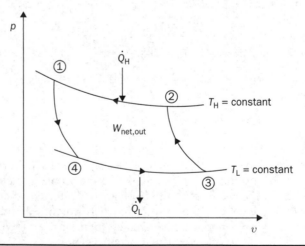

Figure 3.9   Carnot cycle of a heat engine.

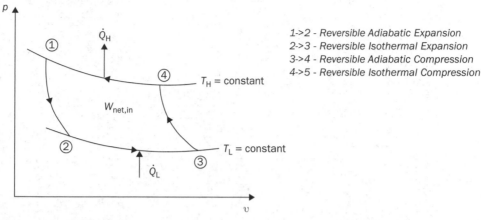

FIGURE **3.10**   Reversed Carnot cycle for a refrigerator.

(iv)

The statement by Clausius was mentioned, i.e., one cannot run a fridge or AC or heat pump without work-input. Another, more subtle observation is the *Clausius inequality for cyclic processes* [see Eq. (3.27) and Sec. 3.5.1]. In a nutshell, focusing on the cyclic heat transfer of, say, a heat engine, and recalling that for a C-HE, $Q_H/T_H - Q_L/T_L = 0$, then mathematically we have, for a *real* HE:

$$\oint \left[\frac{\delta Q}{T}\right]_{real} = \left(\frac{Q_H}{T_H} - \frac{Q_L}{T_L}\right)_{rev} - \frac{Q_{loss}}{T_L} < 0 \qquad (3.27)$$

Clearly, there are two inequalities, i.e., Eq. (3.3) [or Eq. (3.6)] vs. Eq. (3.26), both expressing entropy generation/changes due to irreversibilities causing energy losses and hence a reduction of useful work (see Chap. 4).

It should be noted that mechanical engines, e.g., ICEs and turbojet engines, are not cyclic (but are still called "open cycles") as the fuel–air mixture is just burned, and then the exhaust gases are rejected into the ambient.

3.5   Statements Concerning the Increase-of-Entropy Principle

With the background provided in Sec. 3.4, the underlying theory of the 2nd Law applied to steady closed cycles can now be more readily appreciated.

3.5.1   Clausius Inequality

Consider a Carnot heat engine, e.g., a steam power plant that runs as an internally reversible closed cycle (Fig. 3.11).

Recalling that the absolute temperature scale [see Chap. 1 and Eq. (3.30)] is substance independent, we define for a Carnot cycle (see Fig 3.9 or Fig. 3.11):

$$\eta_{rev} = fct.(T_H, T_L) \equiv 1 - \frac{Q_L}{Q_H} \qquad (3.28)$$

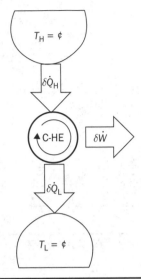

FIGURE 3.11 Differential energy transfer for a Carnot heat engine.

Here, $\eta_{rev} = \eta_{Carnot}$, while T_H and T_L are the high and low reservoir temperatures, respectively. Now defining *the function of* (T_H, T_L) as $1 - \dfrac{T_L}{T_H}$, we can deduce that:

$$\frac{\dot{Q}_H}{T_H} = \frac{\dot{Q}_L}{T_L} \tag{3.29}$$

Equation (3.29) can be employed to: (1) generate an absolute temperature scale T (K) and (2) prove the Clausius inequality, i.e., $\oint \left[\dfrac{\delta Q}{T}\right]_{real} < 0$.

Proof

Setting $T_L \equiv T_{ref} \equiv 273.16$ K, i.e., at 0°C, Eq. (3.29) yields $\dfrac{Q(T_{ref})}{Q(T)} = \dfrac{273.16}{T}$, which sets up the absolute temperature scale:

$$T = 273.16\frac{Q(T)}{Q\big|_{273.16}} \text{ [K]} \tag{3.30}$$

Considering the Clausius inequality, we start with the Clausius statement Eq. (3.1) applied to a reversible closed cycle:

$$\oint dS \equiv \oint \frac{\delta Q}{T}\bigg|_{int.rev} = 0 \tag{3.31}$$

In conjunction with the 1st Law applied to a closed system, i.e., $\delta Q - \delta W = dE$, we have for any cycle $\oint(\delta Q - \delta W) = \oint dE = 0$, as E_1 of State 1 cycles back to the initial state. Specifically, for a Carnot cycle, Eq. (3.31) is proven in light of Eq. (3.29):

$$\oint dS \equiv \oint \frac{\delta Q}{T}\Bigg|_{\text{Carnot}} = \frac{Q_H}{T_H} - \frac{Q_L}{T_L} = 0 \tag{3.32}$$

Now, for a real HE, $Q_L|_{\text{real}}$ can be split into $Q_L|_{\text{rev}} + Q_L|_{\text{loss}}$ where $Q_L|_{\text{loss}}$ is due to irreversibilities. Hence the cyclic integral reads:

$$\oint \frac{\delta Q}{T}\Bigg|_{\text{C–HE}} = \frac{Q_H}{T_H} - \left(\frac{Q_L}{T_L}\Bigg|_{\text{rev}} + \frac{Q_H}{T_H}\Bigg|_{\text{loss}} \right) = -\frac{Q_H}{T_H}\Bigg|_{\text{loss}} < 0 \tag{3.33}$$

The thermal efficiencies of the components of open or closed cycles, i.e., turbine, compressor (or pump or throttle), and heat exchanger are about 80% to 90%. A modern steam power plant runs typically at 45%, whereas solar power plants operate at less than 20% (https://news.energysage.com/what-are-the-most-efficient-solar-panels-on-the-market/). Thus a lot of entropy is inadvertently generated during cyclic processes.

The next section builds on Sec. 3.4 where basic closed cycles, i.e., HE, R, and HP, as well as their Carnot versions were introduced. Specifically Sec. 3.5.2 outlines how important real-world applications, i.e., the Brayton cycles and the Rankine cycles, are.

3.5.2 Brayton and Rankine Cycles

Brayton cycles of gas mixtures are either open, such as car motors and turbojet engines, or closed, e.g., gas power plants. Rankine cycles are closed and representative of steam power plants.

The Brayton Cycle

Figures 3.12a–b depict the schematic of a power plant with the gas turbine being the key element plus the associated *T-s* diagram. Note Heat Process 4 to 1 does not exist in

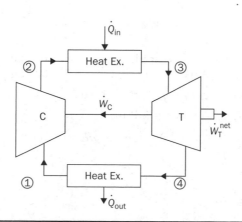

FIGURE 3.12A   Closed Brayton cycle.

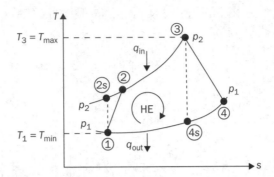

FIGURE 3.12B   Closed Brayton cycle *T-s* diagram with ideal vs. real processes.

open cycles in which the 2-3 heat exchanger is the air–fuel combustor. In order to make a cycle analysis manageable, the following simplifications are assumed:

- The working fluid, i.e., air plus fuel vapor, is air, considered to be an ideal gas.
- All processes are internally reversible, so that process 1-2 is isentropic, 2-3 isobaric, 3-4 isentropic, and 4-1 isobaric again.
- The external combustion process is represented by a fixed heat input, $\dot{Q}_{in} \cong \dot{Q}_{H}$, while the exhaust gases are cooled, i.e., $\dot{Q}_{out} \cong \dot{Q}_{L}$, so that the working fluid returns to the initial state, closing the cycle.

These restrictions are known as the air standard assumptions (ASAs). An important characteristic of all gas power cycles is the *pressure ratio*, $r_p = \dfrac{p_2}{p_1}$, *of the compressor*.

This parameter is used to express the cycle's thermal efficiency:

$$\eta_{Brayton} = \frac{w_{out}}{q_{in}} = 1 - \frac{q_{out}}{q_{in}} = 1 - \frac{c_p(T_4 - T_1)}{c_p(T_3 - T_2)} = 1 - \frac{1}{r_p^{\frac{(n-1)}{n}}} \qquad (3.34a\text{–}d)$$

where $n = k = \dfrac{c_p}{c_v}$ of the gas mixture (or air).

Example 3.14   Consider a simple, closed Brayton cycle for air that operates between the temperature limits of 295 K and 1240 K. The compressor has a pressure ratio of 10 and an efficiency $\eta_{isentr.}^{C} = 0.83$. The turbine in the cycle has an efficiency of $\eta_{isentr.}^{T} = 0.87$. Determine the turbine exit temperature, $T_{T,exit} = T_4$, the net work done by the turbine, w_{net}^{T}, and the thermal efficiency of the cycle, η_{th}.

Sketch

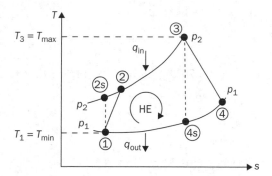

Assumptions

- Steady cycle.
- Air $\cong$ I.G.
- Air-standard assumptions (ASAs) apply.
- $h = h(T)$ only.
- Neglect ΔkE and ΔpE.
- Properties are per $\dot{m}_{air}$.

Method

- Use η_C and η_T to move from isentropic and points to actual temperatures, enthalpies, etc.

- $\eta_C = \dfrac{h_{2s} - h_1}{h_2 - h_1}$.

- $\eta_T = \dfrac{h_3 - h_4}{h_3 - h_{4s}}$.

Solutions

- Pressure ratio $r_p = \dfrac{p_{r2}}{p_{r1}} > p_{r2} = r_p p_{T_1}$.

- Compressors:
  - From air tables for $T_1 = T_{min} = 295\ \text{K} \rightarrow \begin{cases} h_1 = 295.17\ \text{kJ/kg} \\ p_{r1} = 1.3068 \end{cases}$.

    Note: Alternatively, find $h_1 = c_p T_1$ with $c_p \approx 1.006\ \text{kJ/kg·K}$. Clearly, finding accurate c_p-values can be challenging; hence, we use P-T values such as p_r, s^0, etc.

  - Now $p_{r2} = r_p\, p_{r1} := 13.07$; so that $h_{2s} = 570.26\ \dfrac{\text{kJ}}{\text{kg}}$ and $T_{2s} = 564.9\ \text{K}$.

  - From η_C we find $h_2 = h_1 + \dfrac{h_{2s} - h_1}{\eta_C} = 626.60\ \dfrac{\text{kJ}}{\text{kg}}$.

- Turbine:

  - $T_3 = T_{max} = 1240 \text{ K} \begin{cases} h_3 = 1324.93 \text{ kJ/kg} \\ p_{r3} = 272.3 \end{cases}$.

    Hence, $p_{r4} = 27.23$ and so $h_{4s} = 702.07 \dfrac{\text{kJ}}{\text{kg}}$ and $T_{4s} = 689.6 \text{ K}$.

  - From, η_T we determine $h_4 = h_3 - \eta_T (h_3 - h_{4s}) := 783.04 \dfrac{\text{kJ}}{\text{kg}}$.

    Interpolation w. $| h_4 = 783.04$ yields $T_4 = T_{exit} = 764.4 \text{ K}$.

- Cycle:

  - $\eta_{th} = \dfrac{w_{net}}{q_{in}}; \quad w_{net} = q_{in} - q_{out} = h_3 - h_2 - (h_4 - h_1) = 210.4 \dfrac{\text{kJ}}{\text{kg}}$

    where $q_{in} = h_3 - h_2 = 698.3 \dfrac{\text{kJ}}{\text{kg}}$.

  - Finally, $\eta_{th} = \dfrac{210.4}{698.3} = 0.3014$ or 30.1%.

Comments

✓ Using experimental data from the tables, provide accurate results much faster.

✓ The cycle's efficiency is very low, although the compressor and turbine losses are only 65% of the work input and output, respectively.

✓ Find $\eta_{isentropic}^{cycle}$ and compare it to $\eta_{th} = 30.1\%$.

✓ Develop an equation for $\eta_{th}(r_p)$ plot the graph for $2 \leq r_p \leq 20$ and find the maximum η_{th}.

The Rankine Cycle

The Rankine cycle is the heart of steam power plants (see Fig. 3.13). In reality, pressure drops occur across the two heat exchangers, and both turbine and pump generate entropy. However, for an initial analysis, isobaric and isentropic processes are assumed (see ASAs). Thus realism is being introduced via turbine and pump efficiencies, i.e.,

$$\eta_T = \frac{W_{actual}}{W_{isentropic}} \text{ and } \eta_p = \frac{W_{isentropic}}{W_{actual}} .$$

Recall, assuming no heat losses, $\eta_{th} = \dfrac{W_{net}}{q_{in}}$ where $w_{net} = w_{turbine} - w_{pump} = q_{in} - q_{out}$.

Specifically for the isentropic processes:

$$w_p^{is} = h_{2s} - h_1 = v(p_2 - p_1) \tag{3.35a, b}$$

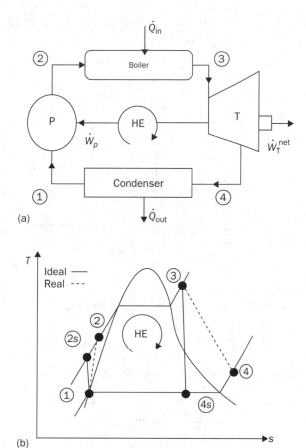

Figure 3.13 Simple Rankine cycle with T-s diagram showing (a) simple Rankine cycle and (b) associated ideal and real T-s diagrams.

and

$$w_T^{is} = h_3 - h_{4s} \tag{3.36}$$

$$q_{in} = h_3 - h_{2s} \tag{3.37}$$

Now, to turn the ideal results into actual work needed (compressor) or work performed (turbine), measured isentropic efficiencies are used. Hence, for both pump and turbine:

$$w_{actual} = \eta_{isentropic} w^{is} \tag{3.38}$$

Example 3.15 Consider an idealized 210-MW steam power plant. As an aside, that's a simple Rankine cycle with isentropic expansion, i.e., the turbine, and isentropic compression, i.e., the pump. The steam enters the turbine at 10 MPa and

500°C and is cooled in the condenser at a pressure of 10 kPa, i.e., the turbine exit condition. Find x_4, $\eta_{th,cycle}$ and $\dot{m}$.

Sketch

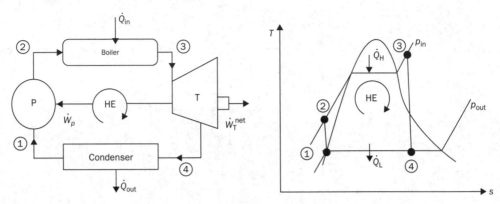

Assumptions

- Steady operation w. | isentropic turbine and pump.
- H_2O two-phase.

Method

- Take each of the four devices separately:

$$w_p = v_1(p_2 - p_1); \ w_T = q_{in} - q_{out}, \text{ where}$$

$$q_{in} = h_3 - h_2 \text{ and } q_{out} = h_4 - h_1$$

$$\eta_{th} = \frac{w_T}{q_{in}} \text{ and } \dot{m} = \frac{\dot{W}_T}{w_T}.$$

- Property Tables.

Solution

- Property Values:

State 1: $h_1 = h_{f \text{ at } p_1} = 191.81\frac{kJ}{kg}; \ v_1 = v_{f \text{ at } p_1} = 0.00101\frac{m^3}{kg}.$

With $w_{pump} = v_1(p_2 - p_1) := 10.09\frac{kJ}{kg} = h_2 - h_1 > h_2 = h_1 + w_{pump} = 201.9\frac{kJ}{kg}.$

State 3: p_3 and $T_3 \rightarrow h_3 = 3375.1\frac{kJ}{kg}$ and $s_3 = 6.5995\frac{kJ}{kg \cdot K}.$

State 4: p_4 and $s_4 = s_3 \rightarrow x_4 = \dfrac{s_4 - s_f}{s_{gf}} = 0.7934.$

Thus, $h_4 = h_f + x_4 h_{fg} = 2089.7\frac{kJ}{kg}.$

- Finally:

$$q_{\text{in}} \sim Q_{\text{H}} = h_3 - h_2 = 3173.2 \frac{\text{kJ}}{\text{kg}};\ q_{\text{out}} \sim Q_{\text{L}} = h_4 - h_1 = 1897.9 \frac{\text{kJ}}{\text{kg}}$$

$$\therefore w_{\text{T}} = q_{\text{in}} - q_{\text{out}} = 1275.4 \frac{\text{kJ}}{\text{kg}}$$

And:

$$\eta_{\text{R-cycle}} = \eta_{\text{th}} = \frac{w_{\text{net}}}{q_{\text{in}}} = 40.2\%$$

And:

$$\dot{m}_{\text{steam}} = \frac{\dot{W}_{\text{T}}}{w_{\text{T}}} = \frac{210{,}000\ \text{kJ/s}}{1275.4\ \text{kJ/kg}} = 164.7\ \text{kg/s}$$

Example 3.16: Simplified Turbojet Engine Analysis

Background Information Gas turbines, because of a high power-to-weight ratio, are key components in jet engines (see Fig. 3.14). The following video link illustrates the working of a turbojet engine: https://www.youtube.com/watch?v=SZ-hCHGoyn4.

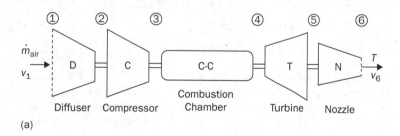

(a)

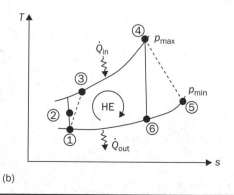

(b)

FIGURE 3.14 Components of a turbojet engine with accompanying *T*-s diagram: (a) components of a turbojet engine; (b) idealized *T*-s diagram.

Note   The *T-s* diagram also indicates entropy generation plus assumed heat rejection (6)–(1). While the turbine power is used to drive the compressor and auxiliary equipment, its high-pressure exit gas mixture is accelerated in the nozzle to generate the airplane thrust.

Specifically, the propulsion power is:

$$\dot{W}_{propulsion} = F v_{aircraft} = \dot{m}\left(v_{exit} - v_{inlet}\right) v_{aircraft}$$

so that

$$\eta_{propulsion} = \frac{\dot{W}_P}{\dot{Q}_{in}}$$

Here F is the thrust of the turbojet, $\dot{m}$ is the gas-mixture (i.e., air) mass flow rate, v_{exit} is the nozzle exit velocity, v_{inlet} is the diffuser inlet velocity, and $v_{aircraft}$ is the plane's speed.

Problem Statement   An aircraft is flying at 280 m/s at an altitude of 9150 m where $p_{amb} = 30$ kPa and $T_{amb} = -32°C$. The compressor pressure ratio $r_p = 12$ with air entering at $\dot{m}_c = 50$ kg/s. The turbine inlet temperature is 1100 K. The heating value of the jet fuel is 42,700 kJ/kg. Assuming $c_p = 1.005$ kJ/kg·K, $n \equiv k = 1.4$ and ideal process conditions, find:

1. The velocity of the exhaust gases.

2. The propulsion power.

3. The rate of fuel consumption.

Solution   Following Newton's 1st Law of Motion, we assume the plane is stationary and the air is approaching with $v_1 = 280$ m/s. The key open cycle components are treated separately, using the process state numbers of the *T-s* diagram. All devices are steady open uniform flow systems. The air-fuel-vapor mixture is approximately an ideal gas.

i. Diffusor:

E-B: $\sum \dot{E}_{in} = \sum \dot{E}_{out}$,   i.e., $h_1 + \dfrac{v_1^2}{2} = h_2 + \dfrac{v_2^2}{2}$ where $h = c_p(T)$ so that

$$c_p(T_2 - T_1) = \frac{v_1^2}{2}(v_2 \approx 0).$$

Hence $T_2 = T_1 + \dfrac{v_1^2}{Rc_p} := 280$ K.

Now, under isentropic conditions, $p_2 = p_1\left(\dfrac{T_2}{T_1}\right)^{\frac{\gamma}{\gamma-1}} := 54.10$ kPa.

ii. Compressor:

Use the given $r_p = \dfrac{p_3}{p_2}$, $p_3 = 649.2$ kPa $:= p_4$, assuming pressure drop across the combustor.

Again, $T_3 = T_2 \left(\dfrac{p_3}{p_2}\right)^{\frac{\gamma-1}{\gamma}} := 569.5$ K.

iii. Turbine:

Assuming that the gas turbine power is solely used to drive the compressor,
$w_T = w_c$ or $h_4 - h_5 = h_3 - h_2$ or $c_p\Delta T_T = c_p\Delta T_C$.

Hence, $T_4 - T_5 = T_3 - T_2$ or $T_5 = T_4 - T_3 + T_2 := 810.5$ K.

iv. Nozzle:

Again, being an adiabatic and reversible device:

$$\sum \dot{E}_{in} = \sum \dot{E}_{out} \text{ or } h_5 + \frac{v_5^2}{2} = h_6 + \frac{v_6^2}{2} \ (v_5 \approx 0)$$

Hence with $h = c_p T$; $c_p(T_5 - T_6) = \dfrac{v_6^2}{2}$, where $T_6 = T_4 \left(\dfrac{p_6}{p_4}\right)^{\frac{\gamma-1}{\gamma}} := 465.5$ K with $T_4 = 1100$ K.

a. Now, solving for $v_6 = v_{exit}$ and recalling that $\dfrac{1\text{ kJ}}{\text{kg}} \cong 10^3 \dfrac{\text{m}^2}{\text{s}^2}$, $v_6 = 832.7\dfrac{\text{m}}{\text{s}}$.

b. $\dot{W}_p = \dot{m}(v_6 - v_1)v_{aircraft} = 50(832.7 - 280)280 = 7738$ kW.

c. $\dot{Q}_{in} = \dot{m}(h_4 - h_3) = \dot{m}c_p(T_4 - T_3) := 26{,}657 \dfrac{\text{kJ}}{\text{s}}$, so that:

$$\dot{m}_{fuel} = \frac{\dot{Q}_{in}}{HV} = \frac{26{,}657}{42{,}700} = 0.6243 \frac{\text{kg}}{\text{s}}$$

Comments

✓ In order to close the cycle theoretically, T_6 has to be reduced to T_1!

✓ The engine is a real fuel guzzler.

✓ Note that $\eta_{prop.} = \dfrac{\dot{W}_p}{\dot{Q}_{in}} := 29\%$.

✓ Check out online videos of jet propulsion cycles.

✓ For a more realistic analysis, assume $\eta_{turbine} = 85\%$ and $\eta_{compression} = 80\%$.

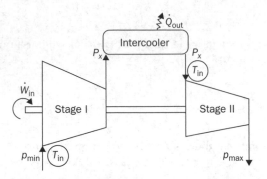

FIGURE 3.15   Schematics of a two-stage compressor system with intercooler.

3.5.3   Multistage Compressors

It should be apparent by now that compressors are (next to gas turbines) key components in open and closed Brayton cycles as well as in turbojet engines. Furthermore, they are stand-alone devices wherever high-pressure air is needed. For example, Fig. 3.15 depicts a two-stage compressor system with intercooling, i.e., via the use of a heat exchanger.

In contrast to pumps that deal with liquids (see Fig. 3.6), compressors (like blowers or fans) deal with gases. While fans produce airflow with small pressure ratios (i.e., $r_p < 2$), compressors achieve r_p-values greater than 10.

The compressor (or flow) work per unit mass is:

$$w_c = \int_1^2 v\, dp \tag{3.39a}$$

where $v(p)$ is given by the polytropic process relationship:

$$v^n p = C \tag{3.39b}$$

As mentioned, the polytropic exponent ranges from $n = 1$ (isothermal process) to $n = k \equiv \dfrac{c_p}{c_v}$ (isentropic process). Clearly, during the compression process, the gas temperature increases and the gas entropy changes. Considering an ideal gas and a reversible process, the solution of Eqs. (3.39a) and (3.39b) read (see Fig. 3.16):

$$w_c = \frac{nR(T_2 - T_1)}{n - 1} = \frac{nRT_1}{n - 1}\left[\left(\frac{p_2}{p_1}\right)^{\frac{n-1}{n}} - 1\right] \tag{3.40}$$

For $n = 1$, the work needed for isothermal compression is:

$$w_c^{\text{isoth}} = C^n \int_1^2 \frac{dp}{p^n} = RT_1 \ln\frac{p_2}{p_1} \tag{3.41}$$

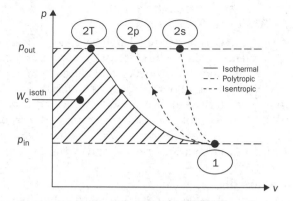

FIGURE 3.16   *p-v* diagram of a compressor for varying processes.

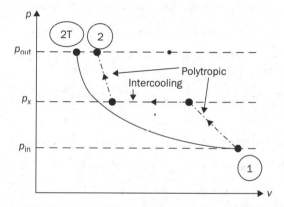

FIGURE 3.17   *p-v* diagram of a multistage compression with an intercooling stage.

It is evident that $w_c^{\text{isothermal}} \cong w_c^{\text{MIN}}$, implying that massive controlled gas cooling would be required. A practical solution is multistage *compression with intercooling*, where Fig. 3.17 depicts a two-stage process as a sample application. Here:

$$w_c = w_c^{\,\text{I}} + w_c^{\,\text{II}} = \frac{nRT_1}{n-1}\left\{\left[\left(\frac{p_2}{p_1}\right)^{\frac{n-1}{n}} - 1\right] + \left[\left(\frac{p_2}{p_x}\right)^{\frac{n-1}{n}} - 1\right]\right\} \qquad (3.42)$$

The result is $p_x = \sqrt{p_1 p_2}$, i.e., the intermediate pressure value (being the geometric mean) that minimizes the total compressor work needed. Clearly, adding compressor stages with multiple intercooling would move the system closer to $w_c^{\text{isothermal}}$.

Clearly, energy saving and hence cost reduction are major challenges in thermal engineering and the material sciences. This chapter and Chap. 4 provide some necessary technical background and incentives to tackle such interdisciplinary challenges, perhaps addressing the topics via course projects.

3.6   Homework Assignments

Again, Part A probes the reader's knowledge base, i.e., a deeper understanding of the concept of entropy, entropy generation, as well as applications in term of ΔS_{system} and S_{gen} calculations for all kinds of basic open and closed systems or devices. In addition, basic questions on cycles round out the Chap. 3 material.

Part B tests the reader's skill level to solve basic 2nd-Law problems. It should be apparent that all Chap. 3 problems are for the most part a repeat of the problems discussed in the previous chapters. However, the additional tasks are finding ΔS_{system} (or S_{gen}) plus the efficiencies for various devices and cycles.

3.6.1   Part A: Insight

Recall that a sketch, some math and descriptions/explanations are required.

1   Explain Eq. (3.1) on a microscopic level and discuss Eq. (3.1) vs. Eq. (3.2).

2   Provide the three balance equations for the transient open system depicted in Fig. 3.2.

3   Explain how the 2nd Law, being an inequality, can be turned into a balance equation.

4   Considering a polytropic process of an ideal gas under isentropic condition, derive a relationship between the pressure ratio and the specific volume ratio.

5   Considering Example 3.2, find S_{gen} in symbolic form. Comment!

6   Why is it important to include $\Delta S_{surr}^{\cdot}$ when computing S_{gen}?

7   Discuss Eq. (3.8) and provide real-world examples on how to reduce S_{gen}.

8   What is the difference between an insulated and an isolated system or device?

9   For an internally reversible process depict $Q = \int_1^2 \delta Q$ in a T-s diagram.

10   Plot polytropic process paths for $0 \le n \le \gamma \equiv k = \dfrac{c_p}{c_v}$ in both p-v and T-s diagrams.

11   Contrast HE-cycles vs. R-cycles vs. HP-cycles; for the latter, show winter vs. summer operation.

12   Why are ICEs and jet engines not operating as true cycles?

13   Explain that, for any HE to work, heat rejection and hence a cold reservoir (known as a TER) are required.

14   Show mathematically and graphically in an h-s diagram that for an adiabatic turbine $P_T = \dot{m}(h_{in} - h_{out})$. How can this equation be used for device improvements?

15   Discuss the different COPs for a refrigerator and a heat pump. What are the typical values, and how can they be improved?

16   Derive Eq. (3.24)!

17   Discuss closed vs. open cycles. For example, how does the open Brayton cycle differ from the jet propulsion cycle?

18 What does the area enclosed by a cycle represent in (i) a T-s diagram and (ii) a p-$\forall$ diagram?

19 Determine, with proper justification, the entropy change of H_2O in a p-c device containing wet steam at 300 kPa after 750 kJ of heat has been transferred.

20 Consider a reversible heat engine receiving heat from a finite body (m, c; initially at T_1) and rejecting waste heat to a second body (same m and c, but initially at T_2). Now, work is produced until the two bodies are at $T_{equilibrium}$. Develop an equation for $T_{eq.}(T_1, T_2)$. Comment!

3.6.2 Part B: Problems

21 Given: A metal wire, i.e., a cylinder 1 mm by 25 mm with a density of 20,000 kg/m³ and a heat capacity of 150 J/kg·K. It is now subjected to an electric current ($I = 1.0$ A and $V = 110$ V). Assuming $T_{initial} = 25°C$, find an expression for the value of the initial temperature rise in K/s. What additional info would be needed to compute $T(t)$ realistically? How would you determine T_{final} and find S_{gen}? Plot an anticipated $T(t)$ graph.

22 Via a supply line, air at 500 kPa and 70°C is pressed into an insulated p-c device, initially with 0.40 m³ containing 1.3 kg of air at 30°C. The process stops when the cylinder volume has increased by 50%. Assuming constant specific heats [i.e., $c_v = 0.718$ kJ/(kg·K) and $c_p = 1.005$ kJ/(kg·K)], develop an equation for T_{final}, solve for T_{final}, and find the entropy generated.

23 An insulated, vertical, spring-loaded p-c device contains R-134a initially with 0.8 m³ at 1.4 MPa and 120°C. A valve is now opened, the (linear) spring unwinds, and fluid escapes until $V_2 = 0.5$ m³ and $p_2 = 0.7$ MPa. Find m_{out} and T_{final} as well as ΔS_{system} and S_{gen}.

24 An adiabatic steam turbine (12.5 MPa and 500°C in; 10 kPa and $x = 0.92$ out with 25 kg/s) drives an insulated air compressor (98 kPa and 295 K in and 1 MPa and 620 K out with 10 kg/s). Determine the net turbine power as well as the entropy generation within the turbine plus compressor. [*Recall*: As $c_p = c_p(T)$ for air, you may want to use the tabulated $s°$-values in Gibbs II.] Graph the two processes!

25 Consider Tank I (0.2 m³, 400 kPa, $x_{steam} = 0.80$) undergoing isentropic expansion to 300 kPa while connected to Tank II, which initially had 3 kg of steam at 200 kPa and 250°C. Note that 600 kJ is lost to an ambient of 0°C.

   i.   Determine the final temperatures in each tank.

   ii. Draw the individual tank process paths in a T-s diagram.

   iii. Compute the total entropy generated.

26 For a (tube-and-shell) heat exchanger, steam enters at 1 MPa and 200°C and leaves as saturated liquid at 1 MPa, while water enters at 2.5 MPa and 50°C and leaves 10°C below the steam exit temperature. Based on reasonable assumptions, determine the steam-to-water mass flow rate ratio as well as the total entropy per unit mass of water generated.

27 Consider a two-stage adiabatic turbine (15 kg/s, 6 MPa, 500°C) where 10% of the exit steam is diverted after the first stage (with an exit pressure of 1.2 MPa) while the rest expands in the second stage to 20 kPa. Determine the turbine power for (i) a reversible process and (ii) a process with 88% isentropic efficiency.

28 Consider a refrigerated water fountain for 20 employees that cools water from 22°C to 8°C at a supply rate of 0.4 L/hr/person and absorbs 45 W from the ambient of 25°C.

    i.   Sketch a design of this system.

   ii.   If the COP of the refrigeration system is 2.9, determine the compressor power [W] needed.

29 Consider a horizontal cylinder with two compartments separated by an adiabatic, frictionless piston. Compartment A contains 0.2 m³ of nitrogen while Compartment B has 0.1 kg of helium—both initially at 20°C and 95 kPa. The cylinder wall plus the helium end are insulated. Now heat from a 500°C source is added to the nitrogen until the helium pressure has reached 120 kPa. Determine the final states of the two gases as well as the amount of heat transferred and entropy generated.

30 Steam enters a two-stage adiabatic turbine at 500°C and 6 MPa with 15 kg/s. After the first stage 10% of the steam at 1.2 MPa is diverted, while the remainder expands in the second stage to 20 kPa. Given an isentropic efficiency of 0.88, find the total entropy change and the actual power output. Show both the ideal and real processes in a *T-s* and an *h-s* diagram.

3.6.3   Part C: Course Projects on Entropy and 2nd-Law Applications

Volunteer C-P Reports plus PowerPoint slide presentations of the results are fine learning experiences. As former participating students reported, such presentations strengthened the knowledge base and skill level, prepared for future report writing in graduate school or on the job, and occasionally impressed recruiters sufficiently to gain job offers. For in-class students, the advantage is mainly a possible course grade improvement.

    As a first step, the Guidelines for Report Writing (App. C) should be read. Possible Chap. 3 projects, some of which may also require a cost-benefit analysis, are:

1 Alternative interpretations and applications of the entropy principle, for example, in power generation, propulsion, climate change, cosmology, a supermarket, a house, etc.

2 Representative sample problem solutions out of *nanothermodynamics*, which deals with size-dependent changes in thermal properties of nanoparticles or nanofilms, including entropy changes/generation. Note that $\varepsilon =$ (surface area)/(system volume) $\sim \ell^2/\ell^3 = \ell^{-1}$ approaches ∞ as the particle diameter (or film thickness) goes to zero.

3 Thermodynamics as well as cost aspects of different energy sources for power generation or propulsion, i.e., fossil fuel vs. renewable *vs.* nuclear. Discuss ideas for power maximization.

4 Low-grade thermal energy harvesting; for example, the organic Rankine cycle (ORC) using an organic fluid with $\eta = 10\%–20\%$, thermoelectric generators (TEGs) based on the Seebeck (power) or Peltier (cooling) effect with $\eta = 4\%–10\%$, superheated CO_2 as the working fluid in Brayton cycles, etc. How can the 2nd-Law efficiency be improved?

5 Thermodynamics and *total cost* analysis of cars powered by ICEs vs. hybrid power vs. electric motor.

CHAPTER 4

Exergy or Minimizing Energy Waste

Exergy, denoted as EX [kJ], quantifies useful (or available) work. Obviously, given an amount of energy supplied to a thermal device, only a certain percentage can be turned into actual work, while the rest, i.e., destroyed exergy, is lost due to irreversibilities (see Chap. 3). Using a turbine as an example, high T and high p steam expands within the device, performing work for which we state:

$$W_{\text{reversible (ideal turbine)}} - W_{\text{useful (real turbine)}} \equiv EX_{\text{destroyed}} \sim S_{\text{gen}} \qquad (4.1)$$

Clearly, minimizing S_{gen} (or $\dot{S}_{\text{gen}}$ on a rate basis) via smart device design and optimal device operation would lead to improved work (or power) output and hence would increase the useful work EX (or $\dot{EX}$). A measure of such device performance would also be the 2nd-Law efficiency:

$$\eta_{\text{II}} = \frac{\text{actual thermal efficiency}}{\text{max possible efficiency}} = \frac{\eta_{\text{th}}}{\eta_{\text{rev}}}$$

$$= \begin{cases} \dfrac{W_{\text{useful}}}{W_{\text{reversible}}} & \text{(e.g., turbines)} \\[2ex] \dfrac{W_{\text{reversible}}}{W_{\text{useful}}} & \text{(e.g., compressors)} \\[2ex] \dfrac{COP_{\text{act}}}{COP_{\text{rev}}} & \text{(e.g., heat pumps)} \end{cases} \qquad (4.2a\text{–}d)$$

Here $\eta_{\text{rev}} = \eta_{\text{Carnot}}$ and $W_{\text{useful}} \hat{=} EX$. Now, the tasks are to develop an exergy balance, following Eqs. (1.10a) and (1.10b), and to determine energy waste, i.e., $EX_{\text{destroyed}} \equiv \|W_{\text{ideal,out}} - W_{\text{real,out}}\| \hat{=} EX_{\text{sink}}$ in Eq. (1.10b).

In summary:

- Exergy is the maximum theoretical work obtainable when processing a system from a given state to the "dead state," i.e., the surrounding at T_0, p_0, v_0, etc., where the system would be at a thermomechanical equilibrium with its environment.

- Exergy is an extensive property and nonconservative because of inherent irreversibilities. Thus:

$$EX = (U - U_0) + p_0 (\forall - \forall_0) - T_0(S - S_0) + \Delta kE + \Delta pE \tag{4.3}$$

The use of Eq. (4.3) on a per unit mass basis can be illustrated with an idealized p-c device with $\forall_{air} = 2450$ cm^3, $p = 7$ bar, and $T = 867°C$. The ambient, i.e., the dead state, is at $T_0 = 27°C$ and $p_0 = 1.013$ bar. To obtain the specific exergy, $ex = EX/m$, it is assumed that air is an ideal gas and that $\Delta ke + \Delta pe$ can be neglected for this closed system. Clearly, we can represent specific exergy, i.e., the maximum useful work per unit mass relative to the ambient (dead) state, as:

$$ex = (u_1 - u_0) + p_0(v_1 - v_0) - T_0(s_1 - s_0)$$

where u_1, v_1, and s_1 indicate the initial state. Applying the IG law and using the property values, we obtain:

$$p_0(v_1 - v_0) = -38.75 \text{ kJ/kg}$$

In addition (from the Property Tables in App. B):

$$u_1 - u_0 = 666.28 \text{ kJ/kg and } T_0(s_1 - s_0) = 258.62 \text{ kJ/kg}$$

Hence, the specific work potential is $ex = 368.9$ kJ/kg.
Now, there are two possible scenarios for the work potential:

1. Opening the cylinder via a valve and letting the gas expand to the dead state, the remaining work potential would be $W_{useful} = 0$, as $ex = 0$.
2. Using the gas to run a turbine will yield $W_{useful} = m_{air} ex$.

4.1   Macroscale Exergy Balance

Formally, Eq. (1.10b) can be rewritten as a general exergy balance for process (or observation time) Δt:

$$\underbrace{\sum EX_{in} - \sum EX_{out}}_{\substack{\text{Net } EX\text{-transfer via} \\ \text{heat, work and mass flow}}} - \underbrace{EX_{destroyed}}_{\substack{\text{due to } S_{gen} \\ \text{(irreversibilities)}}} = \Delta EX|_{system}$$

$$= \underbrace{(EX_{final} - EX_{initial})_{system}}_{\substack{\text{Net } EX\text{-change over } \Delta t \\ \text{inside CV/system/device}}} \tag{4.4}$$

Clearly, exergy is opposite to entropy, as it can only be destroyed, not created; thus the $EX_{destroyed}$ term is negative while S_{gen} is positive in the S-B (see Sec. 3.1).
For steady open uniform flow systems, the right-hand side of Eq. (4.4) is zero, so that on a rate basis:

$$\dot{EX}_{destroyed} = \sum \dot{EX}_{in} - \sum \dot{EX}_{out} \tag{4.5}$$

Figure 4.1 illustrates the connections between the 1st and 2nd Laws governing any system, and interactions with the ambient at $T_0 = T_{surr}$, $p_0 = p_{surr}$, and $v_0 = 0$ [see Eq. (4.4)].

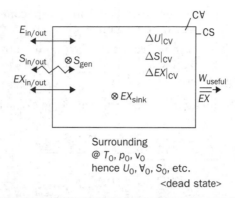

FIGURE 4.1 System schematic listing parameters for the exergy balance.

The ambient (or surrounding of the system) is called the dead state because energy input at T_0, p_0, and v_0 to a device would produce zero work; for example, a turbine would not turn a blade because energy input at the ambient conditions minus output at the same conditions, i.e., T_0, p_0, and v_0, would be zero. Hence exergy can be defined as maximum obtainable work from a process where the system reaches the dead state; i.e., it reaches thermodynamic equilibrium. The ways for obtaining $EX|_{\text{destroyed}}$ $(EX|_d)$ can be threefold:

$$EX|_d = \begin{cases} T_0 S_{\text{gen}}; \ T_0 \,\widehat{=}\, T_{\text{surr}} \\ \text{from } EX - \text{balance (4.4) or (4.10)} \\ \|W_{\text{rev.OUT}} - W_{\text{real.OUT}}\| \end{cases} \quad (4.6a\text{–}c)$$

Specific forms of exergy transfer are due to heat supplied and work done as well as net energy delivered via mass transfer. Thus:

$$EX_{\text{heat}} = \eta_{\text{rev}} Q = \left(1 - \frac{T_0}{T}\right) Q; \quad (4.7)$$

$$EX_{\text{work}} = \begin{cases} W - W_{\text{surr}} \ \text{for boundary work} \\ W \ \text{for other work forms} \end{cases} \quad (4.8)$$

and

$$EX_{\text{mass}} = m\left[(h - h_0) - T_0(s - s_0) + \frac{v^2}{2} + gz\right] \quad (4.9)$$

As for the previous balances, Eqs. (4.7) to (4.9) describe the three different modes of transfer, here for energy. Looking ahead, for Eq. (4.10) we sum up all IN–OUT contributors to generate a realistic net EX, i.e., $\Delta EX_{\text{system}}$. Specifically:

- EX_{heat} of Eq. (4.7) is the maximal percentage of heat supply, i.e., Q at source temperature T, for useful work generation, as $\eta_{\text{rev}} = \eta_{\text{Carnot}}$. Note that typically $T = T_{\text{source}} \gg T_{\text{ambient}} = T_0$.

- EX_{work} of Eq. (4.8) is the direct available work contributor, where, for example, $W_{boundary}$ pushes against the ambient so that $W_{net} = W_{boundary} - W_{surr}$.
- EX_{mass} of Eq. (4.9) provides net energy (minus entropy) input via mass transfer relative to the dead state conditions, assuming v_0 and z_0 to be zero here, i.e., neglecting ΔkE and ΔpE.

Now, summing up the three contributors, Eq. (4.4) can be rewritten as (see Fig. 4.2):

$$\underbrace{\sum\left(1 - \frac{T_0}{T}\right)Q_k}_{EX_{heat} \Rightarrow W_{useful}^{heat}} - \underbrace{\left[W - p_0(\forall_{final} - \forall_{initial})\right]}_{EX_{work} \Rightarrow W_{useful}^{out+boundary}}$$

$$+ \underbrace{\sum_{IN}EX_{mass} - \sum_{OUT}EX_{mass}}_{EX_{mass}^{net} \Rightarrow W_{useful}^{mass}} - \underbrace{EX_{destroyed}}_{=T_0 S_{gen}} = \underbrace{\Delta EX|_{C\forall}}_{\substack{\text{Change of} \\ \text{EX inside C}\forall}} \qquad (4.10)$$

Specifically, the right-hand side of Eq. (4.10) indicates the exergy change within the $C\forall$, i.e.:

$$\Delta EX|_{system} \approx \Delta U + p_0\Delta\forall - T_0\Delta S \qquad (4.11)$$

where $\Delta \triangleq$ final (2) – initial (1). For unit mass and recalling that $\Delta u + p\Delta v = \Delta h$, we obtain:

$$\Delta ex = \frac{\Delta EX}{m} = h_2 - h_1 - T_0(s_2 - s_1) \qquad (4.12)$$

Clearly, for steady open systems, we have:

$$\sum\left(1 - \frac{T_0}{T_k}\right)\dot{Q}_k - \dot{W} + \sum_{IN}\dot{m}(h - T_0 s) - \sum_{OUT}\dot{m}(h - T_0 s) - E\dot{X}_d = 0 \qquad (4.13)$$

It should be noted that, for reversible processes/devices, $E\dot{X}_d \equiv 0$ as no entropy is generated.

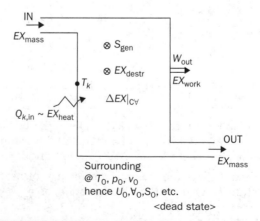

Figure 4.2   Exergy balance application sketch for transient open system.

4.2   Illustrations of the Exergy Balance

Following the deductive learning approach, the general EX balance is first applied to a transient open system for which $EX|_{destroyed}$ has to be found [see Eq. (4.4) or (4.10)]. This example is followed by a steady open system for which the theoretically maximum work output and the associated 2nd-Law efficiency have to be calculated. Additional sample problem solutions, *if done independently*, should deepen the knowledge base and elevate the skill level.

Example 4.1   A container of 0.6 m³ filled with saturated water at 170°C suddenly opens up via a valve and, during process time Δt, releases one-half of the water at approximately constant specific enthalpy and entropy. The temperature in the tank remains constant due to heat supply from a source at 210°C. Determine Q_{in}, S_{gen}, and $EX|_{destroyed}$, assuming the dead state to be at $T_0 = 25°C \triangleq T_{surr}$ and $p_0 = 100$ kPa.

Sketch

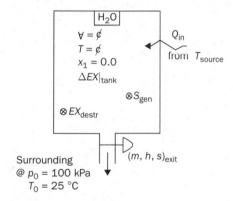

Assumptions

- Transient open system.
- Water is a two-phase fluid.
- Isochoric, isothermal quasi-equilibrium process.
- $W = 0; Q = Q_{in}$.
- $h_{exit} \approx \cancel{c}; s_{exit} \approx \cancel{c}$.
- $\Delta kE = \Delta; pE = 0$.

Method

- M-B: $m_{exit} = m_1 - m_2; m_1 = \forall/v_1$.
- E-B: $Q_{in} - (mh)_{exit} = \Delta U|_{tank}$.
- S-B: $S_{in} - S_{out} + S_{gen} = \Delta S|_{tank}$.
- EX-B: $EX_{in} - EX_{out} - EX|_{destroyed} = \Delta EX|_{tank}; EX|_{destroyed} = T_0 S_{gen}$.

Solution Based on the process description, i.e., $x_1 = 0$, $T_1 = 170°C => p_1 = p_{sat}$ provides the initial condition, while $T_2 = T_1$ with $x_2 > 0$ and $p_2 = p_1 = p_{sat}$ forms the final tank fluid condition.

- M-B: $m_{in} - m_{out} = \Delta m\big|_{tank} => m_{exit} = m_1 - m_2 = \dfrac{1}{2}m_1$.

- E-B: $Q_{in} - (mh)_{exit} = m_2 u_2 - m_1 u_1$.

- Property values:

- $\underbrace{\text{Initial State 1:}}_{(T_1=17°C,\, x_1=0.0)}$ $\begin{cases} u_1 = u_f = 718.2 \text{ kJ/kg} \\ v_1 = v_f = 0.00114 \text{ m}^3/\text{kg} \\ s_1 = s_f = 2.0417 \text{ kJ/(kg·K)} \end{cases}$

- $\underbrace{\text{Exit Conditions}\,(e):}_{(T_e=T_1,\, x_e=x_1)}$ $\begin{cases} h_e = h_f = 719.08 \text{ kJ/kg} \approx ¢ \\ s_e = s_f = s_1 \approx s_2 \approx ¢ \end{cases}$

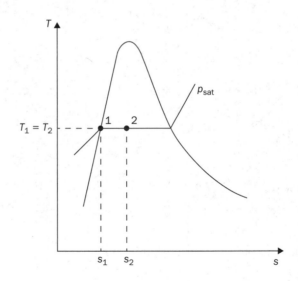

Final State 2: $m_2 = \dfrac{1}{2}m_1 = 0.5\dfrac{\forall}{v_1} = 269.24 \text{ kg} \widehat{=} m_e$. Hence, $v_2 = \dfrac{\forall}{m_2} = 0.00229 \text{ m}^3/\text{kg}$

and $x_2 = \dfrac{v_2 - v_f}{v_{fg}} = 0.004614$ so that with $T_2 = T_1 = 170°C$ and $x_2 = 0.004614$:

$$u_2 = u_f + x_2 u_{fg} = 726.77 \text{ kJ/kg}$$

$$s_2 = s_f + x_2 s_{fg} = 2.063 \text{ kJ/(kg·K)}$$

Now:

$$Q_{in} = (mh)_e + m_2u_2 - m_1u_1 = 2545 \text{ kJ}$$

$$S_{in} - S_{out} + S_{gen} = \Delta S\big|_{tank} \text{ or } \frac{Q}{T_0}\bigg|_{in} - (ms)_e + S_{gen} = (m_2s_2 - m_1s_1)\big|_{tank}$$

So that:

$$S_{gen} = (ms)_2 - (ms)_1 + (ms)_e - \frac{Q_{in}}{T_0}$$

And hence:

$$EX\big|_{destroyed} = T_0S_{gen} = 141.2 \text{ kJ}; \; T_0 = 298 \text{ K}$$

Comments

✓ The exergy balance [see Eq. (4.10)] reduces here to:

$$\left(1 - \frac{T_0}{T_{source}}\right)Q_{in} - 0 + 0 - m_e\big[(h_e - h_o) - T_0(s_e - s_o)\big] - EX\big|_{destr.} = \Delta EX\big|_{tank}$$

✓ As the discharged water vapor mixture doesn't produce any work:

$$EX\big|_{destroyed} = W_{rev,out} - W_{act,out} = W_{rev,out} = 141.2 \text{ kJ}$$

✓ In general, exit conditions would be $s_{exit} = \frac{1}{2}(s_1 + s_2)$ and $h_{exit} = \frac{1}{2}(h_1 + h_2)$.

✓ $\Delta EX\big|_{tank} = EX\big|_{final} - EX\big|_{initial} = \Delta U - T_0\Delta S$ could also be used to estimate $EX\big|_{destroyed}$.

Example 4.2   Water at $T_1 = 150°C$ in an ideal p-c device is heated from its initially saturated liquid State 1 to its final saturated vapor State 2 in an internally reversible process. The dead state, i.e., the surrounding, is at $T_0 = 20°C$ and $p_0 = 100$ kPa. Find, per unit mass, Δex and $ex\big|_{destroyed}$ and show the process in a T-s diagram.

Sketch

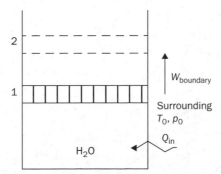

Assumptions

- Transient closed system.
- Water is a two-phase fluid.
- Isobaric ($p = ¢$) process as the piston is frictionless.
- $EX|_{destroyed} = 0$ (internally reversible process).
- ΔkE and ΔpE are negligible.

Method

- Eq. (4.11) is reduced to $\Delta ex = \dfrac{\Delta EX}{m} = \Delta u + p_0 \Delta v - T_0 \Delta s \, ; ex|_{destroyed} = T_0 \Delta s.$

- Eq. (4.5) reads: $\Delta ex = ex|_{in} - ex|_{out} - ex|_{destroyed}.$

- $\dfrac{W}{m} = w = p\Delta v$ and $\dfrac{Q}{m} = T\Delta s.$

Solution   Based on the process description, i.e., saturated water turning into saturated steam at $T_1 = T_2 = 150°C$ and hence $p_1 = p_2 = p_{sat} = 476.16$ kPa. The T-s diagram is:

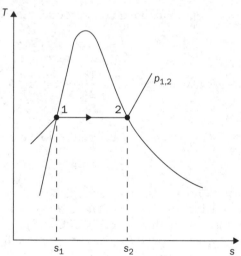

State properties:

- Initial State 1: $\begin{cases} u_1 = 631.8 \text{ kJ/kg} \\ v_1 = 0.0010405 \text{ m}^3/\text{kg} \\ s_1 = 1.8418 \text{ kJ/(kg·K)} \end{cases}$

- Final State 2 : $\begin{cases} u_2 = 2559.5 \text{ kJ/kg} \\ v_2 = 0.3928 \text{ m}^3/\text{kg} \\ s_2 = 6.8379 \text{ kJ/(kg·K)} \end{cases}$

$EX/m = ex$-balance:

$$\Delta ex\big|_{p-c} = ex\big|_2 - ex\big|_1 = \Delta u + p_0 \Delta v - T_0 \Delta s = \Delta ex\big|_{\text{heat in}} - \Delta ex\big|_{\text{work out}} - ex\big|_{\text{destroyed}}.$$

Thus:

$$\Delta ex\big|_{p-c} = u_2 - u_1 + p_0(v_2 - v_1) - T_0(s_2 - s_1) = 1927.82 + 39.17 - 1464.61$$
$$= 502.38 \text{ kJ/kg}$$

and with:

$$\Delta ex\big|_{\text{heat in}} = \left(1 - \frac{T_0}{T_1}\right)\frac{Q}{m} \quad \text{and} \quad \Delta ex\big|_{\text{work out}} = \frac{W}{m} - p_0(v_2 - v_1)$$

where, by definition:

$$W = \int_1^2 p \, d\forall \Rightarrow \frac{W}{m} = p_{\text{sat}}(v_2 - v_1) = 186.38 \text{ kJ/kg}$$

and:

$$Q = \int_1^2 T \, dS = mT \int_1^2 ds \Rightarrow \frac{Q}{m} = T(s_2 - s_1) = 2114.1 \text{ kJ/kg}$$

Finally:

$$\Delta ex\big|_{\text{heat in}} = \left(1 - \frac{293}{423}\right) \times 2114.1 = 649.49 \text{ kJ/kg}$$

while:

$$\Delta ex\big|_{\text{work out}} = \frac{W}{m} - p_0(v_2 - v_1) = 186.38 - 39.17 = 147.21 \text{ kJ/kg}$$

Solving the ex-balance for $ex\big|_{\text{destroyed}}$ yields:

$$ex\big|_{\text{destroyed}} = -\Delta ex\big|_{p-c} + \Delta ex\big|_{\text{heat}} - \Delta ex\big|_{\text{work}}$$
$$= -502.38 + 649.49 - 147.21 = 0 \text{ kJ/kg}$$

Comments

✓ The expansion process, being internally reversible, does not carry any work/energy loss.

✓ Also $ex\big|_{\text{destroyed}} = T_0 s_{\text{gen}}$, where here $s_{\text{gen}} = (s_1 - s_2) + \dfrac{Q}{mT_1} = 0$.

✓ The boundary work $W/m = 186.38$ kJ/kg is determined as part of the *net useful work* because the piston (via Δv) pushes air upward against $p_0 = 100$ kPa.

Example 4.3   Consider a well insulated 5-MW steam turbine with $p_1 = 6$ MPa, $T_1 = 600°C$ and $v_1 = 80$ m/s as the input and with $p_2 = 50$ kPa, $T_2 = 100°C$ and $v_2 = 140$ m/s as the output. Find $P_{rev} \equiv \dot{W}_{rev}$ and η_{II}, assuming the dead state to be at $T_0 = 25°C \triangleq T_{surr}$, $p_0 = 100$ kPa and $v_0 = 0$ m/s.

Sketch

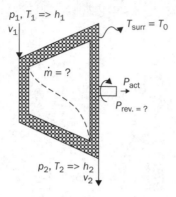

Assumptions

- Steady open system.
- Water is a two-phase fluid.
- $\dot{Q} = 0; \dot{W} = \dot{W}_{actual}$ or $\dot{W}_{reversible}$.
- Quasi-equilibrium expansion process.

Method

- M-B: $\dot{m}_1 = \dot{m}_2 = \dot{m} = ¢$.
- E-B: $\sum \dot{E}_{in} = \sum \dot{E}_{out}$ where $\Delta p \dot{E} = 0$.
- EX-B: $\sum \dot{EX}\big|_{in} = \sum \dot{EX}\big|_{out}$ with $\dot{EX}\big|_{destroyed} = 0$ for $\dot{W}_{rev}$ operation.

Solution   With p_{sat1} and p_{sat2} associated with T_1 and T_2 being greater than p_1 and p_2, the process is in the superheated vapor region.

E-B:

$$\dot{m}\left(h_1 + \frac{v_1^2}{2}\right) = \dot{m}\left(h_2 + \frac{v_2^2}{2}\right) + \dot{W}_{act}$$

$$\therefore \dot{m} = 5.156 \text{ kg/s}$$

Property values:

1.   At p_1 and T_1 => $\begin{cases} h_1 = 3658.8 \text{ kJ/kg} \\ s_1 = 7.1693 \text{ kJ/(kg·K)} \end{cases}$

2. At p_2 and T_2 $=>$ $\begin{cases} h_2 = 2682.4 \text{ kJ/kg} \\ s_2 = 7.6453 \text{ kJ/(kg·K)} \end{cases}$

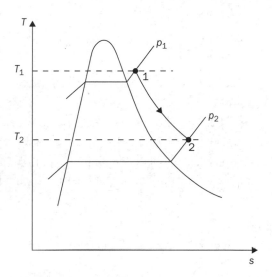

EX-B (see Eq. [4.12]):

$$-\dot{W}_{rev} + m\left[(h_1 - h_2) - T_0(s_1 - s_2) - \Delta ke\right] = 0$$

as $\left.\dot{EX}\right|_{destroyed}$ is zero here.

Solving for $\dot{W}_{rev} = \dot{m}\left[(h_1 - h_2) - T_0(s_1 - s_2) + \frac{1}{2}(v_1^2 - v_2^2)\right] = \dot{W}_{act} - T_0(s_1 - s_2)$.

Hence:

$$\left.\dot{W}_{rev}\right|_{out} = 5000 \text{ kW} - 5.156 \text{ kg/s} \times 298 \text{ K} \times (7.1693 - 7.6953)$$

As expected:

$$\left.\dot{W}_{rev}\right|_{out} = 5808 \text{ kW} > \left.\dot{W}_{act}\right|_{out}$$

Now:

$$\eta_{II} = \frac{\dot{W}_{turbine}}{\left.\dot{W}_{rev}\right|_{out}} = \frac{5000}{5808} = 86.1\%$$

Comments

✓ Alternatively, $\eta_{II} = \dfrac{\eta_{th,turbine}}{\eta_{rev/turbine}} = \dfrac{46\%}{54\%} \times 100\%$.

✓ $\left.\dot{EX}\right|_{destroyed} = \left(\dot{W}_{rev} - \dot{W}_{actual}\right)_{turbine\ output} = 808 \text{ kW} = T_0\dot{S}_{gen}$.

✓ Specifically, the *S-B* yields:

$$\dot{m}(s_1 - s_2) + \dot{S}_{gen} = 0 \text{ or } \dot{S}_{gen} = 5.156(7.6953 - 7.1693)$$

So that:

$$EX\big|_{destroyed} = T_0 \dot{S}_{gen} = 298 \times 2.712 = 808 \text{ kJ/s}$$

4.3 Exergy Concept Applications

From reading Secs. 4.1 and 4.2, it should be transparent that reducing $EX\big|_{destroyed} \triangleq E_{work} = W_{rev} - W_{actual}$ [see Eq. (4.5)] is a key goal. It can be achieved via smart system design and efficient device operation. An equivalent objective would be to increase the 2nd-Law efficiency [see Eq. (4.2)], ultimately to save capital investment cost and fuel cost.

4.3.1 Economical Aspects

For example, considering the cost factors for a steady open system (e.g., a turbine), the *E-B* reads:

$$\dot{Q}_{in} - \dot{Q}_{out} = \dot{W}_{act} + \dot{Q}_{loss} \tag{4.14}$$

Rewritten with $\dot{Q}_{in} - \dot{Q}_{out} = \dot{E}_{net}$ and $\dot{Q}_{loss} \triangleq EX\big|_{destroyed}$, we have:

$$\dot{E}_{net} - \dot{W}_{act} = EX\big|_{destroyed} \tag{4.15}$$

In terms of cost factors, Eq. (4.15) can be expressed as:

$$(C_{investment} + C_{final}) - C_{gain} = C_{loss} \tag{4.16}$$

where $C_{investment} \sim$ design cost and $C_{final} \sim$ operational cost. In order to reduce C_{loss}, two aspects of Eq. (4.16) can be considered:

1. Minimize $C_{investment} + C_{final} = C_{expenditure}$ for a given device performance.

2. Given $C_{expenditure}$, increase C_{gain}, i.e., a better device performance as expressed in $\eta_{th} = \dot{W}_{act}/\dot{W}_{rev}$ or $\eta_{II} = \eta_{th}/\eta_{rev} = W_{useful}/W_{rev}$.

Clearly, combining economic and engineering goals, i.e., (1) and (2), should simultaneously reduce $C_{expenditure}$ and increase C_{gain}.

4.3.2 Maximization of Available Energy Use

As discussed in Sec. 4.1 and stated in Eq. (4.6a), the reduction in useful work potential of available energy can be expressed as:

$$EX\big|_{destr} = T_{ref} S_{gen}^{total} \text{ or } \dot{EX}\big|_{destr} = T_{ref} \dot{S}_{gen}^{total} \tag{4.17a, b}$$

Clearly, in minimizing $\dot{S}_{gen}^{total}$, more useful work can be gained (e.g., via a turbine), or less work is needed to accomplish a task (via compressor or heat pump). Recalling that in engineering analyses, viscous dissipation and heat losses are the major causes of $\dot{S}_{gen}^{total}$, we have:

$$\dot{S}_{gen}^{total} = \dot{S}_{gen}^{friction} + \dot{S}_{gen}^{thermal}$$
(4.18)

as already alluded to with Eq. (3.8).

Viscous dissipation and heat losses are fundamental topics of courses dealing with fluid mechanics and heat transfer and are thus helpful in discussing Eq. (4.18). Nevertheless, for the courageous interested in doing a course project, this problem can be readily approached and solved starting with the triple Prelims: Sketch, Assumptions, and Method (see Sec. 2.1). As you will see, the analysis and specifically, $\dot{S}_{gen}^{friction}$ depends

on the fluid viscosity μ and local velocity gradients squared, e.g., $\left(\dfrac{\partial u}{\partial y}\right)^2$, $\left(\dfrac{\partial v}{\partial x}\right)^2$, etc., while

$\dot{S}_{gen}^{thermal} \sim T^{-2}, k, \nabla^2 T$, where T is the local temperature and k is the fluid conductivity.

Bejan (1996) showed that the volumetric entropy generation rate due to frictional and thermal effects can be expressed as:

$$\dot{S}_{gen}^{total} = \frac{\mu}{T}\varnothing - \frac{1}{T^2}\vec{q} \cdot \nabla T \left[\frac{j}{k \cdot s \cdot m^3}\right]$$
(4.19)

Some important aspects about Eq. (4.19) are as follows:

- $\varnothing$ is the viscous dissipation function proportional to $\mu \nabla v$.

- $\vec{q}$ is Fourier's heat flux vector $(\vec{q} = -k\nabla T)$.

- The dot product of the two vectors $\vec{q}$ and ∇T generates a scalar.

- $\nabla = \hat{i}\dfrac{\partial}{\partial x} + \hat{j}\dfrac{\partial}{\partial y} + \hat{k}\dfrac{\partial}{\partial z}$ is the del operator.

- T is in Kelvin.

For example, for 2-D systems in rectangular coordinates:

$$\dot{S}_{gen}^{total} = \frac{\mu}{T}\left\{2\left[\left(\frac{\partial u}{\partial x}\right)^2 + \left(\frac{\partial v}{\partial y}\right)^2\right] + \left(\frac{\partial u}{\partial y} + \frac{\partial v}{\partial x}\right)^2\right\} + \frac{k}{T^2}\left[\left(\frac{\partial T}{\partial x}\right)^2 + \left(\frac{\partial T}{\partial y}\right)^2\right]$$
(4.20)

Example 4.4 Consider steady laminar fully developed pipe flow, subject to constant wall heating, q_{wall} with fluid inlet temperature T_{in}. Assuming constant fluid properties, show that $u(r)$, the (axial) velocity profile for pipe radius r_o reads:

$$u(r) = u_{max}\left[1 - \left(\frac{r}{r_o}\right)^2\right]; \quad u_{max} = 2u_{average}$$

Neglecting axial temperature changes, the (radial) temperature profile is:

$$T(r) = T_{\text{wall}} - \frac{q_w r_o}{k}\left[\frac{3}{4} - \left(\frac{r}{r_o}\right)^2 + \frac{1}{4}\left(\frac{r}{r_o}\right)^4\right]$$

Sketch

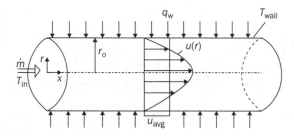

Assumptions

- Thermal Poiseuille flow with $q_w = ¢; T_{\text{in}} < T_w; T_{\text{in}} = T_0 \triangleq T_{\text{ref}}.$
- Thermally fully developed flow.
- Constant fluid properties.

Method

- Reduced form of Eq. (4.18) in cylindrical coordinates.
- Use of given $u(r)$ and $T(r)$.

Solution For the present case, the $\dot{S}_{\text{gen}}^{\text{total}}$-expression, i.e., Eq. (4.19) can be reduced to:

$$\dot{S}_{\text{gen}}^{\text{total}} = \frac{\mu}{T}\left(\frac{\partial u}{\partial r}\right)^2 + \frac{k}{T^2}\left(\frac{\partial T}{\partial r}\right)^2$$

where

$$\frac{\partial T}{\partial r} = \frac{q_w}{k}\left[2\left(\frac{r}{r_o}\right) - \left(\frac{r}{r_o}\right)^3\right]$$

and

$$\frac{\partial u}{\partial r} = -\frac{4u_{\text{avg}}r}{r_o^2}$$

Hence, inserting the u- and T-derivatives into the $\dot{S}_{\text{gen}}^{\text{total}}$ equation yields:

$$\dot{S}_{\text{gen}}^{\text{total}} = \frac{\mu}{T}\frac{16u_{\text{avg}}^2 \cdot r^2}{r_o^4} + \frac{k}{T^2}\left(\frac{q_w}{k}\right)^2\left[2\left(\frac{r}{r_o}\right) - \left(\frac{r}{r_o}\right)^3\right]^2$$

Using $\left(\dfrac{q_w^2}{kT_0^2}\right)$ and $\dfrac{r}{r_o} = R$ for nondimensionalization, we have:

$$\hat{\dot{S}}_{\text{gen}} = \dot{S}_{\text{gen}}^{\text{total}}\left(\frac{kT_0^2}{q_w^2}\right) = \underbrace{\frac{\mu}{T}R^2\left(\frac{16kT_0^2 u_{\text{avg}}^2}{q_w^2 r_o^2}\right)}_{\sim \dot{S}_{\text{gen}}^{\text{friction}}} + \underbrace{\frac{k}{T^2}\left[\frac{4kT_0}{\left(\rho c_p u_{\text{avg}} r_o\right)^2} + \left(\frac{T_0}{T}\right)^2\left(2R - R^3\right)\right]}_{\sim \dot{S}_{\text{gen}}^{\text{thermal}}}.$$

Rewritten, it provides more physical insight as:

$$\hat{\dot{S}}_{\text{gen}} = \underbrace{\frac{16}{Pe^2}\left(\frac{T_0}{T}\right)^2}_{\substack{\sim \dot{S}_{\text{gen}} \text{ due to axial} \\ \text{conduction/convection}}} + \underbrace{\left(2R - R^3\right)^2\left(\frac{T_0^2}{T^2}\right)}_{\sim \dot{S}_{\text{gen}} \text{ due to radial conduction}} + \underbrace{\frac{KT_0}{T}R^2}_{\substack{\sim \dot{S}_{\text{gen}} \text{ due to} \\ \text{fluid friction}}}$$

where the Peclet number $Pe = \dfrac{u_{\text{avg}} \cdot (2r_o)}{D} \triangleq \left(\dfrac{\text{convection}}{\text{conduction}}\right)$, where $T_0 \triangleq T_{\text{in}}$,
$K \equiv \dfrac{16\,kT_0\mu u_{\text{avg}}^2}{q_w^2 r_o^2} = \text{constant}$.

Graph

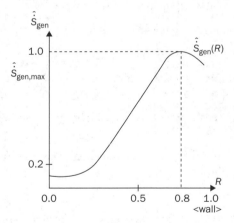

Comments

✓ Along the centerline R = 0, only thermal effects contribute to $\dot{S}_{\text{gen}}^{\text{total}}$, which even vanish for Pe >> 1.

✓ Then, for R > 0, both $\dot{S}_{\text{gen}}^{\text{thermal}} + \dot{S}_{\text{gen}}^{\text{friction}}$ increase with $\hat{\dot{S}}_{\text{gen,max}}$ at R ≈ 0.8.

✓ In the wall region 0.8 < R ≤ 1.0, there is a small decline in $\hat{\dot{S}}_{\text{gen}}$ because of diminished thermal effects, although $\dot{S}_{\text{gen}}^{\text{friction}}$ is highest due to the wall shear stress.

4.4   Homework Assignments

Again, Part A probes the reader's knowledge base, i.e., a deeper understanding of the concept of exergy and the associated balances for mass, energy, and entropy generation.

Part B tests the reader's skill level to solve new applications to the 1st and 2nd Laws of Thermodynamics. It should be apparent that all Chap. 4 problems are for the most part a repeat of the problems discussed in the previous chapters. However, the additional tasks are finding ΔEX_{system} and $EX|_{destroyed}$ plus the efficiencies for various devices and cycles.

4.4.1   Part A: Insight

Recall that a sketch, some math, and descriptions/explanations are required.

1   Considering the 1st Law and the 2nd Law in terms of energy expenditure and exergy produced, compare Student A's lifestyle with Student B's lifestyle in terms of $EX_{destroyed}$. In a given day, Student A sleeps 5 hours, studies 10 hours, watches TV 3 hours, is on the phone with friends for 2 hours, and exercises for 1 hour, while the rest is spent eating, running errands, etc. In contrast, Student B sleeps for 8 hours, studies 5 hours, watches TV 2 hours, interacts with friends for 3 hours, including sports, and spends the rest of the time dreaming, running errands, and eating.

2   For what kind of systems are useful work and actual work the same?

3   For an isentropic process, is the useful work equal to reversible work?

4   How does the 2nd-Law efficiency differ from the thermal efficiency?

5   Is $EX|_{destroyed} = 0$ when $S_{gen} = 0$?

6   In case $COP_I > COP_{II}$ for two refrigerators, will Fridge II also have a higher 2nd-Law efficiency?

7   Depict graphically a representative stationary system in equilibrium when the surrounding is at T_0, p_0, and v_0.

8   Sketch an open transient system receiving Q_k from a TER (thermal energy reservoir) at T_k and work input, but losing heat, Q_0, to the ambient.

9   Provide the exergy balance to Problem 8.

10   Can saturated steam at 1 atm flowing in a pipe with 1 kg/s produce any useful work? If so, what is the maximum work rate?

11   Discuss the pros and cons of Eqs. (4.6a–c).

12   Set up a generalized E-B and S-B in differential for, i.e., $(dU/dt)_{system} = \dots$ and $(dS/dt)_{system} = \dots$, combine both in the form $d(U - T_0 S)/dt$ for the system, and compare the resulting ODE with Eq. (4.10). Identify each term, and comment!

13   Interpret Eq. (4.13) for steady open, *reversible* devices.

14   Give a real-world example to Sec. 4.3.1, i.e., provide actual dollar figures for Eq. (4.16) and ways to maximize C_{gain}.

4.4.2 Part B: Problems

As always, start with the three Prelims, i.e., system Sketch, list of Assumptions, and reduced modeling equations as the Method. Done well, the Prelims solve the larger part of any given problem.

15 Consider a closed heat exchanger (20 L) filled initially with steam at 200°C and 200 kPa, heating a room for a while at a constant 21°C. During this process, the steam temperature decreases to 80°C. Find the heat transferred to the room. Using this heat from the radiator to run a heat engine, what would be the maximum work produced, taking the ambient (i.e., heat sink) to be at 0°C? Secondly, installing a reversible heat pump driven by the ideal HE, what is the heat supplied to the room now? Provide phase diagrams, plot Q_{room} vs. $T_{radiator,final}$, and comment!

16 A tank with an R-134a ($m_{liquid} = 1$ kg, $T = 24°C$), exposed to an ambient of 100 kPa and 24°C, is opened and hence isothermally releases part of the refrigerant until only saturated vapor remains. Determine $EX|_{destroyed}$ two ways.

17 A vessel (0.1 m³ with saturated water at 170°C) discharges half of its total mass while heat from a TER at 210°C is supplied to keep the vessel temperature constant. Assuming the surroundings to be at 25°C and 100 kPa, determine the amount of heat transferred, as well as $W_{reversible}$ and $EX|_{destroyed}$.

18 Consider an insulated p-c device ($p = 300$ kPa, $m_{water} = 15$ kg, and $m_{vapor} = 13$ kg) connected to a steam supply line (400°C and 2 MPa), which stays open until $m_{water} = 0$. Assuming the ambient to be at 25°C and 100 kPa, find m_{in} and $EX|_{destroyed}$.

19 A 4-L pressure cooker, half filled with water and half with steam, operates at 175 kPa (because vapor can escape via a valve), while a hot plate supplies 750 W for 20 min. Assuming ambient conditions of 25°C and 100 kPa, determine how much water remains in the device and the total exergy destruction. Replacing the hot plate with a solar energy source at 180°C supplying the 750 W, what would $EX|_{destroyed}$ be? Comment!

20 Consider a tank filled with 30 kg of nitrogen at 900 K supplying energy to a heat engine that produces work and rejects heat to a constant-pressure system, e.g., a p-c device, containing 15 kg of argon at 300 K. Assuming averaged specific heats, find $W_{HE,max}$ and T_{final} for both N_2 and Ar.

21 A p-c device with R-134a at 60°C and 700 kPa is cooled until only liquid at 20°C exists. Taking 20°C and 100 kPa as the dead state, find $EX_{initial}$ and EX_{final} as well as $EX|_{destroyed}$.

22 Consider a p-c device where the piston rests on stops. The cylinder contains 1.4 kg of R-134a, initially at 100 kPa and 20°C. Heat is supplied from a 150°C source: (i) to move the piston when 120 kPa is reached and (ii) to increase the temperature to 80°C. Taking the ambient to be 25°C and 100 kPa, determine the work done, the heat transferred, the exergy destroyed, and the value of the 2nd-Law efficiency.

23 An adiabatic tank has two equal compartments separated by a massless membrane (i.e., 3 kg of Ar at 70°C and 300 kPa on the left and a vacuum on the right), being surrounded by air of $T_0 = 25°C$. After removing the partition and thermodynamic equilibrium has been reached, what is the exergy destroyed during this expansion process?

24 Initially an insulted p-c device, with ambient conditions of 27°C and 100 kPa, contains 20 L of air at 27°C and 140 kPa. What is the exergy destroyed while an inserted electric heater delivers 100 W for 10 min?

25 An 8-kW compressor (in a 17°C and 100 kPa surrounding) sucks in the ambient air and delivers the air, at a rate of 2.1 kg/min, at 167°C and 600 kPa. Find the increase of the air exergy and the rate of exergy destroyed. *Hint:* Use the tabulated s^0 values for air to determine Δs with Gibbs II. Furthermore, plot (i) the real and ideal compression processes in a T-s diagram, and (ii) T_{exit} plus $E\dot{X}_{destroyed}$ as a function of η_{II} in the range of 75% to 100%.

26 For an adiabatic 5-MW steam turbine (IN: 600°C, 6 MPa, and 80 m/s; OUT: 100°C, 50 kPa, and 140 m/s), calculate the reversible power output plus the 2nd-Law efficiency. What is the percentage of the work potential wasted?

27 Consider an idealized throttling process where the refrigerant R-134a enters at 100°C and 1 MPa, while the ambient is at 30°C. Determine the process work and exergy destroyed for an exit pressure of 0.8 MPa. What is the exit temperature? Provide a T-s diagram, plot $w_{rev,out}$ as a function of p_{exit} (varying from 0.1 MPa to 1 MPa), and comment!

28 Combustion gases of a turbojet engine are accelerated in an adiabatic nozzle from 60 m/s to an exit velocity to be determined. The inlet conditions are 627°C and 230 kPa, while the exit values are 450°C and 70 kPa. Assuming $k = 1.3$ and $c_p = 1.15$ kJ/(kg·°C), also find the change in the gas exergy.

29 A mixing chamber receives dual, same-pressure streams of water (4.6 kg/s, 20°C) and saturated steam (0.19 kg/s) with a discharge of 100% water at 45°C. Taking the ambient to be at 20°C, find the steam inlet temperature, the exergy destruction, and the 2nd-Law efficiency.

30 A counterflow heat exchanger has cold water (1.5 kg/s, 22°C IN, $c_p = 4.18$ kJ/kg·°C) in the inner tube and in the shell pipe oil to be cooled (2 kg/s, 150°C IN and 40°C OUT, $c_p = 2.20$ kJ/ kg·°C). Neglecting energy losses, what are the rates of heat transfer and exergy destruction?

4.4.3   Part C: Aspects of Course Projects on Exergy Maximization

Major tasks in this project area include *realistic* modeling, simulation, and analysis (with cost-benefit estimates) of systems and devices. Then the goal is to achieve exergy optimization via improved device design and operation. Such two-step projects could focus on any real-world system/device discussed in this chapter.

CHAPTER 5

Energy Conversion Cycles

S ection 3.4 introduced idealized closed and open cycles as well as their underlying theory, i.e., the Clausius inequality and the 2nd-Law efficiency. Some sample problem solutions illustrated the basic steps of how to analyze simple cycles. Based on that insight, real-world cycles may be grouped into different categories:

  i. Power plants that provide electricity

  ii. Cooling devices such as refrigerators and air conditioners

  iii. Heat pumps for both heating and cooling

  iv. Propulsion systems to produce torque for cars, trucks, buses, trains, and ships or thrust for planes and rockets

These cyclic devices may differ greatly, considering the following features.

- The type of energy source used for boilers or combustors to generate high-enthalpy working fluids, i.e., fossil fuels, renewables, and nuclear elements.

- The wide range of working fluids, such as water from an elevated reservoir, water-steam mixture, natural gas, refrigerant, kerosene, helium, superheated CO_2, etc.

- The form of cyclic operation: (i) closed cycles such as in Rankine or Brayton power plants, refrigerators, and heat pumps; or (ii) open, i.e., mechanical, "cycles," as in car motors, Brayton gas turbines, and turbojet engines.

- The required system components, which are mainly turbines, compressors, and heat exchangers, but also throttles, diffusors, and nozzles, selected to accomplish a specific task as a function of their working-fluid inlet and outlet conditions.

Again, *Chap. 5 problem solutions rely on the material of the previous chapters, i.e., the in-depth knowledge base acquired and mathematical modeling skills applied.* In summary, Chap. 5 material helps to deepen the understanding of basic engineering thermodynamics and at the same time provides some background information for more advanced courses in thermodynamics, combustion, heat transfer, and HVAC systems. Clearly, for dedicated students Chap. 5 is a starting point to embark on a course project (see App. C).

This chapter begins with a visual summary of the two most important cyclic devices and their associated process diagrams, i.e., the Rankine cycle of power plants and the Brayton cycle of gas-turbine engines. Once fully understood, *solving the ensuing example problems independently* should be straightforward.

Figures 5.1a and b depict a steam-power plant with an extended Rankine cycle and the idealized *T-s* phase diagram, indicating isentropic pumping/expansion and zero-pressure drops across the boiler/reheater and condenser. Switching working fluids, Fig. 5.2 shows the ideal Brayton cycle for a gas-turbine power plant with associated

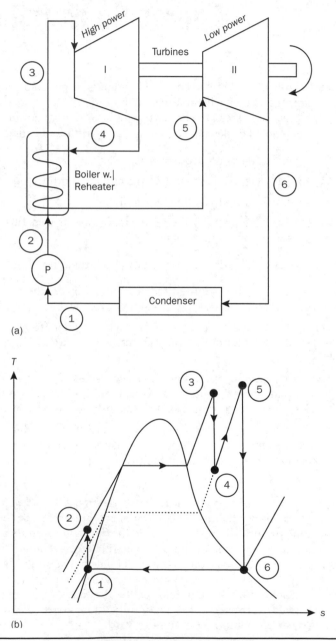

(a)

(b)

FIGURE 5.1 Steam power plant: (a) Rankine cycle with two-stage turbine system plus a reheater; and (b) the idealized *T-s* diagram of the cycle.

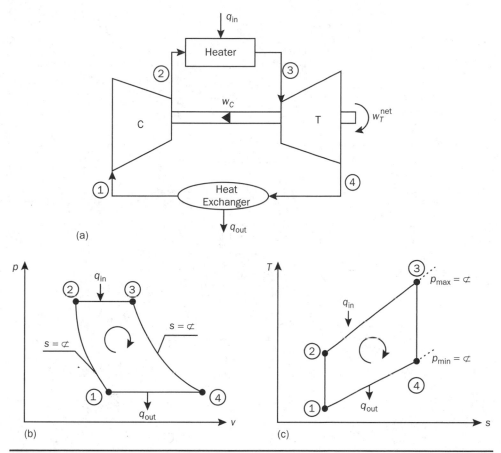

FIGURE 5.2 Gas-turbine engine: (a) Closed Brayton cycle; (b) the idealized *p-v* diagram; and (c) the idealized *T-s* diagram of the cycle.

single-phase *p-v* and *T-s* diagrams. In contrast to these closed cycles, Figs. 5.3 and 5.4 illustrate open "cycles."

To supplement the understanding of these system sketches and process diagrams, video links are provided for each cyclic device:

- Steam Power Plant: https://www.youtube.com/watch?v=AFoovtD3Ap0
- Gas Turbine Engine (Closed Brayton Cycle): https://www.youtube.com/watch?v=_hXXI5oUMFQ&feature=youtu.be&t=436
- Gas Turbine Engine (Open Brayton Cycle): https://youtube/_hXXI5oUMFQ?t=12
- Turbojet Engine: https://www.youtube.com/watch?v=SZ-hCHGoyn4&

In *closed cycles*, the same working fluid circulates through a series of interconnected system components (see cycles in Figs. 5.1 and 5.2). In contrast, the working fluid in *mechanical "cycles"* is constantly supplied and then, after combustion plus power stroke or thrust, the resulting combustion-products are discarded into the ambient (see Figs. 5.3

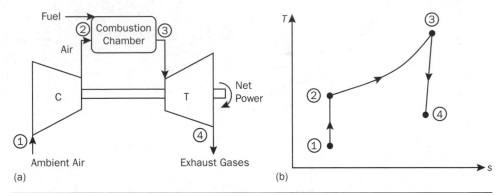

FIGURE 5.3   Gas-turbine engine: (a) open Brayton cycle; and (b) the idealized *T-s* diagram.

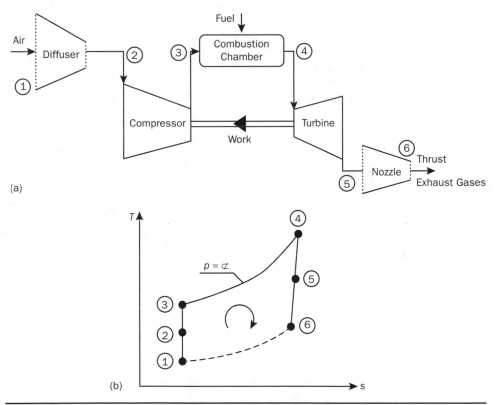

FIGURE 5.4   Open propulsion cycle: (a) turbojet engine; and (b) the idealized *T-s* diagram of the cycle.

and 5.4). As an example, propulsion systems suck in air (i.e., oxygen), compress it, mix it with a fuel, and atomize it. Then, the mixture is combusted in order to generate a power stroke (in ICSs) or expand the product gases in a turbine, which drives the compressor, to ultimately generate thrust via a nozzle (see Fig. 5.4). Still, the performance of all the different systems can be evaluated with specific efficiencies which in general

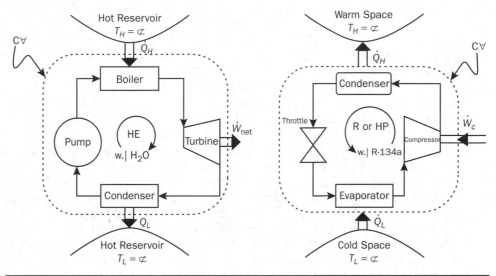

Figure 5.5 Contrasting cyclic systems: heat engine (HE) vs. refrigerator or heat pump (R or HP).

read: $\eta = (desired\ outcome)/(required\ input)$. Regardless of which η-correlation is being used, the result is always less than one in accordance with the Second Law of Thermodynamics (see Chap. 3).

In Fig. 5.5, heat engines (HEs) are compared to refrigerators (Rs) and heat pumps (HPs). Heat engines, e.g., steam power plants, drive generators to produce electricity and refrigerators cool food/drinks and keep ice cream frozen, while heat pumps can heat a place in the winter and in a reverse operation can cool a room in the summer. All cycles are operating between hot and cold constant-temperature reservoirs that are very different entities concerning specific functions for Rs vs. HPs. Regarding two-phase working fluids, HEs typically use H_2O, while Rs and HPs normally employ R-134a. Arguably, the most important components in HEs are the boiler and turbine, while for Rs and HPs they are the compressor and the throttle (also known as the expansion valve).

Recall for these three cycles:

- *E-B:* $\sum \dot{E}_{in} = \sum \dot{E}_{out}$

Here,

- $\dot{W}_{turbine} = \dot{Q}_H - \dot{Q}_L$ and $\dot{W}_{compressor} = \dot{W}_C = \dot{Q}_H - \dot{Q}_L$, where $\dot{W}_{net} = \dot{W}_T - \dot{W}_C$

The thermal efficiency (for HEs) and coefficient of performance (COP for R & HP) are:

- *HE:* $\eta_{th} = \dfrac{desired\ output}{required\ input} = \dfrac{\dot{W}_{net}}{\dot{Q}_H} = 1 - \dfrac{\dot{Q}_L}{\dot{Q}_H}$

- *HP:* $COP_{HP} = \dfrac{\dot{Q}_H}{\dot{W}_C} = \dfrac{\dot{W}_{net}}{\dot{Q}_H} = \dfrac{1}{1 - (\dot{Q}_L/\dot{Q}_H)}$ (Note the two modes of HP-cycle operation for heating in the winter and cooling in the summer by switching the reservoir designation)

$$\bullet \; R: \text{COP}_R = \frac{\dot{Q}_L}{\dot{W}_{\text{in}}} = \frac{\dot{Q}_L}{\dot{Q}_H - \dot{Q}_L} = \frac{1}{(\dot{Q}_H/\dot{Q}_L) - 1}$$

5.1 Closed Power Cycles: Heat Engines

Considering a basic, single-turbine Rankine cycle, the focus is on the performance evaluation of the steam turbine. The pump, driven by the turbine, is characterized by its efficiency while the boiler and condenser are heat exchangers with relatively minor pressure drops. Clearly, the turbine exhaust, either still superheated vapor or wet steam, has to be 100% liquefied for the pump in order to achieve cyclic operation and hence produce work. The *T-s* diagram of Fig. 5.6 nicely depicts the process paths and states of the Rankine cycle. Specifically, saturated water is compressed from a low-pressure p_1 to a very high p_2, where a slight increase in pump-entropy is indicated via pump end-states 2s vs. 2. Then, the compressed water is fed into the boiler where, under quasi-isobaric condition, i.e., $p_3 = p_2$, the water is turned into superheated steam of T_3. This implies quite a high enthalpy rate, $\dot{H} \equiv \dot{m}h = \dot{m}[u(T) + pv]$, entering the turbine. Assuming isentropic expansion, the wet steam leaves the turbine with p_{4s} and T_{4s}. More realistically, Δs is generated inside the turbine and hence the turbine endpoint is 4. It implies that the work output $W_{\text{real}} < W_{\text{ideal}}$, being proportional to the change in specific enthalpy Δh. Again, this inequality is a reflection of the 2nd Law.

A brief review of *T-s* phase diagrams is given in the next section. Then, Example 5.1 provides some numerical and physical insight into how to evaluate steam power plants.

5.1.1 Ideal vs. Real Heat Engines

As an extension of the content in Sec. 3.4, the *T-s* diagram for closed cycles and a few sample solutions for HE components are discussed. The Carnot HE is featured in order to have (again) an unattainable goal for best HE performance.

State 1: $p_2 = 20$ kPa; $x_1 = 0$

State 2: $p_2 = 6$ MPa

State 3: $p_2 \approx p_3$; $T_3 = 600°C$

State 4s: $p_{4s} = p_1$; $s_{4s} = s_3$

Note: $\eta_{\text{th}} = w_T^{\text{real}}/w_T^{\text{ideal}}$ determines State 4

Development of a T-s Phase Diagram

In general, one starts with a thermodynamic state of the working fluid for which two (2) properties are known. Considering the cycle depicted in Fig. 5.6, that would be State 1, which has to lie on the saturated liquid line as $x_1 = 0$ at the given $p_1 = 20$ kPa. Clearly, the associated enthalpy h_1 of the water entering the pump can be read from the Property Tables (see App. B). For the compression to $p_2 = 6$ MPa, it is observed that State 2 is either straight up for an isentropic pump or, more realistically, with a slight decrease in entropy, indicated by the pump efficiency. In any case, compression processes also increase the fluid temperature. Assuming the net heat-input rate goes to an isobaric boiler, State 3 is given by $p_3 = p_2$ and $T_3 = 600°C$. Again, the associated enthalpy h_3 of the steam entering the turbine can be read from the Property Tables. For an isentropic turbine, i.e., an adiabatic and internally

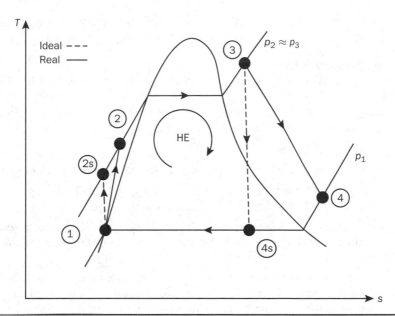

Figure 5.6   Rankine *T-s* phase diagram with ideal and real expansion/compression pathways.

reversible expansion process, State 4*s* appears in the saturated mixture region, given by $s_{4s} = s_3$ and $p_{4s} = p_1$. Again, the associated enthalpy h_{4s} of the wet steam leaving the turbine can be read from the Property Tables. More realistically, the turbine outlet stream is still super-heated vapor, i.e., State 4 is determined by $p_4 = p_1$ and the turbine efficiency $\eta_{th} = (h_3 - h_4)/(h_3 - h_{4s})$. Given η_{th}, h_4 can be determined and hence the real work output.

The prevailing question is: How much can one theoretically improve *any* cycle via smart component design and effective operation? The Carnot cycle (N.L.S. Carnot, 1824) provides a partial answer.

The Carnot Cycle as a Theoretical Performance Measure

By definition, the Carnot cycle connects four reversible processes, i.e., two isentropes and two isotherms. It appears in the *T-s* diagram as a rectangle between $T_{max} = T_H$ and $T_{min} = T_L$, where the enclosed area is equal to the work performed. From the 1st Law:

$$\dot{W}_{max} = \dot{Q}_{in} - \dot{Q}_{out} \text{ or } w_{ideal\,|\,HE} = w_{Carnot} = q_{in} - q_{out} \tag{5.1a–c}$$

so that

$$\eta_{Carnot} = w_{Carnot}/q_{in} = 1 - q_{out}/q_{in} = 1 - T_L(s_2 - s_1)/T_H(s_2 - s_1) \tag{5.2a–c}$$

Hence,

$$\eta_{Carnot} = 1 - T_L/T_H \tag{5.3}$$

Equation (5.3) implies that very high-temperature steam from the boiler entering the turbine and a very low exhaust temperature would maximize the Carnot thermal effi-ciency, and with that the ideal turbine-power produced. Clearly, in practice, there are natural limits to the T_H-level turbine blades can endure and to low T_L-reservoirs, i.e., the type of water or air heat sinks available for the condenser. Nevertheless, the goals of improved HE design and HE operation are to move $\eta_{th,real}$ toward $\eta_{th,Carnot}$.

Sample Problem Solutions

The next steam-power-plant examples provide the approach of how to solve basic Rankine cycles as well as a Carnot cycle plus a Rankine cycle with regeneration. The main tasks are to determine the cycle (i.e., thermal) efficiencies and compare the numbers with insightful discussions.

The procedural steps are as follows:

- Set up the three prelims, i.e., Sketch, Assumptions, and Method.
- The Sketch has to have a cyclic numbering system, starting typically with State 1 being the condenser exit (and hence pump inlet).
- Start the Solution with the *T-s* phase diagram of the cycle, based on the given state values and occasionally state properties obtained from the PTs, *featuring the same numbering as the Sketch.*
- List the two properties for each of the 4+ states and select all needed property values from the Property Tables. Note that $w_{pump} = h_2 - h_1 = v_1(p_2 - p_1)$, used to determine h_2, as State 1 is the pump inlet.
- Evaluate, as usual, all the steady open uniform flow components of the cycle.
- The thermal efficiency of the cycle is then $\eta_{th} = \dot{W}_{net}/\dot{Q}_{in}$ or $\eta_{th} = 1 - q_{out}/q_{in}$.

Example 5.1   Consider a power plant operating between $p_{max} = 6$ MPa and $p_{min} = 50$ kPa. It features $\dot{m} = 20$ kg/s, $T_{turbine,in} = 450°C$, and $\Delta T_{condenser} = 6.3°C$. That odd 6.3°C generates a pump-inlet temperature listed in the Property Tables. Determine $\dot{Q}_{in}$, $P_{turbine,net}$, and $\eta_{cycle} = \eta_{thermal}$.

Sketch

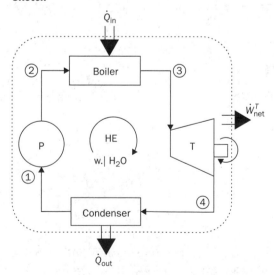

Assumptions

- Steady, closed steam powerplant cycle
- $H_2O \cong$ 2-phase
- Neglect $\Delta p \dot{E}$ and $\Delta k \dot{E}$
- $T_1 = T_{\text{sat}|p_1 = 50\,\text{kPa}}$
- Efficiencies:

$$\eta_T = \frac{\dot{W}^T_{\text{real}}}{\dot{W}^T_{\text{ideal}}}; \eta_{\text{th}} = \frac{\dot{W}_{\text{net}}}{\dot{Q}_{\text{in}}}$$

Method

- Focus on each cyclic component:
    - $w_{\text{pump}} = v_1 \Delta p = h_2 - h_1$
    - $\dot{Q}_{\text{in}} = \dot{m}(h_3 - h_4)$
    - $\dot{W}^T_{\text{net}} = \dot{W}_T - \dot{W}_P$
- Property Tables

Solution

- T-s Diagram

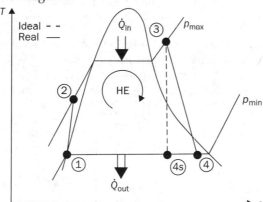

- Property Values

    State 1: p_1 & $T_1 = T_{\text{sat}|50\,\text{kPa}} - 6.3 = 81.3 - 6.3 = 75°C$

    $\therefore h_1 = h_{f|T_1} = 314\,\text{kJ/kg}; v_1 = v_{f|T_1} = 0.001026\,\text{m}^3$

    State 2: $h_2 = h_1 + w_{\text{pump}}; w_p = v_1(p_2 - p_1) = 6.1\,\text{kJ/kg}$

    $\therefore h_2 = 314 + 6.10 = 320.13\,\text{kJ/kg}$

    State 3: p_3 & $T_3 \rightarrow h_3 = 3302.9\,\text{kJ/kg}; s_3 = 6.7219\,\text{kJ/(kg·K)}$

    State 4s: p_4 & $s_{4s} = s_3 \rightarrow x_{4s} = \dfrac{s_{4s} - s_f}{s_{fg}} := 0.866 \rightarrow h_{4s} = h_f + x_{4s}h_{fg} = 2336.4\,\text{kJ/kg}$

    Thus with the given $\eta_T = 0.94$, $h_4 = h_3 - \eta_T(h_3 - h_{4s}) = 2394.4\,\text{kJ/kg}$

- Now,

$$\dot{Q}_{in} = \dot{m}(h_3 - h_2) = 59{,}660 \text{ kW}; \quad \dot{W}_T = \dot{m}(h_3 - h_4) = 18{,}170 \text{ kW}$$

$$\dot{W}_{pump} = \dot{m}w_p = 122 \text{ kW} \rightarrow \dot{W}_{net}^T = \dot{W}_T - \dot{W}_P = 18{,}050 \text{ kW}$$

- Finally, $n_{th} = n_{cycle} = \dfrac{\dot{W}_{net}^T}{\dot{Q}_{in}} = 0.303$

Comments

✓ The isentropic turbine discharges wet steam, i.e., $x_{4s} = 0.866$; then, moving to the more realistic State 4 with $\eta_T = w_T/w_{isentropic}$, the turbine exhaust is still in the saturated mixture region. That endpoint is in contrast to the illustration in Fig. 5.6.

✓ Here the actual specific turbine work is:

- $w_T = h_3 - h_4 = q_{in} - q_{out} = (h_3 - h_2) - (h_4 - h_2)$.

✓ The basic Rankine cycle can be modified in different ways in order to improve η_{cycle}. The unattainable goal is provided with Carnot-cycle results (see Example 5.2). Specifically, one example would be *steam-reheat* after the first-stage (high-pressure) turbine, which is coupled to a low-pressure turbine. Clearly, multi-stage turbines with several feed-water heaters would be advantages (see Sec. 5.1.2). However, capital investment plus operation costs have to be considered in an overall cost-benefit analysis.

Example 5.2   Compare the thermal efficiencies for an idealized Rankine cycle with that of a Carnot cycle. Both have a turbine-inlet pressure of 5 MPa and a condenser pressure of 50 kPa. While the Rankine cycle has saturated liquid as the condenser exit state, the Carnot cycle has saturated liquid entering the boiler.

Sketches

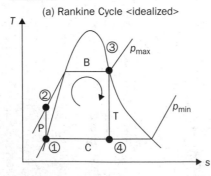

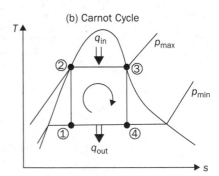

Assumptions

- Steady operations
- $H_2O \cong$ 2-phase
- Isentropic Rankine pump and turbine
- States 3 and 4, i.e., $w_T = h_3 - h_4$, are the same for both cycles

Method

- $\eta_{th} = 1 - \dfrac{q_{out}}{q_{in}}$ for both

- $w_{net} = q_{in} - q_{out}$ for both
- $w_p = v_1(p_2 - p_1)$ for the Rankine cycle
- Property Tables

Solution

(a) *Rankine Cycle:*

States 1 and 2: $h_1 = h_{f|50\,kPa} = 340.54$ kJ/kg; $v_1 = v_{f|50\,kPa} = 0.00103$ m³/kg

∴ $w_p = v_1(p_2 - p_1) = 5.10$ kJ/kg; so that $h_2 = h_1 + w_p = 345.64$ kJ/kg

State 3: $p_3 = 5$ MPa & $x_3 = 1.0 \rightarrow h_3 = 2794.2$ kJ/kg; $s_3 = 5.9737$ kJ/(kg·K)

State 4: $p_4 = 50$ kPa & $s_4 = s_3 \rightarrow x_4 = \dfrac{\left(s_4 - s_f\right)}{s_{fg}} = 0.7409$; ∴ $h_4 = 2071.2$ kJ/kg

Now, $q_{in} = h_3 - h_2 = 2448.6$ kJ/kg; $q_{out} = h_4 - h_1 = 1730.7$ kJ/kg; so that

$$w_T = q_{in} - q_{out} = 717.9 \text{ kJ/kg; and hence, } \eta_{th} = \frac{w_T}{q_{in}} = 0.2932$$

(b) *Carnot Cycle:*

State 3 is the same: $h_3 = 2794.2$ kJ/kg; $s_3 = 5.9737$ kJ/(kg·K); $T_3 = 263.9°C$

so that State 2 with $T_2 = T_3 = 263.9°C$ and $x_2 = 1.0 > h_2 = 1154.5$ kJ/kg;

$s_2 = 2.920$ kJ/(kg·K)

State 1: $p_1 = 50$ kPa & $s_1 = s_2 \rightarrow x_1 = 0.284$ so that $h_1 = h_f + x_1 h_{fg} = 989.05$ kJ/kg

Again,

$q_{in} = h_3 - h_2 = 1639.7$ kJ/kg and $q_{out} = h_4 - h_1 = 1082.2$ kJ/kg

so that

$$w^T_{net} = q_{in} - q_{out} = 557.4 \text{ kJ/kg}$$

and hence,

$$\eta_{th} = \eta_{max} = 1 - \frac{q_{out}}{q_{in}} = 0.34.$$

Comments

✓ Assuming isentropic operation for the Rankine cycle, its efficiency is quite close to $\eta_{max} = \eta_{Carnot}$, i.e., 29.3% vs. 34%.

✓ Having the cycles' T-s diagrams up front, the solution steps become a bit more straightforward.

5.1.2   Ways to Increase Efficiencies of HEs

The turbine is the key component of a cyclic power plant, where

$$P_T = \dot{W}_T = \left[\dot{m}\Delta h - \dot{Q}_{loss}\right] \tag{5.4}$$

Clearly, the power-output can be improved by increasing $\dot{m}$ and $(h_{in} - h_{out})$, while reducing $\dot{Q}_{loss}$. Focusing on Δh, the goals are to elevate T_{in} and p_{in} values which generate $h_{in,max}$ and to lower the condenser pressure being in the saturated mixture region with $T_{out} = T_{sat}$, implying $h_{out,min}$.

A practical way to gain a higher turbine efficiency is the introduction of an open (or closed) feed water heater (FWH), also called regenerator or heat exchanger. Considering a two-stage, i.e., high- and low-pressure, turbine, steam is extracted at an intermediate pressure and temperature. That steam portion heats up the (feed) water from the condenser, which is then supplied to the boiler via a second pump. In open FWHs, heat exchange occurs by direct mixing of the steam from the turbine with the water from the condenser. In contrast, a closed FWH keeps the two H_2O streams separate as in tube-and-shell heat exchangers. For example, Fig. 5.7a depicts the schematics of a steam power plant with an open-loop FWH, while Fig. 5.7b shows the associated T-s diagram. However, this cycle operates under the assumptions of isentropic expansion (turbine) and compression (pumps) and no heat losses elsewhere.

In any case, a new energy balance is required for the FWH to determine the necessary fraction, y, of steam extracted from the turbine. Specifically,

$$y = (\dot{m}_{extracted})/(\dot{m}_{turbine,in}) \tag{5.5}$$

Example 5.3: Efficiency Comparison between a Basic Rankine Cycle and One with Regeneration
Consider an idealized steam power plant operating on a Rankine cycle with and without an open FWH. In both cases, the turbine inlet conditions are 600°C and 15 MPa, while the exiting steam, i.e., the condenser pressure, is at 10 kPa. For regeneration, some steam is extracted at 1.2 MPa to enter the open FWH, i.e., a mixing chamber, where the hot steam portion from the turbine heats up the cold condenser output before being pumped to the boiler. Compare the efficiencies of the two cycles and comment.

Sketch

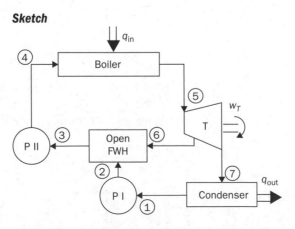

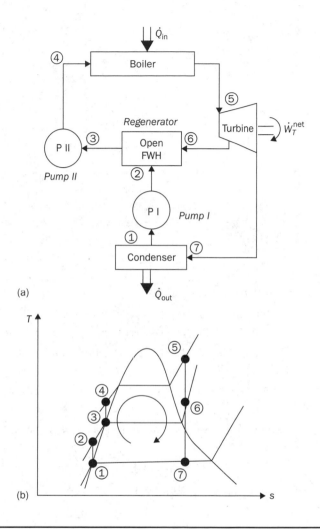

FIGURE 5.7   (a) Rankine power plant with regenerator; and (b) idealized *T-s* phase diagram.

Note   Clearly, in the basic Rankine cycle the FWH plus Pump II and associated streams (6) and (3) do not exist.

Assumptions

- Steady cyclic operation with isentropic turbine and Pumps I and II processes with no heat losses
- $\Delta p\dot{E}$ and $\Delta k\dot{E}$ are negligible

Method

- Together with Property Table values, analyze each steady open uniform flow device, i.e., $w = \Delta h$ and $q = \Delta h$
- For the FWH: $\sum \dot{m}h_{in} = \dot{m}h_{out}$
- Efficiency $\eta_{th} = 1 - q_{out}/q_{in}$

Solution

(a) *Simple Rankine Cycle*

- *T-s* diagram

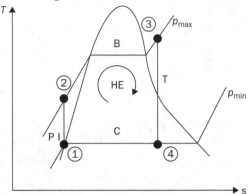

- Property values

State 1 is the same for both cases:

$p_1 = 10$ kPa & $x_1 = 0 \rightarrow h_1 = 191.8$ kJ/kg; $v_1 = 0.00101$ m³/kg

State 2: $p_2 = 15$ MPa & $s_2 = s_1$, still needs $h_2 = h_1 + w_p = h_1 + v_1(p_2 - p_1) = 206.95$ kJ/kg

State 3: $p_3 = p_2$ & $T_3\,600°C \rightarrow h_3 = 35,831.1$ kJ/kg and $s_3 = 6.6796$ kJ/(kg·K)

State 4: $p_4 = p_1$ & $s_4 = s_3 \rightarrow h_4 = 21,152.3$ kJ/kg as $x_4 = 0.804$

Now, knowing h_1 to h_4 we can compute $q_{in} \cong q_{boiler}$ and $q_{out} \cong q_{condenser}$, leading to $\eta_{th} = 1 - q_{out}/q_{in}$:

$$q_{in} = h_3 - h_2 = 3376.2 \text{ kJ/kg and } q_{out} = h_4 - h_1 = 1923.5 \text{ kJ/kg}$$

so that for the basic Rankine cycle:

$$\eta_{th} = 0.43$$

(b) *Rankine Cycle with Regeneration*

- *T-s* diagram

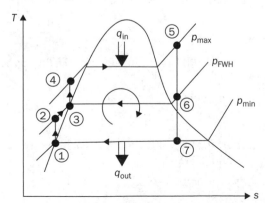

- Property Values

  State 1: $h_1 = h_{f|p_1} = 191.8$ kJ/kg and $v_1 = v_{f|p_1} = 0.00101$ m³/kg

  State 2: $p_2 = p_{FWH} = 1.2$ MPa; $s_2 = s_1$ and $h_2 = h_1 + w_{PI} = h_1 + v_1(p_2 - p_1) = 193.01$ kJ/kg

  State 3: $p_3 = p_2$ & $x_3 = 0 \rightarrow \begin{cases} v_3 = 0.00138 \\ h_3 = 783.33 \end{cases}$

  State 4: $p_4 = 15$ MPa and as $s_4 = s_3$

  $w_{PII} = v_3(p_4 - p_3) = 15.7$ kJ/kg

  so that $h_4 = h_3 + w_{PII} = 814.03$ kJ/kg

  State 5: $p_5 = p_{max} = 15$ MPa & $T_5 = 600°$C

  $\therefore h_s = 3583.1$ kJ/kg and $s_5 = 6.6796$ kJ/(kg·K)

  State 6: $p_6 = p_{FWH} = 1.2$ MPa & $s_6 = s_5 \rightarrow h_6 = 2860.2$ kJ/kg w. | $T_6 = 281.4°$C

  State 7: $p_7 = p_{min} = 10$ kPa and $s_7 = s_5 \rightarrow x_7 = \dfrac{s_7 - s_f}{s_{fg}} = 0.804$

  so that $h_7 = h_f + x_7 h_{fg} = 2115.3$ kJ/kg

Now, while $q_{in} = h_5 - h_4$, $q_{out} = (1 - y)(h_7 - h_1)$ where $y \equiv \dot{m}_6/\dot{m}_5$ is the fraction of steam extracted from the turbine so that $(1 - y)$ is the portion of steam liquified in the condenser. Thus, from the E-B for the open FWH (see Sketch):

$$\dot{m}_6 h_6 + \dot{m}_2 h_2 = \dot{m}_3 h_3$$

where

$$\dot{m}_2 = \dot{m}_1 = \dot{m}_7 = \dot{m}_5 - \dot{m}_6 \text{ and } \dot{m}_3 = \dot{m}_6 + \dot{m}_1 = \dot{m}_5 - \dot{m}_1 + \dot{m}_1 = \dot{m}_5$$

Hence

$$\dot{m}_6 h_6 + (\dot{m}_5 - \dot{m}_6)h_2 = \dot{m}_5 h_3 \Big|\frac{1}{\dot{m}_5} \text{ yields}$$

$$y h_6 + (1 - y)h_2 = h_3 \text{ or } y = \frac{h_3 - h_2}{h_6 - h_2} = 0.227$$

Finally,

$$q_{in} = h_3 - h_4 = 2769.1 \text{ kJ/kg} \quad \text{and} \quad q_{out} = (1 - y)(h_7 - h_1) = 1486.9 \text{ kJ/kg}$$

so that

$$\eta_{th} = 1 - \frac{q_{out}}{q_{in}} = 0.463$$

Comments

✓ The addition of the open FWH improves the Rankine cycle with regeneration by only 3.3%; thus, a cost-benefit (C-B) analysis would help to decide if the FWH should be included.

✓ Specifically, it could determine the number of years it would take to recover initial capital and maintenance costs.

✓ If the simple Rankine cycle could be replaced by a Carnot cycle operating between $T_{max} = 873$ K and $T_{min} = 318.8$ K, $\eta_{Carnot} = 63.5\%$.

Clearly, similar to the multi-stage compressor idea (see Sec. 3.5.3), incorporating several turbine stages and the associated FWHs would push up a cycle's efficiency. An alternative way to improve the thermal efficiency is the operation of a *combined gas-vapor power cycle*. There, an open gas Brayton cycle (see Secs. 3.4 and 5.2) operates on top of a closed steam power (Rankine) cycle, connected by a heat exchanger. However, in light of greater efficiencies, initial capital cost as well as operational and maintenance cost have to be evaluated via a cost-benefit analysis. The following links provide videos and a review of how to perform a C-B analysis of a steam power plant:

• www.LearnEngineering.org (or go to "device" video on YouTube)

• https://www.researchgate.net/publication/329966779_Thermodynamic_and_economic_analysis_of_performance_evaluation_of_all_the_thermal_power_plants_A_review

5.1.3   Brayton Gas-Power Engine

Expanding on the open Brayton cycle of Fig. 5.3 in order to understand the achievement of thermal efficiency improvement, Fig. 5.8a depicts a two-stage compressor with intercooling, driven by a two-stage turbine system that features extra gas reheating and regeneration. The associated *T-s* diagram of the gas-turbine engine with idealized pathways is shown in Fig. 5.8b, i.e., isentropic compression and expansion, plus isobaric heat exchangers without heat losses. As discussed in Sec. 3.5.3, the compressor work needed is minimized when the intermediate pressure $p_2 = p_3 = \sqrt{p_1 \cdot p_4}$ is achieved via intercooling. The combustor, assisted by the regenerator and high-pressure air supplied by the compressor system, provides the high enthalpy to drive the turbine system. One should keep in mind that in order to improve the turbine efficiency, the reheater requires an extra heat source and the intercooler requires an extra heat sink.

Examples 5.4 and 5.5 illustrate the performances of open Brayton cycles, i.e., gas-turbine engines, with different working fluids and approaches of determining the 2nd-Law efficiency as well as exergy destruction.

Example 5.4   Consider a two-stage air-turbine power plant with intercooler and reheater (see Fig. 5.8a), operating between $p_{max} = 1.2$ MPa and $p_{min} = 100$ kPa. The inlet temperature for compressor CI is 300 K and for CII 350 K, while for the turbines $T_{TI} = 1400$ K and $T_{TII} = 1300$ K. The compressor and turbine systems have an isentropic efficiency of 0.80, while the efficiency of the regenerator is 0.75.

Find the cycle's thermal and 2nd-Law efficiencies as well as exergies at State 6 and State 10.

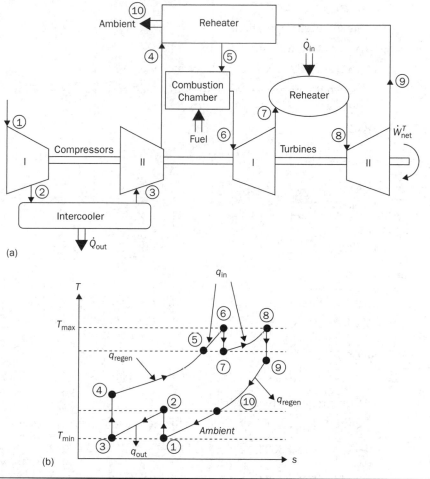

FIGURE 5.8 (a) Two-stage open Brayton cycle with intercooler, reheater, and regenerator; and (b) idealized *T-s* phase diagram for gas-turbine engine.

Sketch

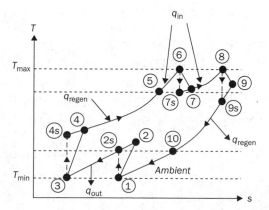

Assumptions

- Steady cyclic operation
- Air $\cong$ I.G. under air-standard conditions
- $\Delta p\dot{E}$ and $\Delta k\dot{E}$ are negligible
- Optimal compressor and turbine stages and hence reheating intercooling pressure $p_{\text{intercooler}} = \sqrt{p_1 p_4}$

Methods

- $\eta_{\text{compr.}} = \dfrac{W_{\text{ideal}}}{W_{\text{real}}}$

- $\eta_{\text{turb.}} = \dfrac{W_{\text{real}}}{W_{\text{ideal}}}$

- $\eta_{\text{th}} = \dfrac{w_{\text{net}}}{q_{\text{in}}}$

- $\eta_{\text{II}} = \dfrac{\eta_{\text{th}}}{\eta_{\text{Carnot}}}$

- $EX = \Delta h - T_o \Delta s$

Solution

- $p_2 = p_3 = \sqrt{p_1 p_4} = \sqrt{100 * 1200} = 346.4$ kPA

- Compression Processes

  State 1: $T_1 = 300$ K $\rightarrow h_1 = 300.43$ kJ/kg as $h = h(T)$ only for I.G.s

  $$T_1 = 300 \text{ K \& } p_1 = 100 \text{ kPa} \rightarrow s_1 = 5.7054 \text{ kJ/kg·K}$$

  State 2: $p_2 = 346.4$ kPa & $s_2 = s_1 \rightarrow h_{2s} = 428.70$ kJ/kg

  State 3: $T_3 = 350$ K & $p_3 = p_2 \rightarrow h_3 = 350.78$ kJ/kg and $s_3 = 5.5040$ kJ/kg·K

  State 4: $p_4 = 1200$ kPa & $s_{4s} = s_3 \rightarrow h_{4s} = 500.42$ kJ/kg so that,

  $$\eta_C = \frac{h_{4s} - h_3}{h_4 - h_3} = 0.80 \rightarrow h_4 = 527.83 \text{ kJ/kg}$$

- Expansion Processes

  State 6: $T_6 = 1400$ K $\rightarrow h_6 = 1514.9$ kJ/kg & w. $|p_6 = 1200$ kPa $\rightarrow s_6 = 6.6574$ kJ/kg·K

  State 7: $p_7 = p_2 = 346.4$ kPa & $s_{7s} = s_6 \rightarrow h_{7s} = 1083.9$ kJ/kg

  State 8: $T_8 = 1300$ K & $p_3 = p_2 = 346.4$ kPa $\rightarrow h_6 = 1395.6$ kJ/kg and $s_8 = 6.9196$ kJ/kg·K

  State 9: $p_9 = 100$ kPa & $s_{9s} = s_8 = 6.9196$ kJ/kg·K $\rightarrow h_{9s} = 996$ kJ/kg

  With $\eta_T = \dfrac{h_8 - h_9}{h_8 - h_{9s}} = 0.80 \rightarrow h_9 = 1075.9$ kJ/kg

- Cycle Components

  An E-B for the 2-stage compressor yields:

  $$w_c = h_2 - h_1 + h_4 - h_3 := 347.50 \text{ kJ/kg}$$

  The same for the turbine system yields:

  $$w_T = h_6 - h_7 + h_8 - h_9 := 664.50 \text{ kJ/kg}$$

  so that

  $$w_{T,net} = w_T - w_C = 317 \text{ kJ/kg}$$

  Now, with $q_{in} = h_6 - h_5$ to calculate $\eta_{th} = \dfrac{w_{net}}{q_{in}}$ we need h_5, which we obtain from the regenerator analysis:

  $$\eta_{reg} = \frac{h_9 - h_{10}}{h_9 - h_4} = 0.75 \rightarrow h_{10} = 672.36 \text{ kJ/kg}$$

  so that

  State 10: $p_{10} = 100$ kPa & $h_{10} = 672.36$ kJ/kg $\rightarrow s_{10} = 65,157$ kJ/kg·K

  so that $q_{reg} = h_9 - h_{10} = h_5 - h_4 \rightarrow h_5 = 941.40$ kJ/kg

  and $q_{in} = h_6 - h_5 = 573.54$ kJ/kg and $\eta_{th} = \dfrac{w_{net}}{q_{in}} = 0.553$

- 2nd-Law efficiency

  $$\eta_{II} = \frac{\eta_{th}}{\eta_{max}} ; \quad \eta_{max} = \eta_{Carnot} = 1 - \frac{T_1}{T_6} = 0.786$$

  $$\therefore \eta_{II} = \frac{0.553}{0.786} = 0.704$$

- Exergies:

  Combustion Chamber Exit: $ex_6 = h_6 - h_o - T_o(s_6 - s_o)$

  $T_o = T_{surr} = 300$ K; $h_o = 300.19$ kJ/kg; $s_o = 5.7054$ kJ/kg·K; $p_o = 100$ kPa

  $\therefore ex_6 = 930.7$ kJ/kg

  Regenerator Exit:

  $ex_{10} = h_{10} - h_o - T_o (s_{10} - s_o)$

  $\therefore ex_{10} = 128.8$ kJ/kg

Comments

✓ Clearly, cycle analyses are getting more challenging when multi-stage turbines and compressors as well as heat exchangers, i.e., intercooler and regenerators, are being added.

✓ *Having a detailed T-s diagram upfront helps greatly to move through the solution steps with confidence.*

✓ It is always of interest to compare numerically for key cycle components the specific exergy values with the specific energy input; ultimately, the total specific exergy destroyed (see Example 5.5).

Example 5.5   A basic gas power plant operates as a closed Brayton cycle between 1240 K and 295 K with isentropic efficiencies of 0.83 for the compressor and 0.87 for the turbine.

Determine (i) the turbine exit temperature; (ii) the net power output; (iii) the cycle (i.e., thermal) efficiency; and (iv) the exergy destroyed in each of the four components of the cycle.

Sketch   See Fig. 5.2

T-s Diagram

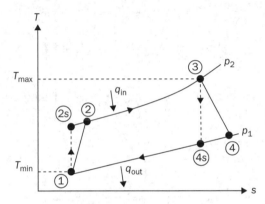

Assumptions

- Steady operation
- Air $\cong$ I.G. with air-standard assumptions
- All devices are steady open uniform flow
- $\Delta p\dot{E}$ and $\Delta k\dot{E}$ are negligible

Method

- $\eta_C = \dfrac{h_{2s} - h_1}{h_2 - h_1}$

- $\eta_T = \dfrac{h_3 - h_4}{h_{3s} - h_{4s}}$

- Pressure ratio, $r_p = p_2/p_1$

- Relative pressure, $r_p = e^{s^\circ/R} \rightarrow \dfrac{p_2}{p_1}\Big|_{s=c} = \dfrac{p_{r2}}{p_{r1}}$

- $\eta_{th} = \dfrac{w_{net}}{q_{in}}$

Solution

- Isentropic process (1) → (2s): $T_1 = 295$ K → $\begin{cases} h_1 = 295.17 \text{ kJ/kg} \\ p_{r1} = 1.3068 \end{cases}$

$\therefore h_{2s} = 570.26$ kJ/kg and $T_{2s} = 564.9$ K via line interpolation

With $\eta_C = \dfrac{h_{2s} - h_1}{h_2 - h_1} = 0.83 \to h_2 = 626.6$ kJ/kg

and (4)

- State 3: $T_3 = 1240$ K → $\begin{cases} h_3 = 1324.93 \text{ kJ/kg} \\ p_{r3} = 272.3 \end{cases}$

$\therefore p_{r4} = \dfrac{p_4}{p_3} p_{r3} = 27.23 \to h_{4s} = 702.07$ kJ/kg and $T_{4s} = 689.6$ K

With $\eta_T = \dfrac{h_3 - h_4}{h_3 - h_{4s}} = 0.87 \to h_4 = 783$ kJ/kg and $T_4 = 764.4$ K

- Cycle Efficiency

$$\eta_{\text{cycle}} = \eta_{\text{th}} = \frac{w_{\text{net}}}{q_{\text{in}}}; \; w_{\text{net}} = q_{\text{in}} - q_{\text{out}}; \; q_{\text{in}} = h_3 - h_2 = 698.3 \text{ kJ/kg}$$

$$q_{\text{out}} = h_4 - h_1 = 487.9 \text{ kJ/kg}$$

$$\therefore \eta_{\text{th}} = \frac{210.4}{698.3} = 0.30$$

- Exergy Destruction

Recall from Sec. 4.1 $EX_d = T_o S_{\text{gen}}$ (or $ex_d \equiv \dfrac{EX_d}{m} = T_o s_{\text{gen}}$)

where $S_{\text{gen}} = \Delta S_{\text{system}} + \Delta S_{\text{surrounding}}$ as given in Chap. 3 by Eq. (3.7). For ideal gases, the needed changes in specific entropy can be computed via the Gibbs relationships, i.e., Eq. (3.13) or (3.14). Thus for any cycle component, ($T_o = T_{\text{surr}} = 295$ K):

$$ex_d = T_o s_{\text{gen}} = T_o \left(s_i - s_j + \frac{q_k}{T_k} \right) = T_o \left(s_i^o - s_j^o - R \ln \frac{p_i}{p_j} + \frac{q_k}{T_k} \right)$$

Specifically,

- $ex_d^{1-2} = T_o(s_2 - s_1) = T_o \left(s_2^o - s_1^o - R \ln \dfrac{p_2}{p_1} \right) := 28.08$ kJ/kg

- $ex_d^{2-3} = T_o \left(s_3 - s_2 + \dfrac{-q_{\text{in}}}{T_H} \right) = T_o \left(s_3^o - s_2^o - 0 + \dfrac{-q_{\text{in}}}{T_H} \right) := 100.3$ kJ/kg

- $ex_d^{3-4} = T_o(s_4 - s_3) = T_o \left(s_4^o - s_3^o - R \ln \dfrac{p_4}{p_3} \right) := 32.86$ kJ/kg

- $ex_d^{4-1} = T_o \left(s_1 - s_4 + \dfrac{q_{\text{out}}}{T_L} \right) = T_o \left(s_1^o - s_4^o - 0 + \dfrac{q_{\text{out}}}{T_L} \right) := 197.9$ kJ/kg

- Using $q_{in} = 698.3$ kJ/kg; $q_{out} = 487.9$ kJ/kg; $s_1^o = 1.68515$ kJ/kg·K w. | $T_1 = 295$ K
  $s_2^o = 2.44117$ kJ/kg·K w. | $h_2 = 626.6$ kJ/kg; $s_3^o = 3.2175$ kJ/kg·K w. | $T_3 = 1240$ K
  and $s_4^o = 2.668$ kJ/kg·K w. | $h_4 = 783$ kJ/kg

Comments

✓ Not surprising, $\eta_{Carnot} = 76\%$ is more than twice the real closed Brayton cycle.

✓ The three numbers to contemplate are $q_{in} = 686.3$ kJ/kg, $w_{net} = 210.4$ kJ/kg, and $\sum ex_d = 359.14$ kJ/kg $= ex_{cycle}$.

✓ As the pressure ratio $r_p = \dfrac{p_2}{p_1}$ is a key system parameter, performance graphs

$\eta_{th}(r_p)$ and $\dot{W}_{net}(r_p)$ should be plotted and discussed for, say, $2 \leq r_p \leq 20$.

5.1.4   Cogeneration

The steam exiting a power-plant turbine may be at a relatively low pressure (i.e., 5–7 atm) but still at a high temperature (i.e., 150–200°C); thus, being often still in the superheated vapor region. Clearly, a lot of energy is being wasted as the condenser rejects heat into the air, lakes, rivers, or oceans to liquefy the exit steam. Many systems or devices near a power plant need process heat for, say, manufacturing or space-heating purposes. Hence, *the energy input to a power plant can be converted, via cogeneration, into electric power as well as process-heat production.* The efficiency of a cogeneration plant, also known as the utilization factor, is defined as:

$$\eta_{cogeneration} = (\dot{W}_{net} + \dot{Q}_{process\,heat})/\dot{Q}_{in} = \dot{Q}_{out}/\dot{Q}_{in} \tag{5.6}$$

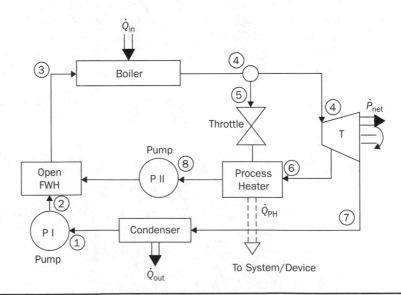

FIGURE 5.9   Cogeneration and regeneration Rankine cycle with adjustable process heat supply.

For example, the Rankine cycle in Fig. 5.7a requires for cogeneration an additional heat exchanger (i.e., the process heater) to supply heat to an external device or system (see Fig. 5.9).

It should be noted that Fig. 5.9 features not only the process heater but also a throttle to control various uses of the steam's energy input. For example, depending on the level of demand for process heat vs. electric power needed, the throttle (an expansion valve) can regulate the amount of steam supplied by the boiler to the process heater (and hence to the turbine) from $\dot{m}_{steam} = 0\%$ to $\dot{m}_{steam} = 100\%$. With respect to Fig. 5.9, the key heat transfer rates and turbine power produced result from basic energy balances. For example,

$$\dot{Q}_{in} = \dot{m}_3(h_4 - h_3);$$

$$\dot{Q}_{out} = \dot{m}_7(h_7 - h_1); \tag{5.7a–c}$$

$$\dot{Q}_{process\ heater} = \dot{m}_5 h_5 + \dot{m}_6 h_6 - \dot{m}_8 h_8$$

5.2 Open Power Cycles: Propulsion Systems

As mentioned, open (or mechanical) cycles have at most one component (e.g., the piston in ICEs) undergoing cyclic movements while the working fluid (e.g., an air-fuels mixture) is being continuously resupplied from the ambient. Examples include the Otto and Diesel motors, the Brayton gas-turbine engine, as well as propulsion systems, such as turbojet engines.

The open Brayton cycle (see Fig. 5.3) was discussed in Sec. 5.1.3 with two examples providing the efficiency numerics for a simple gas-power plant as well as an improved one. Internal combustion engines (ICEs) for automobiles are on their way out, as batteries for electric cars are getting cheaper while battery performance and reliability increase. Still, two-stroke engines for lawn mowers, blowers, chain saws, and motorcycles will endure; so will the diesel engine for trucks, heavy-material movers, ships, and back-up power plants. Using electric airplanes to commute will be another future goal; however, turbojet engines will be needed for large planes flying long distances.

To supplement the understanding of these system sketches and process diagrams, video links are provided for select devices as follows.

Turbojet Engine: https://www.youtube.com/watch?v=KjiUUJdPGX0

Otto Motor: https://youtu.be/fNcZDrfT498?t=24

Diesel Motor: https://www.youtube.com/watch?v=DZt5xU44IfQ

Clearly, thermodynamic analyses as well as cost/benefit evaluations of these open cycles should be explored via *course projects* in comparison to electric and hybrid cars.

5.2.1 Jet Propulsion Engines

Turbojet engines are commonly used for airplane propulsion because of the gas turbine's relatively high power-to-weight ratio. Figures 5.4a and b depict the schematics of a turbojet engine and an idealized *T-s* diagram for computational analysis. Except for the diffusor and nozzle, the main part is the same as the open Brayton cycle (see Fig. 5.3). The front-end decelerates the air relative to the housing, causing a pressure increase from p_1 to p_2 (known as the ram effect). The turbine power is used to drive the compressor as well as auxiliary equipment. The nozzle receives the turbine exhaust gas-mixture, which exits with a very high velocity, causing the desired thrust, $F = \dot{m}\Delta v$. The inlet air plus the combustion-exhaust gases can be treated as an incompressible fluid. However, the maximum local Mach number, Ma = (gas-velocity)/(speed of sound), has to be known *a priori* for proper engine design and operation.

Specifically, $\dot{W}_{propulsion} = Fv_{aircraft} = \dot{m}_{gas}(v_{exit} - v_{inlet})v_{aircraft}$ so that $\eta_{propulsion} = \dot{W}_{prop}/\dot{Q}_{in}$. Here, v_{exit} is the nozzle exit velocity and v_{inlet} is the diffusor inlet velocity.

Example 5.6   A plane at 9150 m altitude ($p_{ambient}$ = 30 kPa and $T_{ambient}$ = –32°C) travels with 280 m/s. Air enters the compressor (r_p = 12) at a mass flow rate of 50 kg/s with c_p = 1.005 kJ/kg·K and k = 1.4. The turbine inlet temperature is 1100 K. The heating value of the jet fuel is 42,700 kJ/kg. Neglecting any kinetic energy impact at both diffuser exit and nozzle inlet, find the velocity of the exhaust gas-mixture, the propulsion power, and the rate of fuel consumption.

Following Newton's 1st Law of Motion, we assume the plane is stationary and the air is approaching with v_1 = 280 m/s. The key open cycle components are treated separately, using the process-state numbers of the *T-s* diagram. All devices are steady open uniform flow systems. The air-fuel-vapor mixture is approx. an ideal gas.

Sketch   See Fig. 5.4a, where the air-inlet velocity is equal to the plane velocity v_1.

T-s Diagram

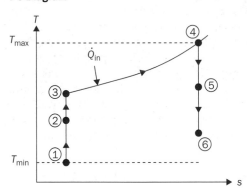

Assumptions

- Steady operation
- Air $\cong$ I.G. subject to air-standard assumptions
- All devices are steady open uniform flow
- Kinetic and potential energies are negligible, except at the diffuser inlet and the nozzle exit
- Gas turbine power is solely used to drive the compressor, i.e., $w_T = w_c$

Method

- $\sum \dot{E}_{in} = \Delta \dot{E}_{out}$

- Pressure ratio, $r_p = p_3/p_2$
- $\Delta h = c_p \Delta T$

Solution

i. *Diffuser*

- E-B: $\sum \dot{E}_{in} = \sum \dot{E}_{out}$, i.e., $h_1 + \dfrac{v_1^2}{2} = h_2 + \dfrac{v_2^2}{2}$

    where $h = c_p T$, so that $c_p(T_2 - T_1) = \dfrac{v_1^2}{2} \langle v_2 \approx 0 \rangle$

    Hence, $T_2 = T_1 \dfrac{v_1^2}{R c_p} := 280$ K

    Now under isentropic conditions, $p_2 = p_1 \left(\dfrac{T_2}{T_1} \right)^{\kappa/(\kappa - 1)} := 54.10$ kPa

ii. *Compressor*

   Using the given $r_p = \dfrac{p_3}{p_2}, p_3 = 649.2$ kPa $= p_4$, assuming no pressure-drop across the combustor.

   Again, $T_3 = T_2 \left(\dfrac{p_3}{p_2} \right)^{\kappa/(\kappa - 1)} := 569.5$ K

iii. *Turbine*

   Assuming that the gas turbine power is solely used to drive the compressor,

$$w_T = w_c \text{ or } h_4 - h_5 = h_3 - h_2 \text{ or } c_p \Delta T_T = c_p \Delta T_c$$

   Hence, $T_4 - T_5 = T_3 - T_2$, or $T_5 = T_4 - T_3 + T_2 := 810.5$ K

iv. *Nozzle*

   Again, being an adiabatic and reversible device:

$$\sum \dot{E}_{in} = \sum \dot{E}_{out}, \text{ or, } h_5 + \dfrac{v_5^2}{2} = h_6 + \dfrac{v_6^2}{2} \langle v_5 \approx 0 \rangle$$

   Hence with $h = c_p T$

$$c_p(T_5 - T_6) = \dfrac{v_6^2}{2}$$

   where $T_6 = T_4 \left(\dfrac{p_6}{p_4} \right)^{(\kappa - 1)/\kappa} := 465.5$ K, $T_4 = 1100$ K

   a. Now, solving for $v_6 = v_{exit}$ and recalling that 1 kJ/kg $\cong 10^3 \text{ m}^2/\text{s}^2$, we obtain: $v_6 = 832.7$ m/s

b. $\dot{W}_p = \dot{m}(v_6 - v_1)v_{plane} = 50(832.7 - 280) * 280 = 7738$ kW

c. $\dot{Q}_{in} = \dot{m}(h_4 - h_3) = \dot{m}c_p(T_4 - T_3) := 26{,}657$ kJ/s

so that $\dot{m}_{fuel} = \dfrac{\dot{Q}_{in}}{HV} = \dfrac{26{,}657}{42{,}700} = 0.6243$ kg/s

Comments

✓ Theoretically, to close the cycle T_6 has to be reduced to T_1!

✓ The engine is a real fuel guzzler.

✓ Note that $\eta_{prop.} = \dfrac{\dot{W}_p}{\dot{Q}_{in}} := 29\%$.

✓ Check out online videos of jet propulsion cycles.

✓ For a more realistic analysis, assume $\eta_{turbine} = 85\%$ and $\eta_{compressor} = 80\%$.

Not surprising, turbojet engines and ICEs, briefly discussed next, are among the favorite course projects selected by dedicated students.

5.2.2 Internal Combustion Engines

ICEs, also known as reciprocating engines, are grouped into spark-ignition (SI) engines, like the Otto motor, and compression-ignition (CI) engines, like the diesel motor. Figures 5.10a and b depict the actual Otto cycle and its idealized processes in p-v and T-s

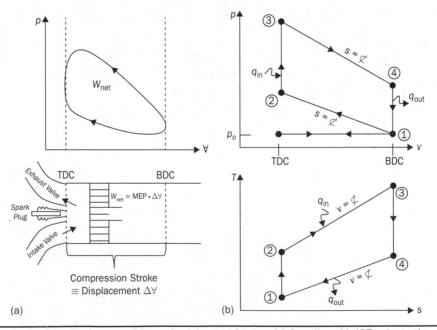

FIGURE 5.10 Four stroke open Otto cycle: (a) approximate $p(\forall)$ function with ICE schematic; and (b) idealized p-v and T-s diagrams.

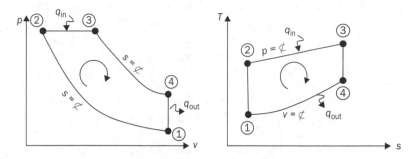

FIGURE 5.11 Idealized *p-v* and *T-s* diagrams for the diesel cycle.

diagrams, while Fig. 5.11 shows the idealized diagrams. The thermal efficiencies of such engines depend largely on the compression ratio $\alpha = \forall_{max}/\forall_{min}$ and the specific heat ratio $\kappa = c_p/c_v$ of the working gas-mixture. The net work-output can be estimated from W_{net} $= p_{mean}\Delta\forall$, where $p_{average}$ MEP is the *mean effective pressure* in the cylinder and $\Delta\forall = \forall_{max} - \forall_{min} = $ TDC – BDC, i.e., "top-dead-center" minus "bottom-dead-center."

Regarding the Otto engine as a steady, closed four-process cycle, $w_{net} = q_{in} - q_{out}$, where $q_{in} = c_v(T_3 - T_2)$ and $q_{out} = c_v(T_4 - T_1)$. Assuming isentropic compression and expansion (see Fig. 5.10b),

$$\eta_{th} = w_{net}/q_{in} = 1 - q_{out}/q_{in} \qquad (5.8a\text{–}c)$$

$$\eta_{thermal}^{Otto} = 1 - \frac{1}{r^{k-1}}$$

where $r = \forall_{max}/\forall_{min}$ is the compression ratio and $k = c_p/c_v$ is the specific heat ratio, typically taken to be around 1.4 for air and the gas mixtures.

When comparing Otto and diesel motor performances for the same compression ratio α, $\eta_{th,Otto} > \eta_{th,Diesel}$. However, typically $\alpha_{Diesel} > \alpha_{Otto}$ and with a more complete combustion diesel motors efficiently deliver more power and are cheaper to build than Otto motors. On the downside, diesel fuel is (in the United States at least) more expensive, and these engines significantly pollute the environment with soot particles and NO_x gases via their exhaust.

5.3   Reversed Closed Cycles: Refrigerators and Heat Pumps

Refrigeration is the cooling of space below the temperature of its surrounding, or the freezing of perishable products. Types of refrigeration cycles include *vapor compression*, which is similar to the reverse Rankine cycle, and *gas refrigeration*, being a reversed Brayton cycle. The historical definition of a cycle's *cooling capacity* is the heat-absorption rate of a metric ton (10^3 kg) of ice, i.e., 3.9 kJ/s, while melting away around > 0°C and 1 atm.

Clearly, selecting the right refrigerant and hence its unique properties, especially $T_{sat}(p_{sat})$, is very important to increase task-specific R-cycle performance. Refrigerants are liquid, liquid-vapor, and vapor substances that circulate through the device, alternately absorbing, transporting, and releasing heat. Most popular is R-134a, an HCFC (hydrochlorofluorocarbon) refrigerant with $T_{sat} = -26°C$ at 1 atm. *When compared to H₂O (i.e., as water, liquid-vapor mixture, or steam) the very different property values for refrigerants have to be kept in mind when analyzing refrigerator and heat pump operations.* Thus, consult very carefully the different Property Tables for R-134a.

Heat pump (HP) systems, capable of both heating and cooling, may use R-134a as well. In the heating mode, heat is pumped from the cold environment to warm, say, a room or a building. The coefficient of performance for a heat pump, for example, for a Carnot HP, $COP_{C-HP} = 1/(1 - T_L/T_H)$, depends the difference in hot and cold reservoir temperatures. Thus, an increasing $\Delta T = T_H - T_L$ decreases the COP_{HP}, so that HPs work best in regions with moderate temperature differences. As HPs are more energy efficient than electric resistance heaters, they are also frequently used to supply hot water for kitchens, showers, and washers.

Figure 5.5 illustrates the different cycle-components, process pathways, and tasks of refrigerators vs. heat pumps.

5.3.1 Refrigeration Systems

As mentioned, refrigerators (Rs) and heat pumps (HPs) use typically R-134a as a two-phase working fluid. Such cycles are also known as vapor-compression systems with possible intercooling to increase efficiency [see Moran et al. (2018), among others]. As with HE-cycles, components of R- and HP-cycles are evaluated individually, assuming either ideal (see Carnot cycle) or real operational conditions. Any ideal cycle-component, say, a turbine or compressor, is typically correlated to the real device via the isentropic thermal efficiency, i.e.,

$$\eta = \dot{W}_{real}/\dot{W}_{ideal} < 1 \tag{5.9}$$

Equation (5.9) is used in the next example to evaluate a realistic R-cycle, where the isentropic compressor efficiency $\eta_{compressor} = 80\%$.

In terms of the elusive Carnot cycles, both Carnot-R and Carnot-HP are Carnot power cycles in reverse. As for the thermal efficiency, the coefficient of performance (COP) is the ratio of "desired output" over "required input" (see Fig. 5.5). So, for the Carnot cycles with $Q_H/Q_L = T_H/T_L$, the associated COPs are:

$$COP_{C-R} = 1/(T_H/T_L - 1) \tag{5.10}$$

$$COP_{C-HP} = 1/(1 - T_L/T_H) \tag{5.11}$$

A key system parameter is the heat-removal rate, $\dot{Q}_L$, required to keep a room/space/compartment cooled at T_L. It should be noted that for effective heat transfer from the condenser (or to the evaporator) finite temperature differences are needed, i.e., $T_H^* = T_H + \Delta T$ while $T_L^* = T_L + \Delta T$.

Example 5.7 An R-134a refrigerator operates: (i) a compressor with 140 kPa and −10°C at the inlet and 800 kPa at the outlet with a mass flow rate of 0.05 kg/s; (ii) a condenser where R-134a is cooled to 26°C and 720 kPa; and then (iii) throttled at 150 kPa and 26°C.

Sketch

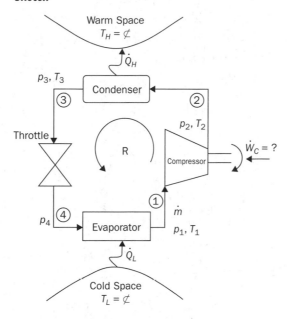

Assumptions

- Steady vapor-compression R-cycle operation with steady open uniform flow components
- Note the slight pressure drop across the condenser
- R-134a $\cong$ 2-phase liquid
- $h_{\text{throttle}} = \not\subset$, i.e., $h_4 = h_3$
- No $\dot{Q}_{\text{loss}}$ and $p = \not\subset$ across the evaporator
- Neglect $\Delta p \dot{E}$ and $\Delta k \dot{E}$
- Take $\dot{Q}_L = \dot{m}(h_1 - h_4) = \dot{Q}_{\text{total}}$

Method

- $\dot{W}_{\text{in}} = \dot{m}(h_2 - h_1)$
- $\eta_C^{\text{isentropic}} = \dfrac{h_{2s} - h_1}{h_2 - h_1}$ and $\text{COP}_R = \dfrac{\dot{Q}_L}{\dot{W}_{\text{in}}}$
- Property Tables

Solution

5.7A   Find the rate of heat removal from the refrigerated space and the needed power input as well as the compressor's isentropic efficiency and the cycle's COP.

- Property Values

  State 1: $p_1 = 0.14$ MPa & $T_1 = -10°C \rightarrow h_1 = 246.4$ kJ/kg

State 2: $p_2 = 0.8$ MPa & $T_2 = 50°C \rightarrow h_2 = 286.7$ kJ/kg

State 3: $p_3 = 0.72$ MPa & $T_3 = 26°C \rightarrow h_3 = h_{f|T_3} = 87.8$ kJ/kg

So that $h_4 = h_3 = 87.8$ kJ/kg

- Work Input and Heat Absorption

$$\dot{W}_{in} = \dot{m}(h_2 - h_1) = 2.02 \text{ kW}$$
$$\dot{W}_L = \dot{m}(h_1 - h_4) = 7.93 \text{ kW}$$
$$\therefore \eta_c = 93.8\% \text{ and COP}_B = 3.93$$

T-s Diagram

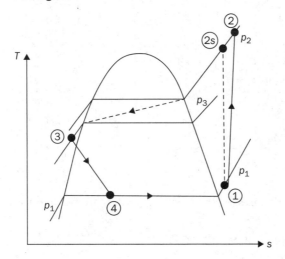

5.7B   Determine the exergy rate destroyed during cyclic operation and the 2nd-Law efficiency of the cycle.

- Recall $\dot{EX}_{destroyed} = T_o\dot{S}_{gen}$, where $T_o = T_H$ for Rs and $\dot{EX}_{cycle} = \sum \dot{EX}_{each\ device} = \dot{W}_{in} - \dot{W}_{in,min} = \dot{W}_{in} - \dot{W}_{Carnot}$

Specifically, $\dot{EX}_{cycle} = \dfrac{\dot{Q}_L}{COP_R} - \dot{Q}_L\left(\dfrac{T_H - T_L}{T_H}\right) := 0.547$ kW

For $\eta_{II}^{cycle} = \dfrac{COP_R}{COP_{C-R}}$, with $COP_{C-R} = \dfrac{1}{T_H/T_L - 1} := 4.38$

$$\eta_{II}^{cycle} = \frac{3.98}{4.38} = 0.896$$

Comments

✓ The realistic compressor-work input being larger than for an isentropic compressor decreases the COP-R only slightly; still, the COP$_{C-R}$ is the highest.

✓ Up front, one has to identify the R-cycle's (or HP-cycle's) T_H and T_L in order to make direct comparisons with COP$_{C-R}$ or COP$_{C-HP}$ results.

✓ As Eqs. (4.6a–c) indicate, there are different ways to calculate the rate of exergy destruction; thus, the Example 5.7B Solution can be augmented with different approaches.

✓ The 2nd-Law efficiency is rather large as COP_R is close to COP_{C-R}.

5.3.2 Heat Pump Systems

Electric motor powered *vapor-compression HPs* are typically used for space heating or cooling in residential and commercial buildings. In the heating mode, HPs extract heat from the atmosphere of >5°C or a body of water. In general, dual-function HPs work best in areas that have a large cooling load and small heating requirements, such as the southern states or Mediterranean countries. Alternatively, thermal-energy driven *absorption HPs* are mainly for industrial applications. Specifically, absorption HPs operate in two loops; one for the absorption medium, e.g., ammonia or lithium-bromide (LiBr), and the other for the refrigerant, usually water. The heat source can be natural gas, propane, solar energy, or geothermally heated water.

Considering a basic vapor-compression HP, its components include (see Figs. 5.5 and 5.12):

- The *reversing valve* controlling the direction of the refrigerant, i.e., changing the HP from the heating to the cooling mode or *vice versa*.

- The *evaporator,* being a heat exchanger with meandering tubes (i.e., the coil), where the refrigerant absorbs heat from a surrounding source (i.e., the air or water reservoir at T_C) and boils to turn into a low-temperature vapor. Via the reversing valve, the vapor stream feeds the *compressor,* which increases the vapor pressure and temperature.

- The *condenser,* another shell-and-tube heat exchanger, liquefies the refrigerant by rejecting heat to the ambient air or a large body of water, i.e., the hot reservoir at T_H.

- The *expansion valve* (or *throttle*) lowers the pressure and hence drops the temperature, so that the refrigerant becomes a low-temperature vapor-liquid mixture that enters the evaporator.

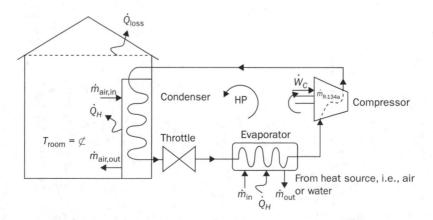

FIGURE 5.12 Residential heat pump system.

- The *plenum* is an air-compartment, which distributes the warm air (winter) or cool air (summer) through the residential/commercial building.

- In order to further visualize this cycle, a video link is provided: https://youtube/G53tTKoakcY?t=55.

While Example 5.8 analyzes the associated HP cycle of Fig. 5.12, Example 5.9 illustrates an HP cycle in cooling mode.

Example 5.8 Consider a residential R-134a heat pump system with a condenser unit in the room supplying a constant heat rate to keep the room at a set temperature. The condenser inlet stream is at 800 kPa and 50°C with a mass-flow rate of 0.022 kg/s that leaves cooled by –3°C at 750 kPa. The refrigerant enters the compressor at 200 kPa and exits at 50°C.

Determine the heat rate generated in the room and the COP of the HP, as well as the results when operating (theoretically) an idealized heat pump.

Sketch

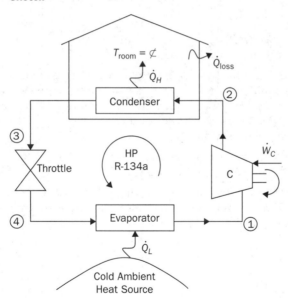

T-s Diagrams

    (i) Real HP-cycle   (ii) Ideal HP-cycle

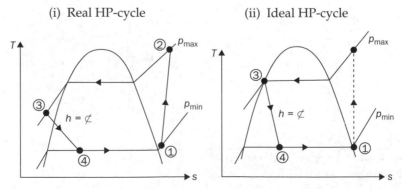

Assumptions

- Steady operation
- Air $\cong$ I.G. subject to the air-standard assumptions
- All devices are steady open uniform flow
- Kinetic and potential energies are negligible

Method

- Property values for all four states
- $\dot{Q}_H = \dot{Q}_{cond.} = \dot{m}(h_2 - h_3)$

- $\dot{W}_C^{in} = \dot{m}(h_2 - h_1)$; $COP_{HP} = \dfrac{\dot{Q}_H}{\dot{W}_C^{in}}$

- Ideal HP-cycle:

  $h_1 = h_{g|p_1}$; $s_1 = s_{g|p_1}$ and $s_{2s} = s_1$

  $h_3 = h_{f|p_2}$ and $h_3 = h_4$; $\dot{Q}_H = \dot{m}(h_{2s} - h_3)$; $COP_{ideal,HP} = \dfrac{h_{2s} - h_3}{h_{2s} - h_1}$

  As $\dot{W}_{ideal}^C = \dot{m}(h_{2s} - h_1)$

Solution

- State 2: $p_2 = 800$ kPa & $T_2 = 50°C \rightarrow h_2 = 286.71$ kJ/kg
- State 3: $p_3 = 750$ kPa & $T_2 = T_{sat|p_3} - 3°C = 26.06°C \rightarrow h_3 = 87.93$
- State 4: $h_4 = h_3 = 87.93$ kJ/kg
- State 1: $p_1 = 200$ kPa & $T_1 = T_{sat|p_1} + 4°C \rightarrow \begin{cases} h_1 = 247.9 \text{ kJ/kg} \\ s_1 = 0.950 \text{ kJ/(kg·K)} \end{cases}$

Hence, $h_2 = h_1 + \eta_C^{is}(h_{2s} - h_1) := 283.48$ kJ/kg

State 2: p_2 & $h_2 \rightarrow T_2 = 54.5°C$ via linear interpolation

State 3: $p_3 = p_2$ & $x_3 = 0.0 \rightarrow h_3 = h_{f|p_2} = 117.8$ kJ/kg $:= h_4$

- Mass flow rate: $\dot{m} = \dfrac{\dot{V}_1}{v_1} := 0.04048$ kg/s with the appropriate conversion factor

- Room heat removal: $\dot{Q}_{ambient} + \dot{Q}_{equipment} + \dot{Q}_{people} \equiv \dot{Q}_L = \dot{m}(h_1 - h_4) = 5.31$ kW

  $\therefore \dot{Q}_{people} = 0.665$ kW

- $COP_{HP} = \dfrac{\dot{Q}_L}{\dot{W}_{in}}$; $\dot{W}_{in} = \dot{m}(h_2 - h_1) := 0.9764$ kW so that

  $COP_{HP} = 5.87$

- Minimum power requirement: $\dot{W}_{in}^{MIN} = \dfrac{\dot{Q}_L}{COP_{C\text{-}HP}}$; $COP_{C\text{-}HP} = \dfrac{1}{T_H/T_L - 1} =$

  37.38

so that $\dot{W}_{in}^{MIN} = 0.1533 \text{ kW} = \dot{m}_{min}(h_2 - h_1)$ so that

$$\dot{m}_{min} = \frac{\dot{W}_{in}^{MIN}}{h_2 - h_1} = 0.00636 \text{ kg/s}$$

Comments

✓ As expected, $\dot{Q}_{ambient}$ is that major heat load the HP has to remove from the room.

✓ Note the isenthalpic process line (3) → (4) in the T-s diagram.

✓ Here, $\dot{m}_{min}$ is only 15.7% of $\dot{m}_{real}$ because of the high Carnot COP, i.e., $\text{COP}_{rev} = 5.87$.

✓ HP-operation is equivalent to an air conditioner.

✓ As expected, the HP-COP$_{ideal}$ is greater than the HP-COP$_{real}$.

✓ Running an equivalent Carnot heat pump, i.e., fully in the saturated mixture region, the $\text{COP}_{C-HP} = 15.9$.

✓ Clearly, with the given $\dot{m}_{R\text{-}134a}$ neither the idealized nor the Carnot HP-system can provide enough heat to maintain the set room temperature in light of the high heat-loss rate.

Example 5.9   An R-134a heat pump keeps a room cool at 26°C while 250 kJ/min enter the room from the outside (being at 34°C), 900 W are generated inside by equipment plus an unknown amount by people in the room. The compressor, at 75% isentropic efficiency, compresses 100 L/min. The condenser operates at 1.2 MPa with saturated liquid at the exit, while the evaporator at 500 kPa has saturated vapor at its outlet.

Find the compressor exit temperature, the COP, and the minimum volumetric flow rate for the same compressor inlet/outlet conditions.

Sketch

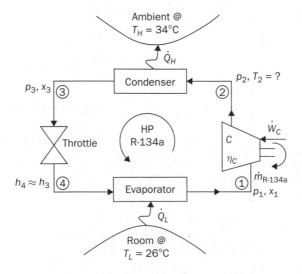

Assumptions

- Steady cyclic operation
- R-134a $\cong$ 2 phase-fluid
- $h_{throttle} = \not{C}$, i.e., $h_4 = h_3$
- Kinetic and potential energies are negligible
- No pressure drop across the heat-exchanger

Method

- $\eta_C^{isentropic} = \dfrac{h_{2s} - h_1}{h_2 - h_1} > h_2 = \cdots$

- $\dot{m}_R = \dfrac{\dot{\forall}_1}{v_1}$

- Property values for all four states

- $\dot{Q}_L = \dot{m}(h_1 - h_4) = \dot{Q}_{ambient} + \dot{Q}_{equipment} + \dot{Q}_{people}$ to keep $T_{room} = T_L = 26°C = $ const.

- $COP_{HP} = \dfrac{\dot{Q}_L}{\dot{W}_{in}}$; $\dot{W}_{in} = \dot{m}(h_2 - h_1)$

- $\dot{W}_{in}^{MIN} = \dfrac{\dot{Q}_L}{COP_{C\text{-}HP}}$; $COP_{C\text{-}HP} = \dfrac{1}{\dfrac{T_H}{T_L} - 1}$

Solution

- Property Values

State 1: $p_1 = 500$ kPa & $x_1 = 1 \rightarrow \begin{cases} h_1 = 259.30 \text{ kJ/kg,} \\ v_1 = 0.04112 \text{ m}^3/\text{kg} \\ s_1 = 0.9242 \text{ kJ/(kg·K)} \end{cases}$

State 2: $p_2 = 1.2$ MPa & $s_{2s} = s_2 \rightarrow h_{2s} = 277.45$ kJ/kg

- Now,

$$\dot{Q}_H = \dot{Q}_{loss}^{house} = \dot{m}(h_2 - h_3) = 4.37 \text{ kW}$$

- Power Input

$$\dot{W}_C = \dot{m}(h_2 - h_1) = 0.854$$

- Hence,

$$COP_{HP} = \frac{\dot{Q}_H}{\dot{W}_C} = 5.12$$

- For the ideal HP-cycle (see the T-s diagram):

State 1: $h_1 = h_{g|p_1} = 244.5$ kJ/kg and $s_1 = s_{g|p_1} = 0.938$ kJ/(kg·K)

State 2: $p_2 = 900$ kPa & $s_{2_s} = s_1 \rightarrow h_{2_s} = 273.3$ kJ/kg

State 3: $h_3 = h_{f|p_2} = 95.48$ kJ/kg $= h_4$

- Thus,

$$\text{COP}_{\text{ideal,HP}} = \frac{h_{2_s} - h_3}{h_{2_s} - h_1} = \frac{\dot{Q}_H^{\text{ideal}}}{\dot{W}_C^{\text{ideal}}} = 6.18$$

Comments

✓ In contrast to Example 5.8, here the heat pump runs as a cooling cycle.

✓ As expected, the ideal $\text{COP}_{\text{HP}} > \text{COP}_{\text{HP}}$ of the real cycle.

✓ Again, an isentropic efficiency has to be given to relate ideal cycles to real cycles.

5.4   Course Projects on Power Cycles

As mentioned in the chapter introduction, doing an independent course project (C-P) and writing a course-project report (C-PR), illustrated via a PowerPoint slide presentation, is a very valuable experience. C-P topics are listed or suggested throughout the text. Accomplishing that for (or without) extra course credit is an excellent exercise to prepare for future job-related tasks or assignments to be encountered in graduate school. The most favorite topics include closed or open cycles as well as 2nd-Law applications. Key features of a C-PR are outlined in App. C; these also apply for the most part to drafting a work report, master's thesis, or journal manuscript.

Equation Sheets for Homework Sets and Closed-Book Tests and Exams

Thermodynamics I deals with the transfer and conversion of different forms of energy, i.e., via heat, work, and mass flow, in different devices (or systems) via different processes using different fluids. In addition to the conservation laws of energy and mass, entropy generation must be checked as well. Any device (or system) interacts with its surrounding during energy and mass transfer. Here we deal with simple compressible substances (solids or fluids), fully determined by only two (2) property values, which undergo quasi-equilibrium processes.

A.1   Laws

- *First Law:* ΣE_{total} = constant. For example, when applied to a closed system such as a piston-cylinder device: $Q_{in} - W_{out} = \Delta E_{system} \approx \Delta U_{system} \equiv \Delta U_{cv}$ during process time Δt

- *Second Law:* Due to process irreversibilities, $S_{gen} = \Delta S_{total} = \Delta S_{system} + \Delta S_{surr} > 0$

- *Third Law:* The entropy of a perfect crystal at 0 K is zero; although, 0 K is not realistically attainable. However, it establishes the absolute temperature scale, where 0 K = −273.15°C

A.2   Balances for Systems/Devices (i.e., Control Volumes $C\forall \equiv CV$)

A.2.1   Mass Balance

Transient Uniform Mass Flow

$$\sum \dot{m}_{in} - \sum \dot{m}_{out} = \left(\frac{\Delta m}{\Delta t} \right)_{cv}$$

During process time Δt (Transient Open)

$$\boxed{\sum m_{in} - \sum m_{out} = (\Delta m)_{cv} = (m_f - m_i)}$$

$f \equiv$ final, $i \equiv$ initial

A.2.2   Energy Balance

Transient Uniform Flow $\sum \dot{E}_{in} - \sum \dot{E}_{out} = \left(\dfrac{\Delta E}{\Delta t}\right)_{cv} \approx \left(\dfrac{\Delta U}{\Delta t}\right)_{cv}$

During process time Δt
(Transient Open)

$$\boxed{\sum E_{in} - \sum E_{out} \approx (\Delta U)_{cv} = \left[(mu)_f - (mu)_i\right]_{cv}}$$

Here $\sum \dot{E}_{in/out} = \sum \dot{Q} + \sum \dot{W} + \sum \dot{H} + \sum \dot{E}_{kin} + \sum \dot{E}_{pot}$

and $\dot{X} = \dot{m}x$ or specific property $x = \dfrac{x}{m}$; $\dot{E}_{kin} = \dfrac{1}{2}\dot{m}v^2$; $\dot{E}_{pot} = \dot{m}gz$;

$\dot{H} = \dot{m}h = \dot{m}(u + pv)$

Note   On a differential basis, $dh = du + d(pv)$

A.2.3 · Entropy Balance

Transient Uniform Flow

$$\boxed{\sum \dot{S}_{in} - \sum \dot{S}_{out} + \dfrac{\sum \dot{Q}}{T_k} + \sum \dot{S}_{gen} = \left(\dfrac{\Delta s}{\Delta t}\right)_{cv}}$$

During process time Δt (Transient Open) $\sum S_{in} - \sum S_{out} + \dfrac{\sum Q}{T_k} + \sum S_{gen} = (\Delta S)_{cv}$

where

$$\boxed{(\Delta S)_{cv} = \left[(ms)_f - (ms)_i\right]_{cv}}$$

Note   Work performed does not generate entropy but can indirectly cause entropy changes in the substance and hence a system or device.

A.2.4   Balances for Steady Open Systems

During steady-state processes, $(\Delta m)_{cv}$, $(\Delta E)_{cv}$, and $(\Delta S)_{cv}$ are zero. Hence, on a steady rate basis:

$$\boxed{\sum \dot{m}_{in} = \sum \dot{m}_{out}}; \quad \boxed{\sum \dot{E}_{in} = \sum \dot{E}_{out}}; \quad \boxed{\dot{S}_{gen} = \sum \dot{S}_{out} - \sum \dot{S}_{in} - \dfrac{\sum \dot{Q}}{T_k}}$$

A.3   Work

A.3.1   Definitions

$$\delta W = F\,ds$$

$$\sum W = \sum W_{mech} + \sum W_{electrical}$$

$$\sum W_{mech} = \sum W_{boundary} + \sum W_{spring} + \sum W_{shaft} + \sum W_{flow} \cdots \qquad W_{electrical} = V_e I \Delta t$$

With $F_{spring} = k_s S$, $W_{spring} = \displaystyle\int_{s_1}^{s_2} k_s \cdot s\, ds.$

If $k_s = \cancel{c}$ $\qquad W_{spring} = \dfrac{1}{2}k_s(s_2^2 - s_1^2)$ where k_s is the spring constant

$W_{shaft/stir} = 2\pi N\, d\tau$ where N is shaft revolution number and τ is the torque

Importantly: $W_{boundary} = \int_{V_1}^{V_2} p\, dV$ for p-c devices and $W_{flow} = \int_{p_1}^{p_2} V\, dp$ for compressor

A.3.2   Flow Work

Flow work is due to net pressure moving fluid volume V into or out of a system/device

$$\delta W = V\, dp$$

Note Inexact differentials are being used for path-dependent quantities such as δW and δQ

- For polytropic compression, $pV^n = C$. Thus, $W = \int_1^2 \left(\dfrac{C}{p}\right)^{\frac{1}{n}} dp$

  For ideal gases, $W_{poly,compr} = \dfrac{n}{n-1} RT_1 \left[\left(\dfrac{p_2}{p_1}\right)^{\frac{n-1}{n}} - 1\right]; n \neq 1$

- For isothermal compression with $n-1$, $pV = mRT = C$ of an ideal gas:

$$W_{isothermal,compr} = RT_1\, \ln\frac{p_2}{p_1}$$

A.3.3   Boundary Work

Boundary work for polytropic processes of gases: $V^n = C$ or $p(V) = \dfrac{C}{V^n}$; $1 \leq n \leq k = \dfrac{c_p}{c_v}$

$$W_b = C\int_1^2 \frac{dV}{V^n} = \frac{p_2 V_2 - p_1 V_1}{1-n} \text{ for } n \neq 1$$

- For an isothermal process with $n \equiv 1$, $pV = mRT = C$ of an ideal gas:

$$W_b = p_1 V_1 \ln\frac{V_2}{V_1} = p_2 V_2 \ln\frac{V_2}{V_1}$$

- For an ideal gas undergoing an isentropic process with constant specific heats:

$$n = k = \frac{c_p}{c_v}$$

$$W = \frac{mR(T_2 - T_1)}{1-n} \text{ for } n \neq 1 \qquad W = mRT_1 \ln\frac{V_2}{V_1} \text{ for } n \equiv 1$$

Note   For $p = c$, $W = p\,\Delta V$, so that $\Delta H = \Delta U + p\,\Delta V$, as in piston-cylinder devices; thus, 3 unknowns are reduced to 2.

A.4   Systems and Processes

A.4.1   Types of Systems

Open Systems: A steady uniform flow device (turbine, compressor, etc.) or a transient system, e.g., a tank containing a supply line or outlet valve.

Closed Systems: Transient (e.g., p-c device or tank), i.e., no mass flow.

Closed and Open Cycles: Heat engine, heat pump, refrigerator, air conditioner, Carnot cycle, Rankine cycle, Brayton cycle; ICEs such as diesel and Otto engines.

A.4.2   Types of Processes

Heat Exchanger vs. Multiple Stream Mixing: The process of transfer of thermal energy from one stream of matter (solid or fluid) to another by radiation, convection, or conduction to affect a change in temperature of the two streams. If heat is transferred without streams coming in direct contact, then the device is called a heat exchanger. If heat is transferred with streams coming in direct contact, then it is called a mixing chamber. Also, this process may or may not see a phase-change of the stream.

Thus, heat transferred: $\boxed{Q = mc_p \Delta T}$

Expansion vs. Compression: Change in system/fluid volume is referred to as expansion or compression. If the volume increases, then it is expansion (as in turbines) and if the volume decreases then it is compression (as in compressors).

Process categories include:

Adiabatic/Isocaloric Process: A process that occurs with no heat transfer/exchange because of perfect insulation.
$\qquad Q_{in/out} = 0$

Isothermal Process: A process that occurs at constant temperature, typically achieved by device/fluid cooling or heating.
$\qquad T = ¢$

Isochoric Process: A process that occurs at constant volume, which means that boundary work done by the system is zero, e.g., a rigid tank.
$\qquad \forall = ¢$

Isobaric Process: A process that occurs at constant pressure. An example would be an ideal piston-cylinder arrangement (p-c device), where only the piston weight exerts the constant pressure.
$\qquad p = ¢$

Isentropic Process: An ideal, reversible process that is adiabatic; hence, $S_{gen} = 0$. An isentropic process is unattainable in reality.
$\qquad S = ¢$
$\qquad Q_{in/out} = 0$

Isenthalpic Process: No change in enthalpy (e.g., a throttle).
$\qquad h = ¢$

Polytropic Process: Any process that obeys $p\forall^n = C$, i.e., $p_1 \forall_1^n = p_2 \forall_2^n$,
$\qquad p\forall^n = C$

where n is the polytropic index and is always a real number.

Hence, with given specific volumes: $\dfrac{p_2}{p_1} = \left(\dfrac{v_1}{v_2}\right)^n$

For an ideal gas: $\dfrac{T_2}{T_1} = \left(\dfrac{v_1}{v_2}\right)^{n-1}$ and $\dfrac{T_2}{T_1} = \left(\dfrac{p_2}{p_1}\right)^{\frac{n-1}{n}}$

1. If $n = 0$, then $p\forall^0 = p = \mathcal{c}$ and it is an isobaric process (constant pressure)

2. If $n = 1$, then for an ideal gas $p\forall = \mathcal{c}$ and it is an isothermal process (constant T)

3. If $n = k = \dfrac{c_p}{c_v}$, then, for an ideal gas, it is an adiabatic process (no heat transferred)

   Note $1 < k < 2$, where $k = \dfrac{c_p}{c_v} = \dfrac{c_v + R}{c_v} = 1 + \dfrac{R}{c_v} > 1$

4. If $n = \infty$, then it is an isochoric process (constant volume)

A.5   Gibbs Relations

Gibbs equations are used to calculate entropy changes in fluids and solids. Recall that entropy is a path-independent state function. Clausius expressed that $\delta Q_{rev} = T\,ds$. For a basic p-c device:

$$\delta Q - \delta W = dU \text{ so that per unit mass: } T\,ds - p\,dv = du$$

Therefore, *Gibbs I: $ds = \dfrac{du}{T} + \dfrac{p}{T}\,dv$*   *Gibbs II: $ds = \dfrac{dh}{T} - \dfrac{v}{T}\,dp$*

A.6   Types of Fluids

A.6.1   The Ideal Gas

Ideal Gases: Air, He, N_2, CO_2, etc.

Equation of State for Ideal Gases: $p\forall = mRT$

Property Tables for IG: Air is the most common IG in thermodynamic problems. Hence, there exist extensive property tables for ideal gases (see App. B) that can be used directly for solving problems. Using data tables saves time and gives more precise values.

A.6.2   Ideal Gas Relations

- Equation of State: $p\forall = mRT$; recalling $k \equiv \dfrac{c_p}{c_v}$ and $R = c_p - c_v$

- $dh = c_p\,dT$ and $du = c_v\,dT$ as u and h are functions of T only

- Heat capacities

$$C_v \equiv \left(\frac{\partial u}{\partial T}\right)_v = \left(\frac{du}{dT}\right)_{IG} \qquad C_p \equiv \left(\frac{\partial h}{\partial T}\right)_p = \left(\frac{dh}{dT}\right)_{IG}$$

- Gibbs applications

$$\Delta S = m\left[\int_{T_1}^{T_2} \frac{c_p(T)}{T}\,dT - R\,\ln\left(\frac{p_2}{p_1}\right)\right] = m\left[\int_{T_1}^{T_2} \frac{c_v(T)}{T}\,dT + R\,\ln\left(\frac{\forall_2}{\forall_1}\right)\right]$$

$$\int_{T_1}^{T_2} \frac{c_p(T)}{T}\, dT = s_2^0 - s_1^0 \text{ where } s^0 \equiv \int_0^T c_p(T)\, \frac{dT}{T} \text{ (see Property Tables, App. B)}$$

Now Gibbs II can be rewritten as:

$$s_2 - s_1 = s_2^0 - s_1^0 - R \ln\frac{p_2}{p_1}$$

For isentropic process, $s_2 - s_1 = 0$, so that

$$s_2^0 - s_1^0 = R \ln\frac{p_2}{p_1} \text{ or } \frac{p_2}{p_1} = \exp\left(\frac{s_2^0 - s_1^0}{R}\right) = \frac{\exp\left(\dfrac{s_2^0}{R}\right)}{\exp\left(\dfrac{s_1^0}{R}\right)}$$

Introducing the relative pressure,

$$p_r = \exp\left(\frac{s^0}{R}\right)$$

We have

$$\left(\frac{p_2}{p_1}\right)_{s=c} = \frac{p_{r2}}{p_{r1}}$$

Similarly,

$$\left(\frac{v_2}{v_1}\right)_{s=c} = \frac{v_{r2}}{v_{r1}}$$

A.6.3   Two-Phase Fluids

R-134a: Three states of refrigeration fluid: Compressed liquid, saturated mixture, or superheated vapor

H_2O: Three states of H_2O: Compressed liquid (water), saturated mixture, or superheated vapor (steam)

Note   Use Property Tables to obtain two-phase fluid data.

A.6.4   Phase Changes and Diagrams

Pure Substances: A substance that has a fixed chemical composition throughout is called a pure substance. Just two Property Table values (see App. B) are sufficient to determine all the properties of a simple substance.

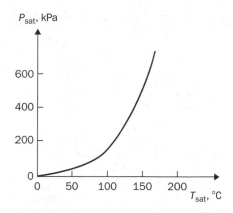

FIGURE A.1 Liquid–vapor saturation of H_2O.

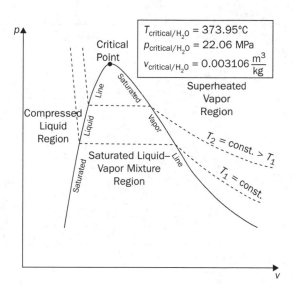

FIGURE A.2 p-v diagram of H_2O.

Phase Change: A pure substance exists in different phases (solid, liquid, or gaseous) depending on its temperature and pressure levels. Transition from one phase to another either by giving away energy or by accepting energy is known as "phase change."

Compressed Liquid: Substance in the liquid phase not about to vaporize.

Superheated Vapor: Substance in the gas phase not about to condense.

Saturation Temperature and Pressure: At a given pressure a substance changes phase at a fixed temperature, called the saturation temperature. Likewise, at a given temperature a substance changes phase at a fixed pressure, called the saturation pressure.

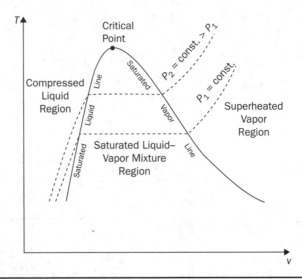

FIGURE A.3   *T-v* diagram of H$_2$O.

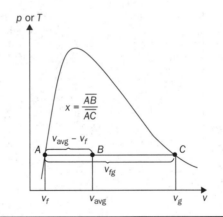

FIGURE A.4   *T-v* or *p-v* diagram of H$_2$O.

Saturated Liquid and Vapor: During a boiling process, both the liquid and the vapor coexist in equilibrium, and under this condition liquid is called saturated liquid and vapor is called saturated vapor.

Critical Point: The state beyond which there is no distinct vaporization process.

Triple Line: All the three phases of a substance coexist in equilibrium at states along the triple line.

Quality: In a saturated vapor-liquid mixture, the mass fraction of vapor is the quality and it is given by

$$x = \frac{m_{\text{vapor}}}{m_{\text{total}}} ; 0 \leq x \leq 1$$

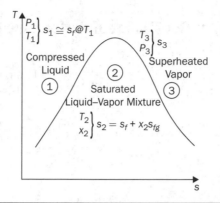

FIGURE A.5   *T*-s diagram of H_2O.

Quality ranges from 0 (saturated liquid) → 1 (saturated vapor). Quality is only significant in the saturated mixture region, and it is used to express average mixture values of any property *y* as given below:

$$y = y_f + xy_{fg}$$

where *f* stands for saturated liquid and *g* for saturated vapor, while *y* could be specific volume (*v*), internal energy (*u*), enthalpy (*h*), or entropy (*s*).

A.7   The Carnot Cycle

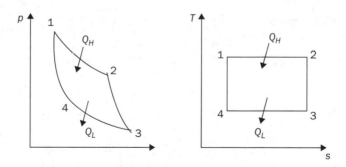

FIGURE A.6   *p*-∀ and *T*-s diagrams of a Carnot cycle.

Note   A Carnot cycle, say, for HEs, Rs, and HPs, consists of two isotherms and two isentropes, generating (theoretically) maximal thermal/cyclic efficiencies.

A.8 Efficiency (η) or Coefficient of Performance (COP)

1st Law for a cycle:

$$W_{net} = Q_H - Q_L \qquad \eta \text{ or COP} = \frac{\text{Desired Output}}{\text{Required Input}} \text{ or } \frac{\text{real}}{\text{ideal}}$$

Isentropic or 2nd-Law efficiencies: $\eta_{T,isentr.} = \dfrac{W_{actual}}{W_{isentr.}} = \eta_{II}^{Turb.}$; $\eta_{C,isentr.} = \dfrac{W_{isentr.}}{W_{actual}} = \eta_{II}^{Compr.}$

1. Heat Engine (HE):

$$\eta_{HE,th} = \frac{W_{net/out}}{Q_{in}} = \frac{Q_H - Q_L}{Q_H} = 1 - \frac{Q_L}{Q_H}$$

- Carnot or Reversible HE:

$$\eta_{HE,th,rev} = \frac{W_{net/out}}{Q_{in}} = \frac{Q_H - Q_L}{Q_H} = 1 - \frac{Q_L}{Q_H} = 1 - \frac{T_L}{T_H}$$

Note $\left(\dfrac{Q_L}{Q_H}\right)_{Carnot} \equiv \dfrac{T_L}{T_H}$

2. Refrigerator (R):

$$COP_{R,th} = \frac{Q_L}{W_{net/in}} = \frac{Q_L}{Q_H - Q_L} = \frac{1}{\dfrac{Q_H}{Q_L} - 1}$$

- Carnot or Reversible R:

$$COP_{R,th,rev} = \frac{Q_L}{W_{net/in}} = \frac{Q_L}{Q_H - Q_L} = \frac{1}{\dfrac{Q_H}{Q_L} - 1} = \frac{1}{\dfrac{T_H}{T_L} - 1}$$

3. Heat Pump (HP):

$$COP_{HP,th} = \frac{Q_H}{W_{net/in}} = \frac{Q_H}{Q_H - Q_L} = \frac{1}{1 - \dfrac{Q_L}{Q_H}}$$

- Carnot or Reversible HP:

$$COP_{HP,th,rev} = \frac{Q_H}{W_{net/in}} = \frac{Q_H}{Q_H - Q_L} = \frac{1}{1 - \dfrac{Q_L}{Q_H}} = \frac{1}{1 - \dfrac{T_L}{T_H}}$$

4. Relation: $COP_{HP} = COP_R + 1$

A.9   Exergy (Useful Work Potential of Energy)

A.9.1   Transient Uniform Flow Equations

Transient Uniform Flow

$$\left(\frac{\Delta EX}{\Delta t}\right)_{system} = \left[\sum \dot{EX}_f - \sum \dot{EX}_i\right]_{CV} = \sum \dot{EX}_{in} - \sum \dot{EX}_{out} - \sum \dot{EX}_{destroyed}$$

During process time Δt

$$\boxed{\Delta EX_{system} = \left[\sum EX_f - \sum EX_i\right]_{CV} = \sum EX_{in} - \sum EX_{out} - \sum EX_{destroyed}}$$

Steady open system $\Delta EX_{system} = 0$; thus, $\sum \dot{EX}_{destroyed} = \sum \dot{EX}_{in} - \sum \dot{EX}_{out}$

A.9.2   Applications

Closed system

$$\boxed{\frac{\Delta EX_{system}}{m} = (u - u_0) + p_0(v - v_0) - T_0(s - s_0)}$$

Open Transient System

$$\boxed{\frac{\Delta EX_{system}}{m} = (h - h_0) - T_0(s - s_0)}$$

Where subscript 0 indicates the dead state. Typically, $T_0 \triangleq T_{amb} = 25°C$ and $p_0 = 1$ atm. Importance of finding Exergy Destruction for system improvement and/or design: $EX_{destroyed} = T_{ref} S_{gen} > 0$; thus, by minimizing $S_{gen} = \Delta S_{system} + \Delta S_{surr}$ we can increase useful work! Typically, $T_{ref} \triangleq T_{amb}$.

A.10   Representative Examples

A.10.1   Solution FORMAT

A proper FORMAT is a simple and effective way to start solving thermodynamics problems. One starts with *three (3) preliminary steps (Sketch, Assumptions, and Method)* followed by the actual solution procedure. A general *Sketch* for a transient open system is shown below. Every system can be modeled as a simplification of a transient open system. *Assumptions* should include the details of the problem that need to be assumed to solve the problem, e.g., type of system, the type of substance, energies that are neglected, etc. *Method* should include reduced equations, need for property tables, laws, or balances that must be used to solve the problem. Each problem should be solved using FORMAT to ensure accurate solutions and communicability to fellow engineers in industry.

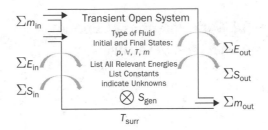

A.10.2 Transient Open System

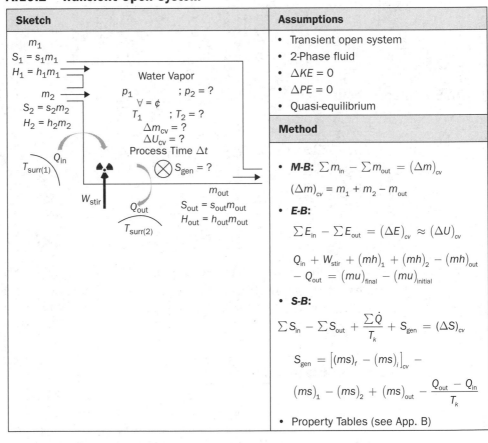

Sketch	Assumptions
	• Transient open system
	• 2-Phase fluid
	• $\Delta KE = 0$
	• $\Delta PE = 0$
	• Quasi-equilibrium

Method

• **M-B:** $\sum m_{in} - \sum m_{out} = (\Delta m)_{cv}$

$(\Delta m)_{cv} = m_1 + m_2 - m_{out}$

• **E-B:**

$$\sum E_{in} - \sum E_{out} = (\Delta E)_{cv} \approx (\Delta U)_{cv}$$

$$Q_{in} + W_{stir} + (mh)_1 + (mh)_2 - (mh)_{out} - Q_{out} = (mu)_{final} - (mu)_{initial}$$

• **S-B:**

$$\sum S_{in} - \sum S_{out} + \frac{\sum \dot{Q}}{T_k} + S_{gen} = (\Delta S)_{cv}$$

$$S_{gen} = \left[(ms)_f - (ms)_i\right]_{cv} - (ms)_1 - (ms)_2 + (ms)_{out} - \frac{Q_{out} - Q_{in}}{T_k}$$

• Property Tables (see App. B)

A.10.3   Closed System: p-c Device

Sketch	Assumptions	Method
Ideal Gas $p_1 = \cancel{c}$; $m = \cancel{c}$ $\forall_1 = ?$; $\forall_2 = ?$ T_1 ; $T_2 = ?$ $\Delta U_{cv} = ?$ $\otimes S_{gen} = ?$ -1 -2 $\downarrow W_b = ?$	• Transient closed system • Ideal gas • $\Delta KE = 0$ • $\Delta PE = 0$ • Adiabatic, i.e., $Q = 0$ • $W_b = p\Delta\forall$	• IG: $p\forall = mRT$ • **M-B:** $$\sum m_{in} - \sum m_{out} = (\Delta m)_{cv} = 0;$$ $$m_1 = m_2$$ • **E-B:** $$\sum E_{in} - \sum E_{out} = (\Delta E)_{cv} \approx (\Delta U)_{cv}$$ $$W_b = m(\Delta u)$$ • **S-B:** $$\sum S_{in} - \sum S_{out} + \frac{\sum \dot{Q}}{T_k} + S_{gen} = (\Delta S)_{cv}$$ $$(\Delta S)_{cv} = \left[(ms)_f - (ms)_i \right]_{cv} = S_{gen}$$ • Property Tables (see App. B)

A.10.4   Steady Open System: Turbine

Sketch	Assumptions	Method
① $\dot{m}, h_1 \begin{cases} T_1 \\ P_1 \end{cases}$ cv $\dot{m}_{steam}$ $\Delta E\vert_{cv} = 0$ $\dot{W}_T = P_T$ ② $\dot{m}, h_2 \begin{cases} T_2 \\ P_2 \end{cases}$	• Steady open system • Water vapor • $\Delta KE \approx 0$ • $\Delta PE = 0$ • Adiabatic; $\dot{Q} = 0$	• **M-B:** $$\sum \dot{m}_{in} - \sum \dot{m}_{out} = (\Delta\dot{m})_{cv}$$ $$(\Delta\dot{m})_{cv} = 0 \text{ for all steady}$$ state devices $$\dot{m}_{steam} = \cancel{c}$$ • **E-B:** $$\sum \dot{E}_{in} - \sum \dot{E}_{out} = (\Delta\dot{E})_{cv} \approx$$ $$(\Delta\dot{U})_{cv} = 0$$ $$\dot{W}_T = \dot{m}(\Delta h) \text{ as } \dot{Q} = 0$$ • **S-B:** $$\sum \dot{S}_{in} - \sum \dot{S}_{out} + \frac{\sum \dot{Q}}{T_k} + \dot{S}_{gen} =$$ $$(\Delta\dot{S})_{cv} = 0$$ $$\dot{S}_{gen} = (\dot{m}s)_2 - (\dot{m}s)_1$$ • Property Tables (see App. B)

A.10.5   Other Common Steady Open Systems

Compressor

Nozzle

Mixing Chamber

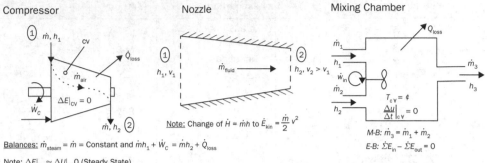

Balances: $\dot{m}_{steam} = \dot{m} =$ Constant and $\dot{m}h_1 + \dot{W}_C = \dot{m}h_2 + \dot{Q}_{loss}$

Note: $\Delta E|_{CV} \approx \Delta U|_{CV} 0$ (Steady State)

Note: Change of $\dot{H} = \dot{m}h$ to $\dot{E}_{kin} = \frac{\dot{m}}{2}v^2$

M-B: $\dot{m}_3 = \dot{m}_1 + \dot{m}_2$

E-B: $\dot{\Sigma E}_{in} - \dot{\Sigma E}_{out} = 0$

Throttle

Heat Exchanger

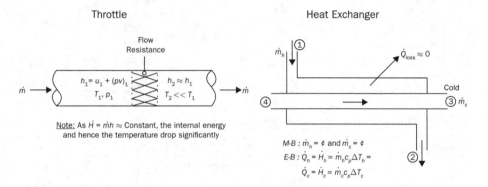

Note: As $\dot{H} = \dot{m}h \approx$ Constant, the internal energy and hence the temperature drop significantly

M-B : $\dot{m}_h = \phi$ and $\dot{m}_c = \phi$

E-B : $\dot{Q}_h = \dot{H}_h = \dot{m}_h c_p \Delta T_h =$

$\dot{Q}_c = \dot{H}_c = \dot{m}_c c_p \Delta T_c$

A.10.6   HE Cycle and R Cycle (or HP)

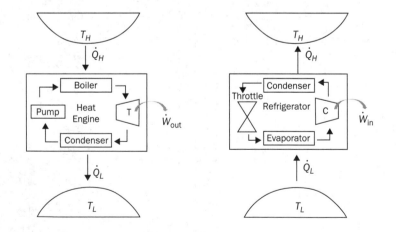

A.11 Linear Interpolation Formula

In the Property Tables, property values are specified for discrete temperature or pressure. For example,

Temperature	Pressure	Property Value
T_1	p_1	X_1
T_2	p_2	X_2

If the value of the property at a temperature or pressure that lies in between the printed pressure and temperature values is required, linear interpolation should be used. With the sample formulas below one can compute values at the new temperature or pressure.

At temperature T: $X = X_1 + \dfrac{(X_2 - X_1)}{T_2 - T_1}(T - T_1)$; for T between T_1 and T_2

At pressure p: $X = X_1 + \dfrac{(X_2 - X_1)}{p_2 - p_1}(p - p_1)$; for p between p_1 and p_2

A.12 Nomenclature

COP   Coefficient of performance [1]

C_p Specific heat at $p = ¢ \left[\dfrac{kJ}{kg \cdot K}\right]$

C_v Specific heat at $v = ¢ \left[\dfrac{kJ}{kg \cdot K}\right]$

E Energy [kJ]

$\dot{E}$ Energy rate $\left[\dfrac{kJ}{s}\right] = [kW]$

EX Exergy [kJ]

h Specific enthalpy $\left[\dfrac{kJ}{kg}\right]$

H Enthalpy [kJ]

I Electric current [A]

k Adiabatic index [1]

k_s Spring constant $\left[\dfrac{kg}{s^2}\right]$

m Mass [kg]

$\dot{m}$ Mass flow rate $\left[\dfrac{kg}{s}\right]$

M Molecular weight $\left[\dfrac{\text{kg}}{\text{kmol}}\right]$

n Polytropic index [1]

η Efficiency [1]

N Revolutions per min [RPM]

ρ Density $\left[\dfrac{\text{kg}}{\text{m}^3}\right]$

p Pressure [kPa]

P Power $\left[\dfrac{\text{kJ}}{\text{s}} = \text{kW}\right]$

Q Heat [kJ]

$\dot{Q}$ Heat flow rate $\left[\dfrac{\text{kJ}}{\text{s}}\right]$

R Gas constant $\left[\dfrac{\text{kJ}}{\text{kg·K}}\right]$

R_u Universal gas constant $\left[\dfrac{\text{kJ}}{\text{kmol·K}}\right]$

s Specific entropy $\left[\dfrac{\text{kJ}}{\text{kg·K}}\right]$

$s^0 = s^0(T)$ only

S Entropy $\left[\dfrac{\text{kJ}}{\text{K}}\right]$

$\dot{S}$ Entropy rate $\left[\dfrac{\text{kJ}}{\text{s·K}}\right]$

τ Torque [N·m]

t Time [s]

T Temperature [K or °C]

u Specific internal energy $\left[\dfrac{\text{kJ}}{\text{kg}}\right]$

U Internal energy [kJ]

ν Specific volume $\left[\dfrac{\text{m}^3}{\text{kg}}\right]$

v Velocity $\left[\dfrac{m}{s}\right]$

V_e Voltage [V]

$\forall$ Volume [m³]

W Work [kJ $\cong 10^3$ N·m]

$\dot{W}$ Work rate = Power $\left[\dfrac{kJ}{s} = kW\right]$

APPENDIX B

Property Tables with Conversion Factors

Substance	Formula	Molar mass M [kg/kmol]	Gas constant R [kJ/kg·K*]	Temperature [K]	Pressure [MPa]	Volume [m³/kmol]
Air	—	28.97	0.2870	132.5	3.77	0.0883
Ammonia	NH_3	17.03	0.4882	405.5	11.28	0.0724
Argon	Ar	39.948	0.2081	151	4.86	0.0749
Benzene	C_6H_6	78.115	0.1064	562	4.92	0.2603
Bromine	Br_2	159.808	0.0520	584	10.34	0.1355
n-Butane	C_4H_{10}	58.124	0.1430	425.2	3.80	0.2547
Carbon dioxide	CO_2	44.01	0.1889	304.2	7.39	0.0943
Carbon monoxide	CO	28.011	0.2968	133	3.50	0.0930
Carbon tetrachloride	CCl_4	153.82	0.05405	556.4	4.56	0.2759
Chlorine	Cl_2	70.906	0.1173	417	7.71	0.1242
Chloroform	$CHCl_3$	119.38	0.06964	536.6	5.47	0.2403
Dichlorodi-fluoromethane (R-12)	CCl_2F_2	120.91	0.06876	384.7	4.01	0.2179
Dichloro-fluoromethane (R-21)	$CHCl_2F$	102.92	0.08078	451.7	5.17	0.1973
Ethane	C_2H_6	30.070	0.2765	305.5	4.48	0.1480
Ethyl alcohol	C_2H_5OH	46.07	0.1805	516	6.38	0.1673
Ethylene	C_2H_4	28.054	0.2964	282.4	5.12	0.1242
Helium	He	4.003	2.0769	5.3	0.23	0.0578
n-Hexane	C_6H_{14}	86.179	0.09647	507.9	3.03	0.3677
Hydrogen (normal)	H_2	2.016	4.1240	33.3	1.30	0.0649
Krypton	Kr	83.80	0.09921	209.4	5.50	0.0924
Methane	CH_4	16.043	0.5182	191.1	4.64	0.0993
Methyl alcohol	CH_3OH	32.042	0.2595	513.2	7.95	0.1180
Methyl chloride	CH_3Cl	50.488	0.1647	416.3	6.68	0.1430
Neon	Ne	20.183	0.4119	44.5	2.73	0.0417

TABLE B.1 Molar Mass, Gas Constant, and Critical-Point Properties

Substance	Formula	Molar mass M [kg/kmol]	Gas constant R [kJ/kg·K*]	Temperature [K]	Pressure [MPa]	Volume [m³/kmol]
Nitrogen	N_2	28.013	0.2968	126.2	3.39	0.0899
Nitrous oxide	N_2O	44.013	0.1889	309.7	7.27	0.0961
Oxygen	O_2	31.999	0.2598	154.8	5.08	0.0780
Propane	C_3H_8	44.097	0.1885	370	4.26	0.1998
Propylene	C_3H_6	42.081	0.1976	365	4.62	0.1810
Sulfur dioxide	SO_2	64.063	0.1298	430.7	7.88	0.1217
Tetrafluoroethane (R-134a)	CF_3CH_2F	102.03	0.08149	374.2	4.059	0.1993
Trichloro-fluoromethane (R-11)	CCl_3F	137.37	0.06052	471.2	4.38	0.2478
Water	H_2O	18.015	0.4615	647.1	22.06	0.0560
Xenon	Xe	131.30	0.06332	289.8	5.88	0.1186

*The unit kJ/kg·K is equivalent to kPa·m³/kg·K. The gas constant is calculated from $R = R_u/M$, where $R_u = 8.31447$ kJ/kmol·K and M is the molar mass.
Source: K. A. Kobe and R. E. Lynn, Jr., *Chemical Review* 52 (1953), pp. 117–236; and *ASHRAE, Handbook of Fundamentals* (Atlanta, GA: American Society of Heating, Refrigerating and Air-Conditioning Engineers, Inc., 1993), pp. 16.4 and 36.1.

TABLE B.1 Molar Mass, Gas Constant, and Critical-Point Properties (*Concluded*)

Gas	Formula	Gas constant R [kJ/kg·K]	c_p [kJ/kg·K]	c_v [kJ/kg·K]	k
Air	—	0.2870	1.005	0.718	1.400
Argon	Ar	0.2081	0.5203	0.3122	1.667
Butane	C_4H_{10}	0.1433	1.7164	1.5734	1.091
Carbon dioxide	CO_2	0.1889	0.846	0.657	1.289
Carbon monoxide	CO	0.2968	1.040	0.744	1.400
Ethane	C_2H_6	0.2765	1.7662	1.4897	1.186
Ethylene	C_2H_4	0.2964	1.5482	1.2518	1.237
Helium	He	2.0769	5.1926	3.1156	1.667
Hydrogen	H_2	4.1240	14.307	10.183	1.405
Methane	CH_4	0.5182	2.2537	1.7354	1.299
Neon	Ne	0.4119	1.0299	0.6179	1.667
Nitrogen	N_2	0.2968	1.039	0.743	1.400
Octane	C_8H_{18}	0.0729	1.7113	1.6385	1.044
Oxygen	O_2	0.2598	0.918	0.658	1.395
Propane	C_3H_8	0.1885	1.6794	1.4909	1.126
Steam	H_2O	0.4615	1.8723	1.4108	1.327

Note: The unit kJ/kg·K is equivalent to kJ/kg·°C.
Source: *Chemical and Process Thermodynamics* 3/E by Kyle, B. G., © 2000. Adapted by permission of Pearson Education, Inc., Upper Saddle River, NJ.

TABLE B.2 Ideal-Gas Specific Heats of Various Common Gases: (a) At 300 K

Temperature [K]	Air			Carbon dioxide, CO_2			Carbon monoxide, CO		
	c_p [kJ/kg·K]	c_v [kJ/kg·K]	k	c_p [kJ/kg·K]	c_v [kJ/kg·K]	k	c_p [kJ/kg·K]	c_v [kJ/kg·K]	k
250	1.003	0.716	1.401	0.791	0.602	1.314	1.039	0.743	1.400
300	1.005	0.718	1.400	0.846	0.657	1.288	1.040	0.744	1.399
350	1.008	0.721	1.398	0.895	0.706	1.268	1.043	0.746	1.398
400	1.013	0.726	1.395	0.939	0.750	1.252	1.047	0.751	1.395
450	1.020	0.733	1.391	0.978	0.790	1.239	1.054	0.757	1.392
500	1.029	0.742	1.387	1.014	0.825	1.229	1.063	0.767	1.387
550	1.040	0.753	1.381	1.046	0.857	1.220	1.075	0.778	1.382
600	1.051	0.764	1.376	1.075	0.886	1.213	1.087	0.790	1.376
650	1.063	0.776	1.370	1.102	0.913	1.207	1.100	0.803	1.370
700	1.075	0.788	1.364	1.126	0.937	1.202	1.113	0.816	1.364
750	1.087	0.800	1.359	1.148	0.959	1.197	1.126	0.829	1.358
800	1.099	0.812	1.354	1.169	0.980	1.193	1.139	0.842	1.353
900	1.121	0.834	1.344	1.204	1.015	1.186	1.163	0.866	1.343
1000	1.142	0.855	1.336	1.234	1.045	1.181	1.185	0.888	1.335

Temperature [K]	Hydrogen, H_2			Nitrogen, N_2			Oxygen, O_2		
	c_p [kJ/kg·K]	c_v [kJ/kg·K]	k	c_p [kJ/kg·K]	c_v [kJ/kg·K]	k	c_p [kJ/kg·K]	c_v [kJ/kg·K]	k
250	14.051	9.927	1.416	1.039	0.742	1.400	0.913	0.653	1.398
300	14.307	10.183	1.405	1.039	0.743	1.400	0.918	0.658	1.395
350	14.427	10.302	1.400	1.041	0.744	1.399	0.928	0.668	1.389
400	14.476	10.352	1.398	1.044	0.747	1.397	0.941	0.681	1.382
450	14.501	10.377	1.398	1.049	0.752	1.395	0.956	0.696	1.373
500	14.513	10.389	1.397	1.056	0.759	1.391	0.972	0.712	1.365
550	14.530	10.405	1.396	1.065	0.768	1.387	0.988	0.728	1.358
600	14.546	10.422	1.396	1.075	0.778	1.382	1.003	0.743	1.350
650	14.571	10.447	1.395	1.086	0.789	1.376	1.017	0.758	1.343
700	14.604	10.480	1.394	1.098	0.801	1.371	1.031	0.771	1.337
750	14.645	10.521	1.392	1.110	0.813	1.365	1.043	0.783	1.332
800	14.695	10.570	1.390	1.121	0.825	1.360	1.054	0.794	1.327
900	14.822	10.698	1.385	1.145	0.849	1.349	1.074	0.814	1.319
1000	14.983	10.859	1.380	1.167	0.870	1.341	1.090	0.830	1.313

Source: Kenneth Wark, *Thermodynamics,* 4th ed. (New York: McGraw-Hill, 1983), p. 783, Table A–4M. Originally published in *Tables of Thermal Properties of Gases,* NBS Circular 564, 1955.

TABLE B.2   Ideal-Gas Specific Heats of Various Common Gases (*Continued*): (b) At Various Temperatures

Substance	Formula	a	b	c	d	Temperature range [K]	% error	
							Max.	Avg.
Nitrogen	N_2	28.90	-0.1571×10^{-2}	0.8081×10^{-5}	-2.873×10^{-9}	273–1800	0.59	0.34
Oxygen	O_2	25.48	1.520×10^{-2}	-0.7155×10^{-5}	1.312×10^{-9}	273–1800	1.19	0.28
Air	—	28.11	0.1967×10^{-2}	0.4802×10^{-5}	-1.966×10^{-9}	273–1800	0.72	0.33
Hydrogen	H_2	29.11	-0.1916×10^{-2}	0.4003×10^{-5}	-0.8704×10^{-9}	273–1800	1.01	0.26
Carbon monoxide	CO	28.16	0.1675×10^{-2}	0.5372×10^{-5}	-2.222×10^{-9}	273–1800	0.89	0.37
Carbon dioxide	CO_2	22.26	5.981×10^{-2}	-3.501×10^{-5}	7.469×10^{-9}	273–1800	0.67	0.22
Water vapor	H_2O	32.24	0.1923×10^{-2}	1.055×10^{-5}	-3.595×10^{-9}	273–1800	0.53	0.24
Nitric oxide	NO	29.34	-0.09395×10^{-2}	0.9747×10^{-5}	-4.187×10^{-9}	273–1500	0.97	0.36
Nitrous oxide	N_2O	24.11	5.8632×10^{-2}	-3.562×10^{-5}	10.58×10^{-9}	273–1500	0.59	0.26
Nitrogen dioxide	NO_2	22.9	5.715×10^{-2}	-3.52×10^{-5}	7.87×10^{-9}	273–1500	0.46	0.18
Ammonia	NH_3	27.568	2.5630×10^{-2}	0.99072×10^{-5}	-6.6909×10^{-9}	273–1500	0.91	0.36
Sulfur	S_2	27.21	2.218×10^{-2}	-1.628×10^{-5}	3.986×10^{-9}	273–1800	0.99	0.38
Sulfur dioxide	SO_2	25.78	5.795×10^{-2}	-3.812×10^{-5}	8.612×10^{-9}	273–1800	0.45	0.24
Sulfur trioxide	SO_3	16.40	14.58×10^{-2}	-11.20×10^{-5}	32.42×10^{-9}	273–1300	0.29	0.13
Acetylene	C_2H_2	21.8	9.2143×10^{-2}	-6.527×10^{-5}	18.21×10^{-9}	273–1500	1.46	0.59
Benzene	C_6H_6	-36.22	48.475×10^{-2}	-31.57×10^{-5}	77.62×10^{-9}	273–1500	0.34	0.20
Methanol	CH_4O	19.0	9.152×10^{-2}	-1.22×10^{-5}	-8.039×10^{-9}	273–1000	0.18	0.08
Ethanol	C_2H_6O	19.9	20.96×10^{-2}	-10.38×10^{-5}	20.05×10^{-9}	273–1500	0.40	0.22
Hydrogen chloride	HCl	30.33	-0.7620×10^{-2}	1.327×10^{-5}	-4.338×10^{-9}	273–1500	0.22	0.08
Methane	CH_4	19.89	5.024×10^{-2}	1.269×10^{-5}	-11.01×10^{-9}	273–1500	1.33	0.57
Ethane	C_2H_6	6.900	17.27×10^{-2}	-6.406×10^{-5}	7.285×10^{-9}	273–1500	0.83	0.28
Propane	C_3H_8	-4.04	30.48×10^{-2}	-15.72×10^{-5}	31.74×10^{-9}	273–1500	0.40	0.12
n-Butane	C_4H_{10}	3.96	37.15×10^{-2}	-18.34×10^{-5}	35.00×10^{-9}	273–1500	0.54	0.24
i-Butane	C_4H_{10}	-7.913	41.60×10^{-2}	-23.01×10^{-5}	49.91×10^{-9}	273–1500	0.25	0.13
n-Pentane	C_5H_{12}	6.774	45.43×10^{-2}	-22.46×10^{-5}	42.29×10^{-9}	273–1500	0.56	0.21
n-Hexane	C_6H_{14}	6.938	55.22×10^{-2}	-28.65×10^{-5}	57.69×10^{-9}	273–1500	0.72	0.20
Ethylene	C_2H_4	3.95	15.64×10^{-2}	-8.344×10^{-5}	17.67×10^{-9}	273–1500	0.54	0.13
Propylene	C_3H_6	3.15	23.83×10^{-2}	-12.18×10^{-5}	24.62×10^{-9}	273–1500	0.73	0.17

Source: B. G. Kyle, *Chemical and Process Thermodynamics* (Englewood Cliffs, NJ: Prentice-Hall, 1984). Used with permission.

Table B.2 Ideal-Gas Specific Heats of Various Common Gases (*Concluded*): (c) As a Function of Temperature [$c_p = a + bT + cT^2 + dT^3$ (T in K, c_p in kJ/kmol·K)]

Substance	Boiling data at 1 atm		Freezing data		Liquid properties		
	Normal boiling point [°C]	Latent heat of vaporization h_{fg} [kJ/kg]	Freezing point [°C]	Latent heat of fusion h_{if} [kJ/kg]	Temperature [°C]	Density ρ [kg/m³]	Specific heat c_p [kJ/kg·K]
Ammonia	−33.3	1357	−77.7	322.4	−33.3	682	4.43
					−20	665	4.52
					0	639	4.60
					25	602	4.80
Argon	−185.9	161.6	−189.3	28	−185.6	1394	1.14
Benzene	80.2	394	5.5	126	20	879	1.72
Brine (20% sodium chloride by mass)	103.9	—	−17.4	—	20	1150	3.11
n-Butane	−0.5	385.2	−138.5	80.3	−0.5	601	2.31
Carbon dioxide	−78.4*	230.5 (at 0°C)	−56.6		0	298	0.59
Ethanol	78.2	838.3	−114.2	109	25	783	2.46
Ethyl alcohol	78.6	855	−156	108	20	789	2.84
Ethylene glycol	198.1	800.1	−10.8	181.1	20	1109	2.84
Glycerine	179.9	974	18.9	200.6	20	1261	2.32
Helium	−268.9	22.8	—	—	−268.9	146.2	22.8
Hydrogen	−252.8	445.7	−259.2	59.5	−252.8	70.7	10.0
Isobutane	−11.7	367.1	−160	105.7	−11.7	593.8	2.28
Kerosene	204–293	251	−24.9	—	20	820	2.00
Mercury	356.7	294.7	−38.9	11.4	25	13,560	0.139
Methane	−161.5	510.4	−182.2	58.4	−161.5	423	3.49
					−100	301	5.79
Methanol	64.5	1100	−97.7	99.2	25	787	2.55
Nitrogen	−195.8	198.6	−210	25.3	−195.8	809	2.06

Table B.3 Properties of Common Liquids, Solids, and Foods: (a) Liquids

| Substance | Boiling data at 1 atm | | Freezing data | | Liquid properties | | |
	Normal boiling point [°C]	Latent heat of vaporization h_{fg} [kJ/kg]	Freezing point [°C]	Latent heat of fusion h_{if} [kJ/kg]	Temperature [°C]	Density ρ [kg/m³]	Specific heat c_p [kJ/kg·K]
Octane	124.8	306.3	−57.5	180.7	20	703	2.10
Oil (light)					25	910	1.80
Oxygen	−183	212.7	−218.8	13.7	−183	1141	1.71
Petroleum	—	230–384			20	640	2.0
Propane	−42.1	427.8	−187.7	80.0	−42.1	581	2.25
					0	529	2.53
					50	449	3.13
Refrigerant-134a	−26.1	217.0	−96.6	—	−50	1443	1.23
					−26.1	1374	1.27
					0	1295	1.34
					25	1207	1.43
Water	100	2257	0.0	333.7	0	1000	4.22
					25	997	4.18
					50	988	4.18
					75	975	4.19
					100	958	4.22

*Sublimation temperature. (At pressures below the triple-point pressure of 518 kPa, carbon dioxide exists as a solid or gas. Also, the freezing-point temperature of carbon dioxide is the triple-point temperature of −56.5°C.)

TABLE B.3 Properties of Common Liquids, Solids, and Foods (*Continued*): (a) Liquids

Metals

Substance	Density ρ [kg/m³]	Specific heat c_p [kJ/kg·K]
Aluminum		
200 K		0.797
250 K		0.859
300 K	2700	0.902
350 K		0.929
400 K		0.949
450 K		0.973
500 K		0.997
Bronze (76% Cu, 2% Zn, 2% Al)	8,280	0.400
Brass, yellow (65% Cu, 35% Zn)	8310	0.400
Copper		
−173°C		0.254
−100°C		0.342
−50°C		0.367
0°C		0.381
27°C	8900	0.386
100°C		0.393
200°C		0.403
Iron	7840	0.45
Lead	11,310	0.128
Magnesium	1730	1.030
Nickel	8890	0.440
Silver	10,470	0.235
Steel, mild	7830	0.500
Tungsten	19,400	0.130

Nonmetals

Substance	Density ρ [kg/m³]	Specific heat c_p [kJ/kg·K]
Asphalt	2110	0.920
Brick, common	1922	0.79
Brick, fireclay (500°C)	2300	0.960
Concrete	2300	0.653
Clay	1000	0.920
Diamond	2420	0.616
Glass, window	2700	0.800
Glass, pyrex	2230	0.840
Graphite	2500	0.711
Granite	2700	1.017
Gypsum or plaster board	800	1.09
Ice		
200 K		1.56
220 K		1.71
240 K		1.86
260 K		2.01
273 K	921	2.11
Limestone	1650	0.909
Marble	2600	0.880
Plywood (Douglas fir)	545	1.21
Rubber (soft)	1100	1.840
Rubber (hard)	1150	2.009
Sand	1520	0.800
Stone	1500	0.800
Woods, hard (maple, oak, etc.)	721	1.26
Woods, soft (fir, pine, etc.)	513	1.38

TABLE B.3 Properties of Common Liquids, Solids, and Foods (*Continued*): (b) Solids (Values Are for Room Temperature Unless Indicated Otherwise)

Table B.3 Properties of Common Liquids, Solids, and Foods (Concluded): (c) Foods

Food	Water content, % (mass)	Freezing point [°C]	Specific heat [kJ/kg·K] Above freezing	Below freezing	Latent heat of fusion [kJ/kg]
Apples	84	-1.1	3.65	1.90	281
Bananas	75	-0.8	3.35	1.78	251
Beef round	67	—	3.08	1.68	224
Broccoli	90	-0.6	3.86	1.97	301
Butter	16	—	—	1.04	53
Cheese, swiss	39	-10.0	2.15	1.33	130
Cherries	80	-1.8	3.52	1.85	267
Chicken	74	-2.8	3.32	1.77	247
Corn, sweet	74	-0.6	3.32	1.77	247
Eggs, whole	74	-0.6	3.32	1.77	247
Ice cream	63	-5.6	2.95	1.63	210

Food	Water content, % (mass)	Freezing point [°C]	Specific heat [kJ/kg·K] Above freezing	Below freezing	Late heat of fusion [kJ/kg]
Lettuce	95	-0.2	4.02	2.04	317
Milk, whole	88	-0.6	3.79	1.95	294
Oranges	87	-0.8	3.75	1.94	291
Potatoes	78	-0.6	3.45	1.82	261
Salmon fish	64	-2.2	2.98	1.65	214
Shrimp	83	-2.2	3.62	1.89	277
Spinach	93	-0.3	3.96	2.01	311
Strawberries	90	-0.8	3.86	1.97	301
Tomatoes, ripe	94	-0.5	3.99	2.02	314
Turkey	64	—	2.98	1.65	214
Watermelon	93	-0.4	3.96	2.01	311

Source: Values are obtained from various handbooks and other sources or are calculated. Water content and freezing-point data of foods are from *ASHRAE, Handbook of Fundamentals,* SI version (Atlanta, GA: American Society of Heating, Refrigerating and Air-Conditioning Engineers, Inc., 1993), Chapter 30, Table 1. Freezing point is the temperature at which freezing starts for fruits and vegetables, and the average freezing temperature for other foods.

Temp. T [°C]	Sat. press. P_{sat} [kPa]	Specific volume [m³/kg]		Internal energy [kJ/kg]			Enthalpy [kJ/kg]			Entropy [kJ/kg·K]		
		Sat. liquid v_f	Sat. vapor v_g	Sat. liquid u_f	Evap. u_{fg}	Sat. vapor u_g	Sat. liquid h_f	Evap. h_{fg}	Sat. vapor h_g	Sat. liquid s_f	Evap. s_{fg}	Sat. vapor s_g
0.01	0.6117	0.001000	206.00	0.000	2374.9	2374.9	0.001	2500.9	2500.9	0.0000	9.1556	9.1556
5	0.8725	0.001000	147.03	21.019	2360.8	2381.8	21.020	2489.1	2510.1	0.0763	8.9487	9.0249
10	1.2281	0.001000	106.32	42.020	2346.6	2388.7	42.022	2477.2	2519.2	0.1511	8.7488	8.8999
15	1.7057	0.001001	77.885	62.980	2332.5	2395.5	62.982	2465.4	2528.3	0.2245	8.5559	8.7803
20	2.3392	0.001002	57.762	83.913	2318.4	2402.3	83.915	2453.5	2537.4	0.2965	8.3696	8.6661
25	3.1698	0.001003	43.340	104.83	2304.3	2409.1	104.83	2441.7	2546.5	0.3672	8.1895	8.5567
30	4.2469	0.001004	32.879	125.73	2290.2	2415.9	125.74	2429.8	2555.6	0.4368	8.0152	8.4520
35	5.6291	0.001006	25.205	146.63	2276.0	2422.7	146.64	2417.9	2564.6	0.5051	7.8466	8.3517
40	7.3851	0.001008	19.515	167.53	2261.9	2429.4	167.53	2406.0	2573.5	0.5724	7.6832	8.2556
45	9.5953	0.001010	15.251	188.43	2247.7	2436.1	188.44	2394.0	2582.4	0.6386	7.5247	8.1633
50	12.352	0.001012	12.026	209.33	2233.4	2442.7	209.34	2382.0	2591.3	0.7038	7.3710	8.0748
55	15.763	0.001015	9.5639	230.24	2219.1	2449.3	230.26	2369.8	2600.1	0.7680	7.2218	7.9898
60	19.947	0.001017	7.6670	251.16	2204.7	2455.9	251.18	2357.7	2608.8	0.8313	7.0769	7.9082
65	25.043	0.001020	6.1935	272.09	2190.3	2462.4	272.12	2345.4	2617.5	0.8937	6.9360	7.8296
70	31.202	0.001023	5.0396	293.04	2175.8	2468.9	293.07	2333.0	2626.1	0.9551	6.7989	7.7540
75	38.597	0.001026	4.1291	313.99	2161.3	2475.3	314.03	2320.6	2634.6	1.0158	6.6655	7.6812
80	47.416	0.001029	3.4053	334.97	2146.6	2481.6	335.02	2308.0	2643.0	1.0756	6.5355	7.6111
85	57.868	0.001032	2.8261	355.96	2131.9	2487.8	356.02	2295.3	2651.4	1.1346	6.4089	7.5435
90	70.183	0.001036	2.3593	376.97	2117.0	2494.0	377.04	2282.5	2659.6	1.1929	6.2853	7.4782
95	84.609	0.001040	1.9808	398.00	2102.0	2500.1	398.09	2269.6	2667.6	1.2504	6.1647	7.4151
100	101.42	0.001043	1.6720	419.06	2087.0	2506.0	419.17	2256.4	2675.6	1.3072	6.0470	7.3542
105	120.90	0.001047	1.4186	440.15	2071.8	2511.9	440.28	2243.1	2683.4	1.3634	5.9319	7.2952
110	143.38	0.001052	1.2094	461.27	2056.4	2517.7	461.42	2229.7	2691.1	1.4188	5.8193	7.2382

Table B.4 Saturated Water—Temperature

Temp. T [°C]	Sat. press. P_{sat} [kPa]	Specific volume [m³/kg]		Internal energy [kJ/kg]			Enthalpy [kJ/kg]			Entropy [kJ/kg·K]		
		Sat. liquid V_f	Sat. vapor V_g	Sat. liquid U_f	Evap. U_{fg}	Sat. vapor U_g	Sat. liquid h_f	Evap. h_{fg}	Sat. vapor h_g	Sat. liquid S_f	Evap. S_{fg}	Sat. vapor S_g
115	169.18	0.001056	1.0360	482.42	2040.9	2523.3	482.59	2216.0	2698.6	1.4737	5.7092	7.1829
120	198.67	0.001060	0.89133	503.60	2025.3	2528.9	503.81	2202.1	2706.0	1.5279	5.6013	7.1292
125	232.23	0.001065	0.77012	524.83	2009.5	2534.3	525.07	2188.1	2713.1	1.5816	5.4956	7.0771
130	270.28	0.001070	0.66808	546.10	1993.4	2539.5	546.38	2173.7	2720.1	1.6346	5.3919	7.0265
135	313.22	0.001075	0.58179	567.41	1977.3	2544.7	567.75	2159.1	2726.9	1.6872	5.2901	6.9773
140	361.53	0.001080	0.50850	588.77	1960.9	2549.6	589.16	2144.3	2733.5	1.7392	5.1901	6.9294
145	415.68	0.001085	0.44600	610.19	1944.2	2554.4	610.64	2129.2	2739.8	1.7908	5.0919	6.8827
150	476.16	0.001091	0.39248	631.66	1927.4	2559.1	632.18	2113.8	2745.9	1.8418	4.9953	6.8371
155	543.49	0.001096	0.34648	653.19	1910.3	2563.5	653.79	2098.0	2751.8	1.8924	4.9002	6.7927
160	618.23	0.001102	0.30680	674.79	1893.0	2567.8	675.47	2082.0	2757.5	1.9426	4.8066	6.7492
165	700.93	0.001108	0.27244	696.46	1875.4	2571.9	697.24	2065.6	2762.8	1.9923	4.7143	6.7067
170	792.18	0.001114	0.24260	718.20	1857.5	2575.7	719.08	2048.8	2767.9	2.0417	4.6233	6.6650
175	892.60	0.001121	0.21659	740.02	1839.4	2579.4	741.02	2031.7	2772.7	2.0906	4.5335	6.6242
180	1002.8	0.001127	0.19384	761.92	1820.9	2582.8	763.05	2014.2	2777.2	2.1392	4.4448	6.5841
185	1123.5	0.001134	0.17390	783.91	1802.1	2586.0	785.19	1996.2	2781.4	2.1875	4.3572	6.5447
190	1255.2	0.001141	0.15636	806.00	1783.0	2589.0	807.43	1977.9	2785.3	2.2355	4.2705	6.5059
195	1398.8	0.001149	0.14089	828.18	1763.6	2591.7	829.78	1959.0	2788.8	2.2831	4.1847	6.4678
200	1554.9	0.001157	0.12721	850.46	1743.7	2594.2	852.26	1939.8	2792.0	2.3305	4.0997	6.4302
205	1724.3	0.001164	0.11508	872.86	1723.5	2596.4	874.87	1920.0	2794.8	2.3776	4.0154	6.3930
210	1907.7	0.001173	0.10429	895.38	1702.9	2598.3	897.61	1899.7	2797.3	2.4245	3.9318	6.3563
215	2105.9	0.001181	0.094680	918.02	1681.9	2599.9	920.50	1878.8	2799.3	2.4712	3.8489	6.3200
220	2319.6	0.001190	0.086094	940.79	1660.5	2601.3	943.55	1857.4	2801.0	2.5176	3.7664	6.2840
225	2549.7	0.001199	0.078405	963.70	1638.6	2602.3	966.76	1835.4	2802.2	2.5639	3.6844	6.2483

TABLE B.4 Saturated Water—Temperature (*Continued*)

Temp. T [°C]	Sat. press. P_{sat} [kPa]	Specific volume [m³/kg]		Internal energy [kJ/kg]			Enthalpy [kJ/kg]			Entropy [kJ/kg·K]		
		Sat. liquid v_f	Sat. vapor v_g	Sat. liquid u_f	Evap. u_{fg}	Sat. vapor u_g	Sat. liquid h_f	Evap. h_{fg}	Sat. vapor h_g	Sat. liquid s_f	Evap. s_{fg}	Sat. vapor s_g
230	2797.1	0.001209	0.071505	986.76	1616.1	2602.9	990.14	1812.8	2802.9	2.6100	3.6028	6.2128
235	3062.6	0.001219	0.065300	1010.0	1593.2	2603.2	1013.7	1789.5	2803.2	2.6560	3.5216	6.1775
240	3347.0	0.001229	0.059707	1033.4	1569.8	2603.1	1037.5	1765.5	2803.0	2.7018	3.4405	6.1424
245	3651.2	0.001240	0.054656	1056.9	1545.7	2602.7	1061.5	1740.8	2802.2	2.7476	3.3596	6.1072
250	3976.2	0.001252	0.050085	1080.7	1521.1	2601.8	1085.7	1715.3	2801.0	2.7933	3.2788	6.0721
255	4322.9	0.001263	0.045941	1104.7	1495.8	2600.5	1110.1	1689.0	2799.1	2.8390	3.1979	6.0369
260	4692.3	0.001276	0.042175	1128.8	1469.9	2598.7	1134.8	1661.8	2796.6	2.8847	3.1169	6.0017
265	5085.3	0.001289	0.038748	1153.3	1443.2	2596.5	1159.8	1633.7	2793.5	2.9304	3.0358	5.9662
270	5503.0	0.001303	0.035622	1177.9	1415.7	2593.7	1185.1	1604.6	2789.7	2.9762	2.9542	5.9305
275	5946.4	0.001317	0.032767	1202.9	1387.4	2590.3	1210.7	1574.5	2785.2	3.0221	2.8723	5.8944
280	6416.6	0.001333	0.030153	1228.2	1358.2	2585.4	1236.7	1543.2	2779.9	3.0681	2.7898	5.8579
285	6914.6	0.001349	0.027755	1253.7	1328.1	2581.8	1263.1	1510.7	2773.7	3.1144	2.7066	5.8210
290	7441.8	0.001366	0.025554	1279.7	1296.9	2576.5	1289.8	1476.9	2766.7	3.1608	2.6225	5.7834
295	7999.0	0.001384	0.023528	1306.0	1264.5	2570.5	1317.1	1441.6	2758.7	3.2076	2.5374	5.7450
300	8587.9	0.001404	0.021659	1332.7	1230.9	2563.6	1344.8	1404.8	2749.6	3.2548	2.4511	5.7059
305	9209.4	0.001425	0.019932	1360.0	1195.9	2555.8	1373.1	1366.3	2739.4	3.3024	2.3633	5.6657
310	9865.0	0.001447	0.018333	1387.7	1159.3	2547.1	1402.0	1325.9	2727.9	3.3506	2.2737	5.6243
315	10,556	0.001472	0.016849	1416.1	1121.1	2537.2	1431.6	1283.4	2715.0	3.3994	2.1821	5.5816
320	11,284	0.001499	0.015470	1445.1	1080.9	2526.0	1462.0	1238.5	2700.6	3.4491	2.0881	5.5372
325	12,051	0.001528	0.014183	1475.0	1038.5	2513.4	1493.4	1191.0	2684.3	3.4998	1.9911	5.4908
330	12,858	0.001560	0.012979	1505.7	993.5	2499.2	1525.8	1140.3	2666.0	3.5516	1.8906	5.4422
335	13,707	0.001597	0.011848	1537.5	945.5	2483.0	1559.4	1086.0	2645.4	3.6050	1.7857	5.3907
340	14,601	0.001638	0.010783	1570.7	893.8	2464.5	1594.6	1027.4	2622.0	3.6602	1.6756	5.3358

TABLE B.4 Saturated Water—Temperature (Continued)

Temp. T [°C]	Sat. press. P_{sat} [kPa]	Specific volume [m³/kg]		Internal energy [kJ/kg]			Enthalpy [kJ/kg]			Entropy [kJ/kg·K]		
		Sat. liquid v_f	Sat. vapor v_g	Sat. liquid u_f	Evap. u_{fg}	Sat. vapor u_g	Sat. liquid h_f	Evap. h_{fg}	Sat. vapor h_g	Sat. liquid s_f	Evap. s_{fg}	Sat. vapor s_g
345	15,541	0.001685	0.009772	1605.5	837.7	2443.2	1631.7	963.4	2595.1	3.7179	1.5585	5.2765
350	16,529	0.001741	0.008806	1642.4	775.9	2418.3	1671.2	892.7	2563.9	3.7788	1.4326	5.2114
355	17,570	0.001808	0.007872	1682.2	706.4	2388.6	1714.0	812.9	2526.9	3.8442	1.2942	5.1384
360	18,666	0.001895	0.006950	1726.2	625.7	2351.9	1761.5	720.1	2481.6	3.9165	1.1373	5.0537
365	19,822	0.002015	0.006009	1777.2	526.4	2303.6	1817.2	605.5	2422.7	4.0004	0.9489	4.9493
370	21,044	0.002217	0.004953	1844.5	385.6	2230.1	1891.2	443.1	2334.3	4.1119	0.6890	4.8009
373.95	22,064	0.003106	0.003106	2015.7	0	2015.7	2084.3	0	2084.3	4.4070	0	4.4070

Source: Tables B.4 through B.8 are generated using the Engineering Equation Solver (EES) software developed by S. A. Klein and F. L. Alvarado. The routine used in calculations is the highly accurate Steam_IAPWS, which incorporates the 1995 Formulation for the Thermodynamic Properties of Ordinary Water Substance for General and Scientific Use, issued by The International Association for the Properties of Water and Steam (IAPWS). This formulation replaces the 1984 formulation of Haar, Gallagher, and Kell (*NBS/NRC Steam Tables,* Hemisphere Publishing Co., 1984), which is also available in EES as the routine STEAM. The new formulation is based on the correlations of Saul and Wagner (*J. Phys. Chem. Ref. Data,* 16, 893, 1987) with modifications to adjust to the International Temperature Scale of 1990. The modifications are described by Wagner and Pruss (*J. Phys. Chem. Ref. Data,* 22, 783, 1993). The properties of ice are based on Hyland and Wexler, "Formulations for the Thermodynamic Properties of the Saturated Phases of H₂O from 173.15 K to 473.15 K," *ASHRAE Trans.,* Part 2A, Paper 2793, 1983.

TABLE B.4 Saturated Water—Temperature *(Concluded)*

Press. P [kPa]	Sat. temp. T_{sat} [°C]	Specific volume [m³/kg]		Internal energy [kJ/kg]			Enthalpy [kJ/kg]			Entropy [kJ/kg·K]		
		Sat. liquid V_f	Sat. vapor V_g	Sat. liquid U_f	Evap. U_{fg}	Sat. vapor U_g	Sat. liquid h_f	Evap. h_{fg}	Sat. vapor h_g	Sat. liquid s_f	Evap. s_{fg}	Sat. vapor s_g
1.0	6.97	0.001000	129.19	29.302	2355.2	2384.5	29.303	2484.4	2513.7	0.1059	8.8690	8.9749
1.5	13.02	0.001001	87.964	54.686	2338.1	2392.8	54.688	2470.1	2524.7	0.1956	8.6314	8.8270
2.0	17.50	0.001001	66.990	73.431	2325.5	2398.9	73.433	2459.5	2532.9	0.2606	8.4621	8.7227
2.5	21.08	0.001002	54.242	88.422	2315.4	2403.8	88.424	2451.0	2539.4	0.3118	8.3302	8.6421
3.0	24.08	0.001003	45.654	100.98	2306.9	2407.9	100.98	2443.9	2544.8	0.3543	8.2222	8.5765
4.0	28.96	0.001004	34.791	121.39	2293.1	2414.5	121.39	2432.3	2553.7	0.4224	8.0510	8.4734
5.0	32.87	0.001005	28.185	137.75	2282.1	2419.8	137.75	2423.0	2560.7	0.4762	7.9176	8.3938
7.5	40.29	0.001008	19.233	168.74	2261.1	2429.8	168.75	2405.3	2574.0	0.5763	7.6738	8.2501
10	45.81	0.001010	14.670	191.79	2245.4	2437.2	191.81	2392.1	2583.9	0.6492	7.4996	8.1488
15	53.97	0.001014	10.020	225.93	2222.1	2448.0	225.94	2372.3	2598.3	0.7549	7.2522	8.0071
20	60.06	0.001017	7.6481	251.40	2204.6	2456.0	251.42	2357.5	2608.9	0.8320	7.0752	7.9073
25	64.96	0.001020	6.2034	271.93	2190.4	2462.4	271.96	2345.5	2617.5	0.8932	6.9370	7.8302
30	69.09	0.001022	5.2287	289.24	2178.5	2467.7	289.27	2335.3	2624.6	0.9441	6.8234	7.7675
40	75.86	0.001026	3.9933	317.58	2158.8	2476.3	317.62	2318.4	2636.1	1.0261	6.6430	7.6691
50	81.32	0.001030	3.2403	340.49	2142.7	2483.2	340.54	2304.7	2645.2	1.0912	6.5019	7.5931
75	91.76	0.001037	2.2172	384.36	2111.8	2496.1	384.44	2278.0	2662.4	1.2132	6.2426	7.4558
100	99.61	0.001043	1.6941	417.40	2088.2	2505.6	417.51	2257.5	2675.0	1.3028	6.0562	7.3589
101.325	99.97	0.001043	1.6734	418.95	2087.0	2506.0	419.06	2256.5	2675.6	1.3069	6.0476	7.3545
125	105.97	0.001048	1.3750	444.23	2068.8	2513.0	444.36	2240.6	2684.9	1.3741	5.9100	7.2841
150	111.35	0.001053	1.1594	466.97	2052.3	2519.2	467.13	2226.0	2693.1	1.4337	5.7894	7.2231
175	116.04	0.001057	1.0037	486.82	2037.7	2524.5	487.01	2213.1	2700.2	1.4850	5.6865	7.1716
200	120.21	0.001061	0.88578	504.50	2024.6	2529.1	504.71	2201.6	2706.3	1.5302	5.5968	7.1270

TABLE B.5 Saturated Water—Pressure

Table B.5 Saturated Water—Pressure (Continued)

Press. P [kPa]	Sat. temp. T_{sat} [°C]	Specific volume [m³/kg] Sat. liquid v_f	Specific volume Sat. vapor v_g	Internal energy [kJ/kg] Sat. liquid u_f	Internal energy Evap. u_{fg}	Internal energy Sat. vapor u_g	Enthalpy [kJ/kg] Sat. liquid h_f	Enthalpy Evap. h_{fg}	Enthalpy Sat. vapor h_g	Entropy [kJ/kg·K] Sat. liquid s_f	Entropy Evap. s_{fg}	Entropy Sat. vapor s_g
225	123.97	0.001064	0.79329	520.47	2012.7	2533.2	520.71	2191.0	2711.7	1.5706	5.5171	7.0877
250	127.41	0.001067	0.71873	535.08	2001.8	2536.8	535.35	2181.2	2716.5	1.6072	5.4453	7.0525
275	130.58	0.001070	0.65732	548.57	1991.6	2540.1	548.86	2172.0	2720.9	1.6408	5.3800	7.0207
300	133.52	0.001073	0.60582	561.11	1982.1	2543.2	561.43	2163.5	2724.9	1.6717	5.3200	6.9917
325	136.27	0.001076	0.56199	572.84	1973.1	2545.9	573.19	2155.4	2728.6	1.7005	5.2645	6.9650
350	138.86	0.001079	0.52422	583.89	1964.6	2548.5	584.26	2147.7	2732.0	1.7274	5.2128	6.9402
375	141.30	0.001081	0.49133	594.32	1956.6	2550.9	594.73	2140.4	2735.1	1.7526	5.1645	6.9171
400	143.61	0.001084	0.46242	604.22	1948.9	2553.1	604.66	2133.4	2738.1	1.7765	5.1191	6.8955
450	147.90	0.001088	0.41392	622.65	1934.5	2557.1	623.14	2120.3	2743.4	1.8205	5.0356	6.8561
500	151.83	0.001093	0.37483	639.54	1921.2	2560.7	640.09	2108.0	2748.1	1.8604	4.9603	6.8207
550	155.46	0.001097	0.34261	655.16	1908.8	2563.9	655.77	2096.6	2752.4	1.8970	4.8916	6.7886
600	158.83	0.001101	0.31560	669.72	1897.1	2566.8	670.38	2085.8	2756.2	1.9308	4.8285	6.7593
650	161.98	0.001104	0.29260	683.37	1886.1	2569.4	684.08	2075.5	2759.6	1.9623	4.7699	6.7322
700	164.95	0.001108	0.27278	696.23	1875.6	2571.8	697.00	2065.8	2762.8	1.9918	4.7153	6.7071
750	167.75	0.001111	0.25552	708.40	1865.6	2574.0	709.24	2056.4	2765.7	2.0195	4.6642	6.6837
800	170.41	0.001115	0.24035	719.97	1856.1	2576.0	720.87	2047.5	2768.3	2.0457	4.6160	6.6616
850	172.94	0.001118	0.22690	731.00	1846.9	2577.9	731.95	2038.8	2770.8	2.0705	4.5705	6.6409
900	175.35	0.001121	0.21489	741.55	1838.1	2579.6	742.56	2030.5	2773.0	2.0941	4.5273	6.6213
950	177.66	0.001124	0.20411	751.67	1829.6	2581.3	752.74	2022.4	2775.2	2.1166	4.4862	6.6027
1000	179.88	0.001127	0.19436	761.39	1821.4	2582.8	762.51	2014.6	2777.1	2.1381	4.4470	6.5850
1100	184.06	0.001133	0.17745	779.78	1805.7	2585.5	781.03	1999.6	2780.7	2.1785	4.3735	6.5520
1200	187.96	0.001138	0.16326	796.96	1790.9	2587.8	798.33	1985.4	2783.8	2.2159	4.3058	6.5217

Press. P [kPa]	Sat. temp. T_{sat} [°C]	Specific volume [m³/kg]		Internal energy [kJ/kg]			Enthalpy [kJ/kg]			Entropy [kJ/kg·K]		
		Sat. liquid v_f	Sat. vapor v_g	Sat. liquid u_f	Evap. u_{fg}	Sat. vapor u_g	Sat. liquid h_f	Evap. h_{fg}	Sat. vapor h_g	Sat. liquid s_f	Evap. s_{fg}	Sat. vapor s_g
1300	191.60	0.001144	0.15119	813.10	1776.8	2589.9	814.59	1971.9	2786.5	2.2508	4.2428	6.4936
1400	195.04	0.001149	0.14078	828.35	1763.4	2591.8	829.96	1958.9	2788.9	2.2835	4.1840	6.4675
1500	198.29	0.001154	0.13171	842.82	1750.6	2593.4	844.55	1946.4	2791.0	2.3143	4.1287	6.4430
1750	205.72	0.001166	0.11344	876.12	1720.6	2596.7	878.16	1917.1	2795.2	2.3844	4.0033	6.3877
2000	212.38	0.001177	0.099587	906.12	1693.0	2599.1	908.47	1889.8	2798.3	2.4467	3.8923	6.3390
2250	218.41	0.001187	0.088717	933.54	1667.3	2600.9	936.21	1864.3	2800.5	2.5029	3.7926	6.2954
2500	223.95	0.001197	0.079952	958.87	1643.2	2602.1	961.87	1840.1	2801.9	2.5542	3.7016	6.2558
3000	233.85	0.001217	0.066667	1004.6	1598.5	2603.2	1008.3	1794.9	2803.2	2.6454	3.5402	6.1856
3500	242.56	0.001235	0.057061	1045.4	1557.6	2603.0	1049.7	1753.0	2802.7	2.7253	3.3991	6.1244
4000	250.35	0.001252	0.049779	1082.4	1519.3	2601.7	1087.4	1713.5	2800.8	2.7966	3.2731	6.0696
5000	263.94	0.001286	0.039448	1148.1	1448.9	2597.0	1154.5	1639.7	2794.2	2.9207	3.0530	5.9737
6000	275.59	0.001319	0.032449	1205.8	1384.1	2589.9	1213.8	1570.9	2784.6	3.0275	2.8627	5.8902
7000	285.83	0.001352	0.027378	1258.0	1323.0	2581.0	1267.5	1505.2	2772.6	3.1220	2.6927	5.8148
8000	295.01	0.001384	0.023525	1306.0	1264.5	2570.5	1317.1	1441.6	2758.7	3.2077	2.5373	5.7450
9000	303.35	0.001418	0.020489	1350.9	1207.6	2558.5	1363.7	1379.3	2742.9	3.2866	2.3925	5.6791
10,000	311.00	0.001452	0.018028	1393.3	1151.8	2545.2	1407.8	1317.6	2725.5	3.3603	2.2556	5.6159
11,000	318.08	0.001488	0.015988	1433.9	1096.6	2530.4	1450.2	1256.1	2706.3	3.4299	2.1245	5.5544
12,000	324.68	0.001526	0.014264	1473.0	1041.3	2514.3	1491.3	1194.1	2685.4	3.4964	1.9975	5.4939
13,000	330.85	0.001566	0.012781	1511.0	985.5	2496.6	1531.4	1131.3	2662.7	3.5606	1.8730	5.4336
14,000	336.67	0.001610	0.011487	1548.4	928.7	2477.1	1571.0	1067.0	2637.9	3.6232	1.7497	5.3728
15,000	342.16	0.001657	0.010341	1585.5	870.3	2455.7	1610.3	1000.5	2610.8	3.6848	1.6261	5.3108
16,000	347.36	0.001710	0.009312	1622.6	809.4	2432.0	1649.9	931.1	2581.0	3.7461	1.5005	5.2466

TABLE B.5 Saturated Water—Pressure (Continued)

Press. P [kPa]	Sat. temp. T_{sat} [°C]	Specific volume [m³/kg]		Internal energy [kJ/kg]			Enthalpy [kJ/kg]			Entropy [kJ/kg·K]		
		Sat. liquid V_f	Sat. vapor V_g	Sat. liquid U_f	Evap. U_{fg}	Sat. vapor U_g	Sat. liquid h_f	Evap. h_{fg}	Sat. vapor h_g	Sat. liquid s_f	Evap. s_{fg}	Sat. vapor s_g
17,000	352.29	0.001770	0.008374	1660.2	745.1	2405.4	1690.3	857.4	2547.7	3.8082	1.3709	5.1791
18,000	356.99	0.001840	0.007504	1699.1	675.9	2375.0	1732.2	777.8	2510.0	3.8720	1.2343	5.1064
19,000	361.47	0.001926	0.006677	1740.3	598.9	2339.2	1776.8	689.2	2466.0	3.9396	1.0860	5.0256
20,000	365.75	0.002038	0.005862	1785.8	509.0	2294.8	1826.6	585.5	2412.1	4.0146	0.9164	4.9310
21,000	369.83	0.002207	0.004994	1841.6	391.9	2233.5	1888.0	450.4	2338.4	4.1071	0.7005	4.8076
22,000	373.71	0.002703	0.003644	1951.7	140.8	2092.4	2011.1	161.5	2172.6	4.2942	0.2496	4.5439
22,064	373.95	0.003106	0.003106	2015.7	0	2015.7	2084.3	0	2084.3	4.4070	0	4.4070

Table B.5   Saturated Water—Pressure (Concluded)

P = 0.01 MPa (45.81°C)*

T [°C]	v [m³/kg]	u [kJ/kg]	h [kJ/kg]	s [kJ/kg·K]
Sat.†	14.670	2437.2	2583.9	8.1488
50	14.867	2443.3	2592.0	8.1741
100	17.196	2515.5	2687.5	8.4489
150	19.513	2587.9	2783.0	8.6893
200	21.826	2661.4	2879.6	8.9049
250	24.136	2736.1	2977.5	9.1015
300	26.446	2812.3	3076.7	9.2827
400	31.063	2969.3	3280.0	9.6094
500	35.680	3132.9	3489.7	9.8998
600	40.296	3303.3	3706.3	10.1631
700	44.911	3480.8	3929.9	10.4056
800	49.527	3665.4	4160.6	10.6312
900	54.143	3856.9	4398.3	10.8429
1000	58.758	4055.3	4642.8	11.0429
1100	63.373	4260.0	4893.8	11.2326
1200	67.989	4470.9	5150.8	11.4132
1300	72.604	4687.4	5413.4	11.5857

P = 0.05 MPa (81.32°C)

T [°C]	v [m³/kg]	u [kJ/kg]	h [kJ/kg]	s [kJ/kg·K]
Sat.†	3.2403	2483.2	2645.2	7.5931
50				
100	3.4187	2511.5	2682.4	7.6953
150	3.8897	2585.7	2780.2	7.9413
200	4.3562	2660.0	2877.8	8.1592
250	4.8206	2735.1	2976.2	8.3568
300	5.2841	2811.6	3075.8	8.5387
400	6.2094	2968.9	3279.3	8.8659
500	7.1338	3132.6	3489.3	9.1566
600	8.0577	3303.1	3706.0	9.4201
700	8.9813	3480.6	3929.7	9.6626
800	9.9047	3665.2	4160.4	9.8883
900	10.8280	3856.8	4398.2	10.1000
1000	11.7513	4055.2	4642.7	10.3000
1100	12.6745	4259.9	4893.7	10.4897
1200	13.5977	4470.8	5150.7	10.6704
1300	14.5209	4687.3	5413.3	10.8429

P = 0.10 MPa (99.61°C)

T [°C]	v [m³/kg]	u [kJ/kg]	h [kJ/kg]	s [kJ/kg·K]
Sat.†	1.6941	2505.6	2675.0	7.3589
50				
100	1.6959	2506.2	2675.8	7.3611
150	1.9367	2582.9	2776.6	7.6148
200	2.1724	2658.2	2875.5	7.8356
250	2.4062	2733.9	2974.5	8.0346
300	2.6389	2810.7	3074.5	8.2172
400	3.1027	2968.3	3278.6	8.5452
500	3.5655	3132.2	3488.7	8.8362
600	4.0279	3302.8	3705.6	9.0999
700	4.4900	3480.4	3929.4	9.3424
800	4.9519	3665.0	4160.2	9.5682
900	5.4137	3856.7	4398.0	9.7800
1000	5.8755	4055.0	4642.6	9.9800
1100	6.3372	4259.8	4893.6	10.1698
1200	6.7988	4470.7	5150.6	10.3504
1300	7.2605	4687.2	5413.3	10.5229

P = 0.20 MPa (120.21°C)

T [°C]	v [m³/kg]	u [kJ/kg]	h [kJ/kg]	s [kJ/kg·K]
Sat.	0.88578	2529.1	2706.3	7.1270
150	0.95986	2577.1	2769.1	7.2810
200	1.08049	2654.6	2870.7	7.5081

P = 0.30 MPa (133.52°C)

T [°C]	v [m³/kg]	u [kJ/kg]	h [kJ/kg]	s [kJ/kg·K]
Sat.	0.60582	2543.2	2724.9	6.9917
150	0.63402	2571.0	2761.2	7.0792
200	0.71643	2651.0	2865.9	7.3132

P = 0.40 MPa (143.61°C)

T [°C]	v [m³/kg]	u [kJ/kg]	h [kJ/kg]	s [kJ/kg·K]
Sat.	0.46242	2553.1	2738.1	6.8955
150	0.47088	2564.4	2752.8	6.9306
200	0.53434	2647.2	2860.9	7.1723

TABLE B.6   Superheated Water

197

T [°C]	*v* [m³/kg]	*u* [kJ/kg]	*h* [kJ/kg]	*s* [kJ/kg·K]	*v* [m³/kg]	*u* [kJ/kg]	*h* [kJ/kg]	*s* [kJ/kg·K]	*v* [m³/kg]	*u* [kJ/kg]	*h* [kJ/kg]	*s* [kJ/kg·K]
250	1.19890	2731.4	2971.2	7.7100	0.79645	2728.9	2967.9	7.5180	0.59520	2726.4	2964.5	7.3804
300	1.31623	2808.8	3072.1	7.8941	0.87535	2807.0	3069.6	7.7037	0.65489	2805.1	3067.1	7.5677
400	1.54934	2967.2	3277.0	8.2236	1.03155	2966.0	3275.5	8.0347	0.77265	2964.9	3273.9	7.9003
500	1.78142	3131.4	3487.7	8.5153	1.18672	3130.6	3486.6	8.3271	0.88936	3129.8	3485.5	8.1933
600	2.01302	3302.2	3704.8	8.7793	1.34139	3301.6	3704.0	8.5915	1.00558	3301.0	3703.3	8.4580
700	2.24434	3479.9	3928.8	9.0221	1.49580	3479.5	3928.2	8.8345	1.12152	3479.0	3927.6	8.7012
800	2.47550	3664.7	4159.8	9.2479	1.65004	3664.3	4159.3	9.0605	1.23730	3663.9	4158.9	8.9274
900	2.70656	3856.3	4397.7	9.4598	1.80417	3856.0	4397.3	9.2725	1.35298	3855.7	4396.9	9.1394
1000	2.93755	4054.8	4642.3	9.6599	1.95824	4054.5	4642.0	9.4726	1.46859	4054.3	4641.7	9.3396
1100	3.16848	4259.6	4893.3	9.8497	2.11226	4259.4	4893.1	9.6624	1.58414	4259.2	4892.9	9.5295
1200	3.39938	4470.5	5150.4	10.0304	2.26624	4470.3	5150.2	9.8431	1.69966	4470.2	5150.0	9.7102
1300	3.63026	4687.1	5413.1	10.2029	2.42019	4686.9	5413.0	10.0157	1.81516	4686.7	5412.8	9.8828
	$P = 0.50$ MPa (151.83°C)				**$P = 0.60$ MPa (158.83°C)**				**$P = 0.80$ MPa (170.41°C)**			
Sat.	0.37483	2560.7	2748.1	6.8207	0.31560	2566.8	2756.2	6.7593	0.24035	2576.0	2768.3	6.6616
200	0.42503	2643.3	2855.8	7.0610	0.35212	2639.4	2850.6	6.9683	0.26088	2631.1	2839.8	6.8177
250	0.47443	2723.8	2961.0	7.2725	0.39390	2721.2	2957.6	7.1833	0.29321	2715.9	2950.4	7.0402
300	0.52261	2803.3	3064.6	7.4614	0.43442	2801.4	3062.0	7.3740	0.32416	2797.5	3056.9	7.2345
350	0.57015	2883.0	3168.1	7.6346	0.47428	2881.6	3166.1	7.5481	0.35442	2878.6	3162.2	7.4107
400	0.61731	2963.7	3272.4	7.7956	0.51374	2962.5	3270.8	7.7097	0.38429	2960.2	3267.7	7.5735
500	0.71095	3129.0	3484.5	8.0893	0.59200	3128.2	3483.4	8.0041	0.44332	3126.6	3481.3	7.8692
600	0.80409	3300.4	3702.5	8.3544	0.66976	3299.8	3701.7	8.2695	0.50186	3298.7	3700.1	8.1354
700	0.89696	3478.6	3927.0	8.5978	0.74725	3478.1	3926.4	8.5132	0.56011	3477.2	3925.3	8.3794
800	0.98966	3663.6	4158.4	8.8240	0.82457	3663.2	4157.9	8.7395	0.61820	3662.5	4157.0	8.6061

TABLE B.6 Superheated Water (*Continued*)

Superheated Water (Continued)

T [°C]	P = 1.00 MPa (179.88°C)				P = 1.20 MPa (187.96°C)				P = 1.40 MPa (195.04°C)			
	v [m³/kg]	u [kJ/kg]	h [kJ/kg]	s [kJ/kg·K]	v [m³/kg]	u [kJ/kg]	h [kJ/kg]	s [kJ/kg·K]	v [m³/kg]	u [kJ/kg]	h [kJ/kg]	s [kJ/kg·K]
900	1.08227	3855.4	4396.6	9.0362	0.90179	3855.1	4396.2	8.9518	0.67619	3854.5	4395.5	8.8185
1000	1.17480	4054.0	4641.4	9.2364	0.97893	4053.8	4641.1	9.1521	0.73411	4053.3	4640.5	9.0189
1100	1.26728	4259.0	4892.6	9.4263	1.05603	4258.8	4892.4	9.3420	0.79197	4258.3	4891.9	9.2090
1200	1.35972	4470.0	5149.8	9.6071	1.13309	4469.8	5149.6	9.5229	0.84980	4469.4	5149.3	9.3898
1300	1.45214	4686.6	5412.6	9.7797	1.21012	4686.4	5412.5	9.6955	0.90761	4686.1	5412.2	9.5625

T [°C]	P = 1.00 MPa (179.88°C)				P = 1.20 MPa (187.96°C)				P = 1.40 MPa (195.04°C)			
	v [m³/kg]	u [kJ/kg]	h [kJ/kg]	s [kJ/kg·K]	v [m³/kg]	u [kJ/kg]	h [kJ/kg]	s [kJ/kg·K]	v [m³/kg]	u [kJ/kg]	h [kJ/kg]	s [kJ/kg·K]
Sat.	0.19437	2582.8	2777.1	6.5850	0.16326	2587.8	2783.8	6.5217	0.14078	2591.8	2788.9	6.4675
200	0.20602	2622.3	2828.3	6.6956	0.16934	2612.9	2816.1	6.5909	0.14303	2602.7	2803.0	6.4975
250	0.23275	2710.4	2943.1	6.9265	0.19241	2704.7	2935.6	6.8313	0.16356	2698.9	2927.9	6.7488
300	0.25799	2793.7	3051.6	7.1246	0.21386	2789.7	3046.3	7.0335	0.18233	2785.7	3040.9	6.9553
350	0.28250	2875.7	3158.2	7.3029	0.23455	2872.7	3154.2	7.2139	0.20029	2869.7	3150.1	7.1379
400	0.30661	2957.9	3264.5	7.4670	0.25482	2955.5	3261.3	7.3793	0.21782	2953.1	3258.1	7.3046
500	0.35411	3125.0	3479.1	7.7642	0.29464	3123.4	3477.0	7.6779	0.25216	3121.8	3474.8	7.6047
600	0.40111	3297.5	3698.6	8.0311	0.33395	3296.3	3697.0	7.9456	0.28597	3295.1	3695.5	7.8730
700	0.44783	3476.3	3924.1	8.2755	0.37297	3475.3	3922.9	8.1904	0.31951	3474.4	3921.7	8.1183
800	0.49438	3661.7	4156.1	8.5024	0.41184	3661.0	4155.2	8.4176	0.35288	3660.3	4154.3	8.3458
900	0.54083	3853.9	4394.8	8.7150	0.45059	3853.3	4394.0	8.6303	0.38614	3852.7	4393.3	8.5587
1000	0.58721	4052.7	4640.0	8.9155	0.48928	4052.2	4639.4	8.8310	0.41933	4051.7	4638.8	8.7595
1100	0.63354	4257.9	4891.4	9.1057	0.52792	4257.5	4891.0	9.0212	0.45247	4257.0	4890.5	8.9497
1200	0.67983	4469.0	5148.9	9.2866	0.56652	4468.7	5148.5	9.2022	0.48558	4468.3	5148.1	9.1308
1300	0.72610	4685.8	5411.9	9.4593	0.60509	4685.5	5411.6	9.3750	0.51866	4685.1	5411.3	9.3036

T [°C]	P = 1.60 MPa (201.37°C)				P = 1.80 MPa (207.11°C)				P = 2.00 MPa (212.38°C)			
	v [m³/kg]	u [kJ/kg]	h [kJ/kg]	s [kJ/kg·K]	v [m³/kg]	u [kJ/kg]	h [kJ/kg]	s [kJ/kg·K]	v [m³/kg]	u [kJ/kg]	h [kJ/kg]	s [kJ/kg·K]
Sat.	0.12374	2594.8	2792.8	6.4200	0.11037	2597.3	2795.9	6.3775	0.09959	2599.1	2798.3	6.3390
225	0.13293	2645.1	2857.8	6.5537	0.11678	2637.0	2847.2	6.4825	0.10381	2628.5	2836.1	6.4160

TABLE B.6 Superheated Water (Continued)

Table B.6, Superheated Water (Continued)

T [°C]	v [m³/kg]	u [kJ/kg]	h [kJ/kg]	s [kJ/kg·K]	v [m³/kg]	u [kJ/kg]	h [kJ/kg]	s [kJ/kg·K]	v [m³/kg]	u [kJ/kg]	h [kJ/kg]	s [kJ/kg·K]
250	0.14190	2692.9	2919.9	6.6753	0.12502	2686.7	2911.7	6.6088	0.11150	2680.3	2903.3	6.5475
300	0.15866	2781.6	3035.4	6.8864	0.14025	2777.4	3029.9	6.8246	0.12551	2773.2	3024.2	6.7684
350	0.17459	2866.6	3146.0	7.0713	0.15460	2863.6	3141.9	7.0120	0.13860	2860.5	3137.7	6.9583
400	0.19007	2950.8	3254.9	7.2394	0.16849	2948.3	3251.6	7.1814	0.15122	2945.9	3248.4	7.1292
500	0.22029	3120.1	3472.6	7.5410	0.19551	3118.5	3470.4	7.4845	0.17568	3116.9	3468.3	7.4337
600	0.24999	3293.9	3693.9	7.8101	0.22200	3292.7	3692.3	7.7543	0.19962	3291.5	3690.7	7.7043
700	0.27941	3473.5	3920.5	8.0558	0.24822	3472.6	3919.4	8.0005	0.22326	3471.7	3918.2	7.9509
800	0.30865	3659.5	4153.4	8.2834	0.27426	3658.8	4152.4	8.2284	0.24674	3658.0	4151.5	8.1791
900	0.33780	3852.1	4392.6	8.4965	0.30020	3851.5	4391.9	8.4417	0.27012	3850.9	4391.1	8.3925
1000	0.36687	4051.2	4638.2	8.6974	0.32606	4050.7	4637.6	8.6427	0.29342	4050.2	4637.1	8.5936
1100	0.39589	4256.6	4890.0	8.8878	0.35188	4256.2	4889.6	8.8331	0.31667	4255.7	4889.1	8.7842
1200	0.42488	4467.9	5147.7	9.0689	0.37766	4467.6	5147.3	9.0143	0.33989	4467.2	5147.0	8.9654
1300	0.45383	4684.8	5410.9	9.2418	0.40341	4684.5	5410.6	9.1872	0.36308	4684.2	5410.3	9.1384
	P = 2.50 MPa (223.95°C)				P = 3.00 MPa (233.85°C)				P = 3.50 MPa (242.56°C)			
Sat.	0.07995	2602.1	2801.9	6.2558	0.06667	2603.2	2803.2	6.1856	0.05706	2603.0	2802.7	6.1244
225	0.08026	2604.8	2805.5	6.2629								
250	0.08705	2663.3	2880.9	6.4107	0.07063	2644.7	2856.5	6.2893	0.05876	2624.0	2829.7	6.1764
300	0.09894	2762.2	3009.6	6.6459	0.08118	2750.8	2994.3	6.5412	0.06845	2738.8	2978.4	6.4484
350	0.10979	2852.5	3127.0	6.8424	0.09056	2844.4	3116.1	6.7450	0.07680	2836.0	3104.9	6.6601
400	0.12012	2939.8	3240.1	7.0170	0.09938	2933.6	3231.7	6.9235	0.08456	2927.2	3223.2	6.8428
450	0.13015	3026.2	3351.6	7.1768	0.10789	3021.2	3344.9	7.0856	0.09198	3016.1	3338.1	7.0074
500	0.13999	3112.8	3462.8	7.3254	0.11620	3108.6	3457.2	7.2359	0.09919	3104.5	3451.7	7.1593
600	0.15931	3288.5	3686.8	7.5979	0.13245	3285.5	3682.8	7.5103	0.11325	3282.5	3678.9	7.4357

Table B.6 Superheated Water (Continued)

T [°C]	v [m³/kg]	u [kJ/kg]	h [kJ/kg]	s [kJ/kg·K]	v [m³/kg]	u [kJ/kg]	h [kJ/kg]	s [kJ/kg·K]	v [m³/kg]	u [kJ/kg]	h [kJ/kg]	s [kJ/kg·K]
700	0.17835	3469.3	3915.2	7.8455	0.14841	3467.0	3912.2	7.7590	0.12702	3464.7	3909.3	7.6855
800	0.19722	3656.2	4149.2	8.0744	0.16420	3654.3	4146.9	7.9885	0.14061	3652.5	4144.6	7.9156
900	0.21597	3849.4	4389.3	8.2882	0.17988	3847.9	4387.5	8.2028	0.15410	3846.4	4385.7	8.1304
1000	0.23466	4049.0	4635.6	8.4897	0.19549	4047.7	4634.2	8.4045	0.16751	4046.4	4632.7	8.3324
1100	0.25330	4254.7	4887.9	8.6804	0.21105	4253.6	4886.7	8.5955	0.18087	4252.5	4885.6	8.5236
1200	0.27190	4466.3	5146.0	8.8618	0.22658	4465.3	5145.1	8.7771	0.19420	4464.4	5144.1	8.7053
1300	0.29048	4683.4	5409.5	9.0349	0.24207	4682.6	5408.8	8.9502	0.20750	4681.8	5408.0	8.8786
	P = 4.0 MPa (250.35°C)				**P = 4.5 MPa (257.44°C)**				**P = 5.0 MPa (263.94°C)**			
Sat.	0.04978	2601.7	2800.8	6.0696	0.04406	2599.7	2798.0	6.0198	0.03945	2597.0	2794.2	5.9737
275	0.05461	2668.9	2887.3	6.2312	0.04733	2651.4	2864.4	6.1429	0.04144	2632.3	2839.5	6.0571
300	0.05887	2726.2	2961.7	6.3639	0.05138	2713.0	2944.2	6.2854	0.04535	2699.0	2925.7	6.2111
350	0.06647	2827.4	3093.3	6.5843	0.05842	2818.6	3081.5	6.5153	0.05197	2809.5	3069.3	6.4516
400	0.07343	2920.8	3214.5	6.7714	0.06477	2914.2	3205.7	6.7071	0.05784	2907.5	3196.7	6.6483
450	0.08004	3011.0	3331.2	6.9386	0.07076	3005.8	3324.2	6.8770	0.06332	3000.6	3317.2	6.8210
500	0.08644	3100.3	3446.0	7.0922	0.07652	3096.0	3440.4	7.0323	0.06858	3091.8	3434.7	6.9781
600	0.09886	3279.4	3674.9	7.3706	0.08766	3276.4	3670.9	7.3127	0.07870	3273.3	3666.9	7.2605
700	0.11098	3462.4	3906.3	7.6214	0.09850	3460.0	3903.3	7.5647	0.08852	3457.7	3900.3	7.5136
800	0.12292	3650.6	4142.3	7.8523	0.10916	3648.8	4140.0	7.7962	0.09816	3646.9	4137.7	7.7458
900	0.13476	3844.8	4383.9	8.0675	0.11972	3843.3	4382.1	8.0118	0.10769	3841.8	4380.2	7.9619
1000	0.14653	4045.1	4631.2	8.2698	0.13020	4043.9	4629.8	8.2144	0.11715	4042.6	4628.3	8.1648
1100	0.15824	4251.4	4884.4	8.4612	0.14064	4250.4	4883.2	8.4060	0.12655	4249.3	4882.1	8.3566
1200	0.16992	4463.5	5143.2	8.6430	0.15103	4462.6	5142.2	8.5880	0.13592	4461.6	5141.3	8.5388
1300	0.18157	4680.9	5407.2	8.8164	0.16140	4680.1	5406.5	8.7616	0.14527	4679.3	5405.7	8.7124

TABLE B.6   Superheated Water (Continued)

Table B.6 Superheated Water (Continued) — P = 6.0, 7.0, 8.0 MPa

T [°C]	v [m³/kg]	u [kJ/kg]	h [kJ/kg]	s [kJ/kg·K]	v [m³/kg]	u [kJ/kg]	h [kJ/kg]	s [kJ/kg·K]	v [m³/kg]	u [kJ/kg]	h [kJ/kg]	s [kJ/kg·K]
	P = 6.0 MPa (275.59°C)				P = 7.0 MPa (285.83°C)				P = 8.0 MPa (295.01°C)			
Sat.	0.03245	2589.9	2784.6	5.8902	0.027378	2581.0	2772.6	5.8148	0.023525	2570.5	2758.7	5.7450
300	0.03619	2668.4	2885.6	6.0703	0.029492	2633.5	2839.9	5.9337	0.024279	2592.3	2786.5	5.7937
350	0.04225	2790.4	3043.9	6.3357	0.035262	2770.1	3016.9	6.2305	0.029975	2748.3	2988.1	6.1321
400	0.04742	2893.7	3178.3	6.5432	0.039958	2879.5	3159.2	6.4502	0.034344	2864.6	3139.4	6.3658
450	0.05217	2989.9	3302.9	6.7219	0.044187	2979.0	3288.3	6.6353	0.038194	2967.8	3273.3	6.5579
500	0.05667	3083.1	3423.1	6.8826	0.048157	3074.3	3411.4	6.8000	0.041767	3065.4	3399.5	6.7266
550	0.06102	3175.2	3541.3	7.0308	0.051966	3167.9	3531.6	6.9507	0.045172	3160.5	3521.8	6.8800
600	0.06527	3267.2	3658.8	7.1693	0.055665	3261.0	3650.6	7.0910	0.048463	3254.7	3642.4	7.0221
700	0.07355	3453.0	3894.3	7.4247	0.062850	3448.3	3888.3	7.3487	0.054829	3443.6	3882.2	7.2822
800	0.08165	3643.2	4133.1	7.6582	0.069856	3639.5	4128.5	7.5836	0.061011	3635.7	4123.8	7.5185
900	0.08964	3838.8	4376.6	7.8751	0.076750	3835.7	4373.0	7.8014	0.067082	3832.7	4369.3	7.7372
1000	0.09756	4040.1	4625.4	8.0786	0.083571	4037.5	4622.5	8.0055	0.073079	4035.0	4619.6	7.9419
1100	0.10543	4247.1	4879.7	8.2709	0.090341	4245.0	4877.4	8.1982	0.079025	4242.8	4875.0	8.1350
1200	0.11326	4459.8	5139.4	8.4534	0.097075	4457.9	5137.4	8.3810	0.084934	4456.1	5135.5	8.3181
1300	0.12107	4677.7	5404.1	8.6273	0.103781	4676.1	5402.6	8.5551	0.090817	4674.5	5401.0	8.4925

P = 9.0, 10.0, 12.5 MPa

T [°C]	v [m³/kg]	u [kJ/kg]	h [kJ/kg]	s [kJ/kg·K]	v [m³/kg]	u [kJ/kg]	h [kJ/kg]	s [kJ/kg·K]	v [m³/kg]	u [kJ/kg]	h [kJ/kg]	s [kJ/kg·K]
	P = 9.0 MPa (303.35°C)				P = 10.0 MPa (311.00°C)				P = 12.5 MPa (327.81°C)			
Sat.	0.020489	2558.5	2742.9	5.6791	0.018028	2545.2	2725.5	5.6159	0.013496	2505.6	2674.3	5.4638
325	0.023284	2647.6	2857.1	5.8738	0.019877	2611.6	2810.3	5.7596				
350	0.025816	2725.0	2957.3	6.0380	0.022440	2699.6	2924.0	5.9460	0.016138	2624.9	2826.6	5.7130
400	0.029960	2849.2	3118.8	6.2876	0.026436	2833.1	3097.5	6.2141	0.020030	2789.6	3040.0	6.0433
450	0.033524	2956.3	3258.0	6.4872	0.029782	2944.5	3242.4	6.4219	0.023019	2913.7	3201.5	6.2749
500	0.036793	3056.3	3387.4	6.6603	0.032811	3047.0	3375.1	6.5995	0.025630	3023.2	3343.6	6.4651
550	0.039885	3153.0	3512.0	6.8164	0.035655	3145.4	3502.0	6.7585	0.028033	3126.1	3476.5	6.6317

T [°C]	v [m³/kg]	u [kJ/kg]	h [kJ/kg]	s [kJ/kg·K]	v [m³/kg]	u [kJ/kg]	h [kJ/kg]	s [kJ/kg·K]	v [m³/kg]	u [kJ/kg]	h [kJ/kg]	s [kJ/kg·K]
600	0.042861	3248.4	3634.1	6.9605	0.038378	3242.0	3625.8	6.9045	0.030306	3225.8	3604.6	6.7828
650	0.045755	3343.4	3755.2	7.0954	0.041018	3338.0	3748.1	7.0408	0.032491	3324.1	3730.2	6.9227
700	0.048589	3438.8	3876.1	7.2229	0.043597	3434.0	3870.0	7.1693	0.034612	3422.0	3854.6	7.0540
800	0.054132	3632.0	4119.2	7.4606	0.048629	3628.2	4114.5	7.4085	0.038724	3618.8	4102.8	7.2967
900	0.059562	3829.6	4365.7	7.6802	0.053547	3826.5	4362.0	7.6290	0.042720	3818.9	4352.9	7.5195
1000	0.064919	4032.4	4616.7	7.8855	0.058391	4029.9	4613.8	7.8349	0.046641	4023.5	4606.5	7.7269
1100	0.070224	4240.7	4872.7	8.0791	0.063183	4238.5	4870.3	8.0289	0.050510	4233.1	4864.5	7.9220
1200	0.075492	4454.2	5133.6	8.2625	0.067938	4452.4	5131.7	8.2126	0.054342	4447.7	5127.0	8.1065
1300	0.080733	4672.9	5399.5	8.4371	0.072667	4671.3	5398.0	8.3874	0.058147	4667.3	5394.1	8.2819
	P = 15.0 MPa (342.16°C)				P = 17.5 MPa (354.67°C)				P = 20.0 MPa (365.75°C)			
Sat.	0.010341	2455.7	2610.8	5.3108	0.007932	2390.7	2529.5	5.1435	0.005862	2294.8	2412.1	4.9310
350	0.011481	2520.9	2693.1	5.4438								
400	0.015671	2740.6	2975.7	5.8819	0.012463	2684.3	2902.4	5.7211	0.009950	2617.9	2816.9	5.5526
450	0.018477	2880.8	3157.9	6.1434	0.015204	2845.4	3111.4	6.0212	0.012721	2807.3	3061.7	5.9043
500	0.020828	2998.4	3310.8	6.3480	0.017385	2972.4	3276.7	6.2424	0.014793	2945.3	3241.2	6.1446
550	0.022945	3106.2	3450.4	6.5230	0.019305	3085.8	3423.6	6.4266	0.016571	3064.7	3396.2	6.3390
600	0.024921	3209.3	3583.1	6.6796	0.021073	3192.5	3561.3	6.5890	0.018185	3175.3	3539.0	6.5075
650	0.026804	3310.1	3712.1	6.8233	0.022742	3295.8	3693.8	6.7366	0.019695	3281.4	3675.3	6.6593
700	0.028621	3409.8	3839.1	6.9573	0.024342	3397.5	3823.5	6.8735	0.021134	3385.1	3807.8	6.7991
800	0.032121	3609.3	4091.1	7.2037	0.027405	3599.7	4079.3	7.1237	0.023870	3590.1	4067.5	7.0531
900	0.035503	3811.2	4343.7	7.4288	0.030348	3803.5	4334.6	7.3511	0.026484	3795.7	4325.4	7.2829
1000	0.038808	4017.1	4599.2	7.6378	0.033215	4010.7	4592.0	7.5616	0.029020	4004.3	4584.7	7.4950
1100	0.042062	4227.7	4858.6	7.8339	0.036029	4222.3	4852.8	7.7588	0.031504	4216.9	4847.0	7.6933

TABLE B.6 Superheated Water (Continued)

Table B.6 Superheated Water (Continued)

T [°C]	v [m³/kg]	u [kJ/kg]	h [kJ/kg]	s [kJ/kg·K]	v [m³/kg]	u [kJ/kg]	h [kJ/kg]	s [kJ/kg·K]	v [m³/kg]	u [kJ/kg]	h [kJ/kg]	s [kJ/kg·K]
	P = 25.0 MPa				**P = 30.0 MPa**				**P = 35.0 MPa**			
1200	0.045279	4443.1	5122.3	8.0192	0.038806	4438.5	5117.6	7.9449	0.033952	4433.8	5112.9	7.8802
1300	0.048469	4663.3	5390.3	8.1952	0.041556	4659.2	5386.5	8.1215	0.036371	4655.2	5382.7	8.0574
375	0.001978	1799.9	1849.4	4.0345	0.001792	1738.1	1791.9	3.9313	0.001701	1702.8	1762.4	3.8724
400	0.006005	2428.5	2578.7	5.1400	0.002798	2068.9	2152.8	4.4758	0.002105	1914.9	1988.6	4.2144
425	0.007886	2607.8	2805.0	5.4708	0.005299	2452.9	2611.8	5.1473	0.003434	2253.3	2373.5	4.7751
450	0.009176	2721.2	2950.6	5.6759	0.006737	2618.9	2821.0	5.4422	0.004957	2497.5	2671.0	5.1946
500	0.011143	2887.3	3165.9	5.9643	0.008691	2824.0	3084.8	5.7956	0.006933	2755.3	2997.9	5.6331
550	0.012736	3020.8	3339.2	6.1816	0.010175	2974.5	3279.7	6.0403	0.008348	2925.8	3218.0	5.9093
600	0.014140	3140.0	3493.5	6.3637	0.011445	3103.4	3446.8	6.2373	0.009523	3065.6	3399.0	6.1229
650	0.015430	3251.9	3637.7	6.5243	0.012590	3221.7	3599.4	6.4074	0.010565	3190.9	3560.7	6.3030
700	0.016643	3359.9	3776.0	6.6702	0.013654	3334.3	3743.9	6.5599	0.011523	3308.3	3711.6	6.4623
800	0.018922	3570.7	4043.8	6.9322	0.015628	3551.2	4020.0	6.8301	0.013278	3531.6	3996.3	6.7409
900	0.021075	3780.2	4307.1	7.1668	0.017473	3764.6	4288.8	7.0695	0.014904	3749.0	4270.6	6.9853
1000	0.023150	3991.5	4570.2	7.3821	0.019240	3978.6	4555.8	7.2880	0.016450	3965.8	4541.5	7.2069
1100	0.025172	4206.1	4835.4	7.5825	0.020954	4195.2	4823.9	7.4906	0.017942	4184.4	4812.4	7.4118
1200	0.027157	4424.6	5103.5	7.7710	0.022630	4415.3	5094.2	7.6807	0.019398	4406.1	5085.0	7.6034
1300	0.029115	4647.2	5375.1	7.9494	0.024279	4639.2	5367.6	7.8602	0.020827	4631.2	5360.2	7.7841

T [°C]	v [m³/kg]	u [kJ/kg]	h [kJ/kg]	s [kJ/kg·K]	v [m³/kg]	u [kJ/kg]	h [kJ/kg]	s [kJ/kg·K]	v [m³/kg]	u [kJ/kg]	h [kJ/kg]	s [kJ/kg·K]
	P = 40.0 MPa				**P = 50.0 MPa**				**P = 60.0 MPa**			
375	0.001641	1677.0	1742.6	3.8290	0.001560	1638.6	1716.6	3.7642	0.001503	1609.7	1699.9	3.7149
400	0.001911	1855.0	1931.4	4.1145	0.001731	1787.8	1874.4	4.0029	0.001633	1745.2	1843.2	3.9317
425	0.002538	2097.5	2199.0	4.5044	0.002009	1960.3	2060.7	4.2746	0.001816	1892.9	2001.8	4.1630
450	0.003692	2364.2	2511.8	4.9449	0.002487	2160.3	2284.7	4.5896	0.002086	2055.1	2180.2	4.4140

T [°C]	v [m³/kg]	u [kJ/kg]	h [kJ/kg]	s [kJ/kg·K]	v [m³/kg]	u [kJ/kg]	h [kJ/kg]	s [kJ/kg·K]	v [m³/kg]	u [kJ/kg]	h [kJ/kg]	s [kJ/kg·K]
500	0.005623	2681.6	2906.5	5.4744	0.003890	2528.1	2722.6	5.1762	0.002952	2393.2	2570.3	4.9356
550	0.006985	2875.1	3154.4	5.7857	0.005118	2769.5	3025.4	5.5563	0.003955	2664.6	2901.9	5.3517
600	0.008089	3026.8	3350.4	6.0170	0.006108	2947.1	3252.6	5.8245	0.004833	2866.8	3156.8	5.6527
650	0.009053	3159.5	3521.6	6.2078	0.006957	3095.6	3443.5	6.0373	0.005591	3031.3	3366.8	5.8867
700	0.009930	3282.0	3679.2	6.3740	0.007717	3228.7	3614.6	6.2179	0.006265	3175.4	3551.3	6.0814
800	0.011521	3511.8	3972.6	6.6613	0.009073	3472.2	3925.8	6.5225	0.007456	3432.6	3880.0	6.4033
900	0.012980	3733.3	4252.5	6.9107	0.010296	3702.0	4216.8	6.7819	0.008519	3670.9	4182.1	6.6725
1000	0.014360	3952.9	4527.3	7.1355	0.011441	3927.4	4499.4	7.0131	0.009504	3902.0	4472.2	6.9099
1100	0.015686	4173.7	4801.1	7.3425	0.012534	4152.2	4778.9	7.2244	0.010439	4130.9	4757.3	7.1255
1200	0.016976	4396.9	5075.9	7.5357	0.013590	4378.6	5058.1	7.4207	0.011339	4360.5	5040.8	7.3248
1300	0.018239	4623.3	5352.8	7.7175	0.014620	4607.5	5338.5	7.6048	0.012213	4591.8	5324.5	7.5111

*The temperature in parentheses is the saturation temperature at the specified pressure.
†Properties of saturated vapor at the specified pressure.

TABLE B.6 Superheated Water (Concluded)

Table B.7 split: Top section

T [°C]	v [m³/kg]	u [kJ/kg]	h [kJ/kg]	s [kJ/kg·K]	v [m³/kg]	u [kJ/kg]	h [kJ/kg]	s [kJ/kg·K]	v [m³/kg]	u [kJ/kg]	h [kJ/kg]	s [kJ/kg·K]
	P = 5 MPa (263.94°C)				P = 10 MPa (311.00°C)				P = 15 MPa (342.16°C)			
Sat.	0.0012862	1148.1	1154.5	2.9207	0.0014522	1393.3	1407.9	3.3603	0.0016572	1585.5	1610.3	3.6848
0	0.0009977	0.04	5.03	0.0001	0.0009952	0.12	10.07	0.0003	0.0009928	0.18	15.07	0.0004
20	0.0009996	83.61	88.61	0.2954	0.0009973	83.31	93.28	0.2943	0.0009951	83.01	97.93	0.2932
40	0.0010057	166.92	171.95	0.5705	0.0010035	166.33	176.37	0.5685	0.0010013	165.75	180.77	0.5666
60	0.0010149	250.29	255.36	0.8287	0.0010127	249.43	259.55	0.8260	0.0010105	248.58	263.74	0.8234
80	0.0010267	333.82	338.96	1.0723	0.0010244	332.69	342.94	1.0691	0.0010221	331.59	346.92	1.0659
100	0.0010410	417.65	422.85	1.3034	0.0010385	416.23	426.62	1.2996	0.0010361	414.85	430.39	1.2958
120	0.0010576	501.91	507.19	1.5236	0.0010549	500.18	510.73	1.5191	0.0010522	498.50	514.28	1.5148
140	0.0010769	586.80	592.18	1.7344	0.0010738	584.72	595.45	1.7293	0.0010708	582.69	598.75	1.7243
160	0.0010988	672.55	678.04	1.9374	0.0010954	670.06	681.01	1.9316	0.0010920	667.63	684.01	1.9259
180	0.0011240	759.47	765.09	2.1338	0.0011200	756.48	767.68	2.1271	0.0011160	753.58	770.32	2.1206
200	0.0011531	847.92	853.68	2.3251	0.0011482	844.32	855.80	2.3174	0.0011435	840.84	858.00	2.3100
220	0.0011868	938.39	944.32	2.5127	0.0011809	934.01	945.82	2.5037	0.0011752	929.81	947.43	2.4951
240	0.0012268	1031.6	1037.7	2.6983	0.0012192	1026.2	1038.3	2.6876	0.0012121	1021.0	1039.2	2.6774
260	0.0012755	1128.5	1134.9	2.8841	0.0012653	1121.6	1134.3	2.8710	0.0012560	1115.1	1134.0	2.8586
280					0.0013226	1221.8	1235.0	3.0565	0.0013096	1213.4	1233.0	3.0410
300					0.0013980	1329.4	1343.3	3.2488	0.0013783	1317.6	1338.3	3.2279
320									0.0014733	1431.9	1454.0	3.4263
340									0.0016311	1567.9	1592.4	3.6555

Table B.7 split: Bottom section

T [°C]	v [m³/kg]	u [kJ/kg]	h [kJ/kg]	s [kJ/kg·K]	v [m³/kg]	u [kJ/kg]	h [kJ/kg]	s [kJ/kg·K]	v [m³/kg]	u [kJ/kg]	h [kJ/kg]	s [kJ/kg·K]
	P = 20 MPa (365.75°C)				P = 30 MPa				P = 50 MPa			
Sat.	0.0020378	1785.8	1826.6	4.0146								
0	0.0009904	0.23	20.03	0.0005	0.0009857	0.29	29.86	0.0003	0.0009767	0.29	49.13	0.0010
20	0.0009929	82.71	102.57	0.2921	0.0009886	82.11	111.77	0.2897	0.0009805	80.93	129.95	0.2845

TABLE B.7 Compressed Liquid Water

T [°C]	v [m³/kg]	u [kJ/kg]	h [kJ/kg]	s [kJ/kg·K]	v [m³/kg]	u [kJ/kg]	h [kJ/kg]	s [kJ/kg·K]	v [m³/kg]	u [kJ/kg]	h [kJ/kg]	s [kJ/kg·K]
40	0.0009992	165.17	185.16	0.5646	0.0009951	164.05	193.90	0.5607	0.0009872	161.90	211.25	0.5528
60	0.0010084	247.75	267.92	0.8208	0.0010042	246.14	276.26	0.8156	0.0009962	243.08	292.88	0.8055
80	0.0010199	330.50	350.90	1.0627	0.0010155	328.40	358.86	1.0564	0.0010072	324.42	374.78	1.0442
100	0.0010337	413.50	434.17	1.2920	0.0010290	410.87	441.74	1.2847	0.0010201	405.94	456.94	1.2705
120	0.0010496	496.85	517.84	1.5105	0.0010445	493.66	525.00	1.5020	0.0010349	487.69	539.43	1.4859
140	0.0010679	580.71	602.07	1.7194	0.0010623	576.90	608.76	1.7098	0.0010517	569.77	622.36	1.6916
160	0.0010886	665.28	687.05	1.9203	0.0010823	660.74	693.21	1.9094	0.0010704	652.33	705.85	1.8889
180	0.0011122	750.78	773.02	2.1143	0.0011049	745.40	778.55	2.1020	0.0010914	735.49	790.06	2.0790
200	0.0011390	837.49	860.27	2.3027	0.0011304	831.11	865.02	2.2888	0.0011149	819.45	875.19	2.2628
220	0.0011697	925.77	949.16	2.4867	0.0011595	918.15	952.93	2.4707	0.0011412	904.39	961.45	2.4414
240	0.0012053	1016.1	1040.2	2.6676	0.0011927	1006.9	1042.7	2.6491	0.0011708	990.55	1049.1	2.6156
260	0.0012472	1109.0	1134.0	2.8469	0.0012314	1097.8	1134.7	2.8250	0.0012044	1078.2	1138.4	2.7864
280	0.0012978	1205.6	1231.5	3.0265	0.0012770	1191.5	1229.8	3.0001	0.0012430	1167.7	1229.9	2.9547
300	0.0013611	1307.2	1334.4	3.2091	0.0013322	1288.9	1328.9	3.1761	0.0012879	1259.6	1324.0	3.1218
320	0.0014450	1416.6	1445.5	3.3996	0.0014014	1391.7	1433.7	3.3558	0.0013409	1354.3	1421.4	3.2888
340	0.0015693	1540.2	1571.6	3.6086	0.0014932	1502.4	1547.1	3.5438	0.0014049	1452.9	1523.1	3.4575
360	0.0018248	1703.6	1740.1	3.8787	0.0016276	1626.8	1675.6	3.7499	0.0014848	1556.5	1630.7	3.6301
380					0.0018729	1782.0	1838.2	4.0026	0.0015884	1667.1	1746.5	3.8102

TABLE B.7 Compressed Liquid Water (*Concluded*)

207

Specific volume [m³/kg], Internal energy [kJ/kg], Enthalpy [kJ/kg], Entropy [kJ/kg·K]

Temp. T [°C]	Sat. press. P_{sat} [kPa]	Sat. ice v_i	Sat. vapor v_g	Sat. ice u_i	Subl. u_{ig}	Sat. vapor u_g	Sat. ice h_i	Subl. h_{ig}	Sat. vapor h_g	Sat. ice s_i	Subl. s_{ig}	Sat. vapor s_g
0.01	0.61169	0.001091	205.99	-333.40	2707.9	2374.5	-333.40	2833.9	2500.5	-1.2202	10.374	9.154
0	0.61115	0.001091	206.17	-333.43	2707.9	2374.5	-333.43	2833.9	2500.5	-1.2204	10.375	9.154
-2	0.51772	0.001091	241.62	337.63	2709.4	2371.8	-337.63	2834.5	2496.8	-1.2358	10.453	9.218
-4	0.43748	0.001090	283.84	-341.80	2710.8	2369.0	-341.80	2835.0	2493.2	-1.2513	10.533	9.282
-6	0.36873	0.001090	334.27	-345.94	2712.2	2366.2	-345.93	2835.4	2489.5	-1.2667	10.613	9.347
-8	0.30998	0.001090	394.66	-350.04	2713.5	2363.5	-350.04	2835.8	2485.8	-1.2821	10.695	9.413
-10	0.25990	0.001089	467.17	-354.12	2714.8	2360.7	-354.12	2836.2	2482.1	-1.2976	10.778	9.480
-12	0.21732	0.001089	554.47	-358.17	2716.1	2357.9	-358.17	2836.6	2478.4	-1.3130	10.862	9.549
-14	0.18121	0.001088	659.88	-362.18	2717.3	2355.2	-362.18	2836.9	2474.7	-1.3284	10.947	9.618
-16	0.15068	0.001088	787.51	-366.17	2718.6	2352.4	-366.17	2837.2	2471.0	-1.3439	11.033	9.689
-18	0.12492	0.001088	942.51	-370.13	2719.7	2349.6	-370.13	2837.5	2467.3	-1.3593	11.121	9.761
-20	0.10326	0.001087	1131.3	-374.06	2720.9	2346.8	-374.06	2837.7	2463.6	-1.3748	11.209	9.835
-22	0.08510	0.001087	1362.0	-377.95	2722.0	2344.1	-377.95	2837.9	2459.9	-1.3903	11.300	9.909
-24	0.06991	0.001087	1644.7	-381.82	2723.1	2341.3	-381.82	2838.1	2456.2	-1.4057	11.391	9.985
-26	0.05725	0.001087	1992.2	-385.66	2724.2	2338.5	-385.66	2838.2	2452.5	-1.4212	11.484	10.063
-28	0.04673	0.001086	2421.0	-389.47	2725.2	2335.7	-389.47	2838.3	2448.8	-1.4367	11.578	10.141
-30	0.03802	0.001086	2951.7	-393.25	2726.2	2332.9	-393.25	2838.4	2445.1	-1.4521	11.673	10.221
-32	0.03082	0.001086	3610.9	-397.00	2727.2	2330.2	-397.00	2838.4	2441.4	-1.4676	11.770	10.303
-34	0.02490	0.001085	4432.4	-400.72	2728.1	2327.4	-400.72	2838.5	2437.7	-1.4831	11.869	10.386
-36	0.02004	0.001085	5460.1	-404.40	2729.0	2324.6	-404.40	2838.4	2434.0	-1.4986	11.969	10.470
-38	0.01608	0.001085	6750.5	-408.07	2729.9	2321.8	-408.07	2838.4	2430.3	-1.5141	12.071	10.557
-40	0.01285	0.001084	8376.7	-411.70	2730.7	2319.0	-411.70	2838.3	2426.6	-1.5296	12.174	10.644

TABLE B.8 Saturated Ice–Water Vapor

TABLE B.9 Saturated Refrigerant-134a—Temperature

Temp. T [°C]	Sat. press. P_{sat} [kPa]	Specific volume [m³/kg]		Internal energy [kJ/kg]			Enthalpy [kJ/kg]			Entropy [kJ/kg·K]		
		Sat. liquid v_f	Sat. vapor v_g	Sat. liquid u_f	Evap. u_{fg}	Sat. vapor u_g	Sat. liquid h_f	Evap. h_{fg}	Sat. vapor h_g	Sat. liquid s_f	Evap. s_{fg}	Sat. vapor s_g
-40	51.25	0.0007054	0.36081	-0.036	207.40	207.37	0.000	225.86	225.86	0.00000	0.96866	0.96866
-38	56.86	0.0007083	0.32732	2.475	206.04	208.51	2.515	224.61	227.12	0.01072	0.95511	0.96584
-36	62.95	0.0007112	0.29751	4.992	204.67	209.66	5.037	223.35	228.39	0.02138	0.94176	0.96315
-34	69.56	0.0007142	0.27090	7.517	203.29	210.81	7.566	222.09	229.65	0.03199	0.92859	0.96058
-32	76.71	0.0007172	0.24711	10.05	201.91	211.96	10.10	220.81	230.91	0.04253	0.91560	0.95813
-30	84.43	0.0007203	0.22580	12.59	200.52	213.11	12.65	219.52	232.17	0.05301	0.90278	0.95579
-28	92.76	0.0007234	0.20666	15.13	199.12	214.25	15.20	218.22	233.43	0.06344	0.89012	0.95356
-26	101.73	0.0007265	0.18946	17.69	197.72	215.40	17.76	216.92	234.68	0.07382	0.87762	0.95144
-24	111.37	0.0007297	0.17395	20.25	196.30	216.55	20.33	215.59	235.92	0.08414	0.86527	0.94941
-22	121.72	0.0007329	0.15995	22.82	194.88	217.70	22.91	214.26	s237.17	0.09441	0.85307	0.94748
-20	132.82	0.0007362	0.14729	25.39	193.45	218.84	25.49	212.91	238.41	0.10463	0.84101	0.94564
-18	144.69	0.0007396	0.13583	27.98	192.01	219.98	28.09	211.55	239.64	0.11481	0.82908	0.94389
-16	157.38	0.0007430	0.12542	30.57	190.56	221.13	30.69	210.18	240.87	0.12493	0.81729	0.94222
-14	170.93	0.0007464	0.11597	33.17	189.09	222.27	33.30	208.79	242.09	0.13501	0.80561	0.94063
-12	185.37	0.0007499	0.10736	35.78	187.62	223.40	35.92	207.38	243.30	0.14504	0.79406	0.93911
-10	200.74	0.0007535	0.099516	38.40	186.14	224.54	38.55	205.96	244.51	0.15504	0.78263	0.93766
-8	217.08	0.0007571	0.092352	41.03	184.64	225.67	41.19	204.52	245.72	0.16498	0.77130	0.93629
-6	234.44	0.0007608	0.085802	43.66	183.13	226.80	43.84	203.07	246.91	0.17489	0.76008	0.93497
-4	252.85	0.0007646	0.079804	46.31	181.61	227.92	46.50	201.60	248.10	0.18476	0.74896	0.93372
-2	272.36	0.0007684	0.074304	48.96	180.08	229.04	49.17	200.11	249.28	0.19459	0.73794	0.93253
0	293.01	0.0007723	0.069255	51.63	178.53	230.16	51.86	198.60	250.45	0.20439	0.72701	0.93139
2	314.84	0.0007763	0.064612	54.30	176.97	231.27	54.55	197.07	251.61	0.21415	0.71616	0.93031

Temp. T [°C]	Sat. press. P_{sat} [kPa]	Specific volume [m³/kg]		Internal energy [kJ/kg]			Enthalpy [kJ/kg]			Entropy [kJ/kg·K]		
		Sat. liquid v_f	Sat. vapor v_g	Sat. liquid u_f	Evap. u_{fg}	Sat. vapor u_g	Sat. liquid h_f	Evap. h_{fg}	Sat. vapor h_g	Sat. liquid s_f	Evap. s_{fg}	Sat. vapor s_g
4	337.90	0.0007804	0.060338	56.99	175.39	232.38	57.25	195.51	252.77	0.22387	0.70540	0.92927
6	362.23	0.0007845	0.056398	59.68	173.80	233.48	59.97	193.94	253.91	0.23356	0.69471	0.92828
8	387.88	0.0007887	0.052762	62.39	172.19	234.58	62.69	192.35	255.04	0.24323	0.68410	0.92733
10	414.89	0.0007930	0.049403	65.10	170.56	235.67	65.43	190.73	256.16	0.25286	0.67356	0.92641
12	443.31	0.0007975	0.046295	67.83	168.92	236.75	68.18	189.09	257.27	0.26246	0.66308	0.92554
14	473.19	0.0008020	0.043417	70.57	167.26	237.83	70.95	187.42	258.37	0.27204	0.65266	0.92470
16	504.58	0.0008066	0.040748	73.32	165.58	238.90	73.73	185.73	259.46	0.28159	0.64230	0.92389
18	537.52	0.0008113	0.038271	76.08	163.88	239.96	76.52	184.01	260.53	0.29112	0.63198	0.92310
20	572.07	0.0008161	0.035969	78.86	162.16	241.02	79.32	182.27	261.59	0.30063	0.62172	0.92234
22	608.27	0.0008210	0.033828	81.64	160.42	242.06	82.14	180.49	262.64	0.31011	0.61149	0.92160
24	646.18	0.0008261	0.031834	84.44	158.65	243.10	84.98	178.69	263.67	0.31958	0.60130	0.92088
26	685.84	0.0008313	0.029976	87.26	156.87	244.12	87.83	176.85	264.68	0.32903	0.59115	0.92018
28	727.31	0.0008366	0.028242	90.09	155.05	245.14	90.69	174.99	265.68	0.33846	0.58102	0.91948
30	770.64	0.0008421	0.026622	92.93	153.22	246.14	93.58	173.08	266.66	0.34789	0.57091	0.91879
32	815.89	0.0008478	0.025108	95.79	151.35	247.14	96.48	171.14	267.62	0.35730	0.56082	0.91811
34	863.11	0.0008536	0.023691	98.66	149.46	248.12	99.40	169.17	268.57	0.36670	0.55074	0.91743
36	912.35	0.0008595	0.022364	101.55	147.54	249.08	102.33	167.16	269.49	0.37609	0.54066	0.91675
38	963.68	0.0008657	0.021119	104.45	145.58	250.04	105.29	165.10	270.39	0.38548	0.53058	0.91606
40	1017.1	0.0008720	0.019952	107.38	143.60	250.97	108.26	163.00	271.27	0.39486	0.52049	0.91536
42	1072.8	0.0008786	0.018855	110.32	141.58	251.89	111.26	160.86	272.12	0.40425	0.51039	0.91464
44	1130.7	0.0008854	0.017824	113.28	139.52	252.80	114.28	158.67	272.95	0.41363	0.50027	0.91391
46	1191.0	0.0008924	0.016853	116.26	137.42	253.68	117.32	156.43	273.75	0.42302	0.49012	0.91315
48	1253.6	0.0008996	0.015939	119.26	135.29	254.55	120.39	154.14	274.53	0.43242	0.47993	0.91236

TABLE B.9 Saturated Refrigerant-134a—Temperature (*Continued*)

Temp. T [°C]	Sat. press. P_{sat} [kPa]	Specific volume [m³/kg] Sat. liquid v_f	Sat. vapor v_g	Internal energy [kJ/kg] Sat. liquid u_f	Evap. u_{fg}	Sat. vapor u_g	Enthalpy [kJ/kg] Sat. liquid h_f	Evap. h_{fg}	Sat. vapor h_g	Entropy [kJ/kg·K] Sat. liquid s_f	Evap. s_{fg}	Sat. vapor s_g
52	1386.2	0.0009150	0.014265	125.33	130.88	256.21	126.59	149.39	275.98	0.45126	0.45941	0.91067
56	1529.1	0.0009317	0.012771	131.49	126.28	257.77	132.91	144.38	277.30	0.47018	0.43863	0.90880
60	1682.8	0.0009498	0.011434	137.76	121.46	259.22	139.36	139.10	278.46	0.48920	0.41749	0.90669
65	1891.0	0.0009750	0.009950	145.77	115.05	260.82	147.62	132.02	279.64	0.51320	0.39039	0.90359
70	2118.2	0.0010037	0.008642	154.01	108.14	262.15	156.13	124.32	280.46	0.53755	0.36227	0.89982
75	2365.8	0.0010372	0.007480	162.53	100.60	263.13	164.98	115.85	280.82	0.56241	0.33272	0.89512
80	2635.3	0.0010772	0.006436	171.40	92.23	263.63	174.24	106.35	280.59	0.58800	0.30111	0.88912
85	2928.2	0.0011270	0.005486	180.77	82.67	263.44	184.07	95.44	279.51	0.61473	0.26644	0.88117
90	3246.9	0.0011932	0.004599	190.89	71.29	262.18	194.76	82.35	277.11	0.64336	0.22674	0.87010
95	3594.1	0.0012933	0.003726	202.40	56.47	258.87	207.05	65.21	272.26	0.67578	0.17711	0.85289
100	3975.1	0.0015269	0.002630	218.72	29.19	247.91	224.79	33.58	258.37	0.72217	0.08999	0.81215

Source: Tables B.9 through B.11 are generated using the Engineering Equation Solver (EES) software developed by S. A. Klein and F. L. Alvarado. The routine used in calculations is the R134a, which is based on the fundamental equation of state developed by R. Tillner-Roth and H. D. Baehr, "An International Standard Formulation for the Thermodynamic Properties of 1,1,1,2-Tetrafluoroethane (HFC-134a) for temperatures from 170 K to 455 K and Pressures up to 70 MPa," *J. Phys. Chem. Ref. Data*, Vol. 23, No. 5, 1994. The enthalpy and entropy values of saturated liquid are set to zero at −40°C (and −40°F).

TABLE B.9 Saturated Refrigerant-134a—Temperature (*Concluded*)

Press. P [kPa]	Sat. temp. T_{sat}, °C	Specific volume [m³/kg]		Internal energy [kJ/kg]			Enthalpy [kJ/kg]			Entropy [kJ/kg·K]		
		Sat. liquid v_f	Sat. vapor v_g	Sat. liquid u_f	Evap. u_{fg}	Sat. vapor u_g	Sat. liquid h_f	Evap. h_{fg}	Sat. vapor h_g	Sat. liquid s_f	Evap. s_{fg}	Sat. vapor s_g
60	−36.95	0.0007098	0.31121	3.798	205.32	209.12	3.841	223.95	227.79	0.01634	0.94807	0.96441
70	−33.87	0.0007144	0.26929	7.680	203.20	210.88	7.730	222.00	229.73	0.03267	0.92775	0.96042
80	−31.13	0.0007185	0.23753	11.15	201.30	212.46	11.21	220.25	231.46	0.04711	0.90999	0.95710
90	−28.65	0.0007223	0.21263	14.31	199.57	213.88	14.37	218.65	233.02	0.06008	0.89419	0.95427
100	−26.37	0.0007259	0.19254	17.21	197.98	215.19	17.28	217.16	234.44	0.07188	0.87995	0.95183
120	−22.32	0.0007324	0.16212	22.40	195.11	217.51	22.49	214.48	236.97	0.09275	0.85503	0.94779
140	−18.77	0.0007383	0.14014	26.98	192.57	219.54	27.08	212.08	239.16	0.11087	0.83368	0.94456
160	−15.60	0.0007437	0.12348	31.09	190.27	221.35	31.21	209.90	241.11	0.12693	0.81496	0.94190
180	−12.73	0.0007487	0.11041	34.83	188.16	222.99	34.97	207.90	242.86	0.14139	0.79826	0.93965
200	−10.09	0.0007533	0.099867	38.28	186.21	224.48	38.43	206.03	244.46	0.15457	0.78316	0.93773
240	−5.38	0.0007620	0.083897	44.48	182.67	227.14	44.66	202.62	247.28	0.17794	0.75664	0.93458
280	−1.25	0.0007699	0.072352	49.97	179.50	229.46	50.18	199.54	249.72	0.19829	0.73381	0.93210
320	2.46	0.0007772	0.063604	54.92	176.61	231.52	55.16	196.71	251.88	0.21637	0.71369	0.93006
360	5.82	0.0007841	0.056738	59.44	173.94	233.38	59.72	194.08	253.81	0.23270	0.69566	0.92836
400	8.91	0.0007907	0.051201	63.62	171.45	235.07	63.94	191.62	255.55	0.24761	0.67929	0.92691
450	12.46	0.0007985	0.045619	68.45	168.54	237.00	68.81	188.71	257.53	0.26465	0.66069	0.92535
500	15.71	0.0008059	0.041118	72.93	165.82	238.75	73.33	185.98	259.30	0.28023	0.64377	0.92400
550	18.73	0.0008130	0.037408	77.10	163.25	240.35	77.54	183.38	260.92	0.29461	0.62821	0.92282
600	21.55	0.0008199	0.034295	81.02	160.81	241.83	81.51	180.90	262.40	0.30799	0.61378	0.92177
650	24.20	0.0008266	0.031646	84.72	158.48	243.20	85.26	178.51	263.77	0.32051	0.60030	0.92081
700	26.69	0.0008331	0.029361	88.24	156.24	244.48	88.82	176.21	265.03	0.33230	0.58763	0.91994

TABLE B.10   Saturated Refrigerant-134a—Pressure

Press. P [kPa]	Sat. temp. T_{sat}, °C	Specific volume [m³/kg]		Internal energy [kJ/kg]			Enthalpy [kJ/kg]			Entropy [kJ/kg·K]		
		Sat. liquid v_f	Sat. vapor v_g	Sat. liquid u_f	Evap. u_{fg}	Sat. vapor u_g	Sat. liquid h_f	Evap. h_{fg}	Sat. vapor h_g	Sat. liquid s_f	Evap. s_{fg}	Sat. vapor s_g
750	29.06	0.0008395	0.027371	91.59	154.08	245.67	92.22	173.98	266.20	0.34345	0.57567	0.91912
800	31.31	0.0008458	0.025621	94.79	152.00	246.79	95.47	171.82	267.29	0.35404	0.56431	0.91835
850	33.45	0.0008520	0.024069	97.87	149.98	247.85	98.60	169.71	268.31	0.36413	0.55349	0.91762
900	35.51	0.0008580	0.022683	100.83	148.01	248.85	101.61	167.66	269.26	0.37377	0.54315	0.91692
950	37.48	0.0008641	0.021438	103.69	146.10	249.79	104.51	165.64	270.15	0.38301	0.53323	0.91624
1000	39.37	0.0008700	0.020313	106.45	144.23	250.68	107.32	163.67	270.99	0.39189	0.52368	0.91558
1200	46.29	0.0008934	0.016715	116.70	137.11	253.81	117.77	156.10	273.87	0.42441	0.48863	0.91303
1400	52.40	0.0009166	0.014107	125.94	130.43	256.37	127.22	148.90	276.12	0.45315	0.45734	0.91050
1600	57.88	0.0009400	0.012123	134.43	124.04	258.47	135.93	141.93	277.86	0.47911	0.42873	0.90784
1800	62.87	0.0009639	0.010559	142.33	117.83	260.17	144.07	135.11	279.17	0.50294	0.40204	0.90498
2000	67.45	0.0009886	0.009288	149.78	111.73	261.51	151.76	128.33	280.09	0.52509	0.37675	0.90184
2500	77.54	0.0010566	0.006936	166.99	96.47	263.45	169.63	111.16	280.79	0.57531	0.31695	0.89226
3000	86.16	0.0011406	0.005275	183.04	80.22	263.26	186.46	92.63	279.09	0.62118	0.25776	0.87894

TABLE B.10 Saturated Refrigerant-134a—Pressure (Concluded)

Table B.11 Superheated Refrigerant-134a

T [°C]	v [m³/kg]	u [kJ/kg]	h [kJ/kg]	s [kJ/kg·K]	v [m³/kg]	u [kJ/kg]	h [kJ/kg]	s [kJ/kg·K]	v [m³/kg]	u [kJ/kg]	h [kJ/kg]	s [kJ/kg·K]
	P = 0.06 MPa (T_{sat} = 36.95°C)				P = 0.10 MPa (T_{sat} = 26.37°C)				P = 0.14 MPa (T_{sat} = 18.77°C)			
Sat.	0.31121	209.12	227.79	0.9644	0.19254	215.19	234.44	0.9518	0.14014	219.54	239.16	0.9446
−20	0.33608	220.60	240.76	1.0174	0.19841	219.66	239.50	0.9721				
−10	0.35048	227.55	248.58	1.0477	0.20743	226.75	247.49	1.0030	0.14605	225.91	246.36	0.9724
0	0.36476	234.66	256.54	1.0774	0.21630	233.95	255.58	1.0332	0.15263	233.23	254.60	1.0031
10	0.37893	241.92	264.66	1.1066	0.22506	241.30	263.81	1.0628	0.15908	240.66	262.93	1.0331
20	0.39302	249.35	272.94	1.1353	0.23373	248.79	272.17	1.0918	0.16544	248.22	271.38	1.0624
30	0.40705	256.95	281.37	1.1636	0.24233	256.44	280.68	1.1203	0.17172	255.93	279.97	1.0912
40	0.42102	264.71	289.97	1.1915	0.25088	264.25	289.34	1.1484	0.17794	263.79	288.70	1.1195
50	0.43495	272.64	298.74	1.2191	0.25937	272.22	298.16	1.1762	0.18412	271.79	297.57	1.1474
60	0.44883	280.73	307.66	1.2463	0.26783	280.35	307.13	1.2035	0.19025	279.96	306.59	1.1749
70	0.46269	288.99	316.75	1.2732	0.27626	288.64	316.26	1.2305	0.19635	288.28	315.77	1.2020
80	0.47651	297.41	326.00	1.2997	0.28465	297.08	325.55	1.2572	0.20242	296.75	325.09	1.2288
90	0.49032	306.00	335.42	1.3260	0.29303	305.69	334.99	1.2836	0.20847	305.38	334.57	1.2553
100	0.50410	314.74	344.99	1.3520	0.30138	314.46	344.60	1.3096	0.21449	314.17	344.20	1.2814
	P = 0.18 MPa (T_{sat} = 12.73°C)				P = 0.20 MPa (T_{sat} = 10.09°C)				P = 0.24 MPa (T_{sat} = 5.38°C)			
Sat.	0.11041	222.99	242.86	0.9397	0.09987	224.48	244.46	0.9377	0.08390	227.14	247.28	0.9346
−10	0.11189	225.02	245.16	0.9484	0.09991	224.55	244.54	0.9380				
0	0.11722	232.48	253.58	0.9798	0.10481	232.09	253.05	0.9698	0.08617	231.29	251.97	0.9519
10	0.12240	240.00	262.04	1.0102	0.10955	239.67	261.58	1.0004	0.09026	238.98	260.65	0.9831
20	0.12748	247.64	270.59	1.0399	0.11418	247.35	270.18	1.0303	0.09423	246.74	269.36	1.0134
30	0.13248	255.41	279.25	1.0690	0.11874	255.14	278.89	1.0595	0.09812	254.61	278.16	1.0429
40	0.13741	263.31	288.05	1.0975	0.12322	263.08	287.72	1.0882	0.10193	262.59	287.06	1.0718
50	0.14230	271.36	296.98	1.1256	0.12766	271.15	296.68	1.1163	0.10570	270.71	296.08	1.1001
60	0.14715	279.56	306.05	1.1532	0.13206	279.37	305.78	1.1441	0.10942	278.97	305.23	1.1280

Table B.11 Superheated Refrigerant-134a *(Continued)*

T [°C]	v [m³/kg]	u [kJ/kg]	h [kJ/kg]	s [kJ/kg·K]	v [m³/kg]	u [kJ/kg]	h [kJ/kg]	s [kJ/kg·K]	v [m³/kg]	u [kJ/kg]	h [kJ/kg]	s [kJ/kg·K]
70	0.15196	287.91	315.27	1.1805	0.13641	287.73	315.01	1.1714	0.11310	287.36	314.51	1.1554
80	0.15673	296.42	324.63	1.2074	0.14074	296.25	324.40	1.1983	0.11675	295.91	323.93	1.1825
90	0.16149	305.07	334.14	1.2339	0.14504	304.92	333.93	1.2249	0.12038	304.60	333.49	1.2092
100	0.16622	313.88	343.80	1.2602	0.14933	313.74	343.60	1.2512	0.12398	313.44	343.20	1.2356
	P = 0.28 MPa (Tsat = −1.25°C)				P = 0.32 MPa (Tsat = 2.46°C)				P = 0.40 MPa (Tsat = 8.91°C)			
Sat.	0.07235	229.46	249.72	0.9321	0.06360	231.52	251.88	0.9301	0.051201	235.07	255.55	0.9269
0	0.07282	230.44	250.83	0.9362								
10	0.07646	238.27	259.68	0.9680	0.06609	237.54	258.69	0.9544	0.051506	235.97	256.58	0.9305
20	0.07997	246.13	268.52	0.9987	0.06925	245.50	267.66	0.9856	0.054213	244.18	265.86	0.9628
30	0.08338	254.06	277.41	1.0285	0.07231	253.50	276.65	1.0157	0.056796	252.36	275.07	0.9937
40	0.08672	262.10	286.38	1.0576	0.07530	261.60	285.70	1.0451	0.059292	260.58	284.30	1.0236
50	0.09000	270.27	295.47	1.0862	0.07823	269.82	294.85	1.0739	0.061724	268.90	293.59	1.0528
60	0.09324	278.56	304.67	1.1142	0.08111	278.15	304.11	1.1021	0.064104	277.32	302.96	1.0814
70	0.09644	286.99	314.00	1.1418	0.08395	286.62	313.48	1.1298	0.066443	285.86	312.44	1.1094
80	0.09961	295.57	323.46	1.1690	0.08675	295.22	322.98	1.1571	0.068747	294.53	322.02	1.1369
90	0.10275	304.29	333.06	1.1958	0.08953	303.97	332.62	1.1840	0.071023	303.32	331.73	1.1640
100	0.10587	313.15	342.80	1.2222	0.09229	312.86	342.39	1.2105	0.073274	312.26	341.57	1.1907
110	0.10897	322.16	352.68	1.2483	0.09503	321.89	352.30	1.2367	0.075504	321.33	351.53	1.2171
120	0.11205	331.32	362.70	1.2742	0.09775	331.07	362.35	1.2626	0.077717	330.55	361.63	1.2431
130	0.11512	340.63	372.87	1.2997	0.10045	340.39	372.54	1.2882	0.079913	339.90	371.87	1.2688
140	0.11818	350.09	383.18	1.3250	0.10314	349.86	382.87	1.3135	0.082096	349.41	382.24	1.2942
	P = 0.50 MPa (Tsat = 15.71°C)				P = 0.60 MPa (Tsat = 21.55°C)				P = 0.70 MPa (Tsat = 26.69°C)			
Sat.	0.041118	238.75	259.30	0.9240	0.034295	241.83	262.40	0.9218	0.029361	244.48	265.03	0.9199
20	0.042115	242.40	263.46	0.9383								

Superheated Refrigerant-134a (Continued)

T [°C]	P = 0.50 MPa (Tsat = 15.71°C)				P = 0.60 MPa (Tsat = 21.55°C)				P = 0.70 MPa (Tsat = 26.69°C)			
	v [m³/kg]	u [kJ/kg]	h [kJ/kg]	s [kJ/kg·K]	v [m³/kg]	u [kJ/kg]	h [kJ/kg]	s [kJ/kg·K]	v [m³/kg]	u [kJ/kg]	h [kJ/kg]	s [kJ/kg·K]
30	0.044338	250.84	273.01	0.9703	0.035984	249.22	270.81	0.9499	0.029966	247.48	268.45	0.9313
40	0.046456	259.26	282.48	1.0011	0.037865	257.86	280.58	0.9816	0.031696	256.39	278.57	0.9641
50	0.048499	267.72	291.96	1.0309	0.039659	266.48	290.28	1.0121	0.033322	265.20	288.53	0.9954
60	0.050485	276.25	301.50	1.0599	0.041389	275.15	299.98	1.0417	0.034875	274.01	298.42	1.0256
70	0.052427	284.89	311.10	1.0883	0.043069	283.89	309.73	1.0705	0.036373	282.87	308.33	1.0549
80	0.054331	293.64	320.80	1.1162	0.044710	292.73	319.55	1.0987	0.037829	291.80	318.28	1.0835
90	0.056205	302.51	330.61	1.1436	0.046318	301.67	329.46	1.1264	0.039250	300.82	328.29	1.1114
100	0.058053	311.50	340.53	1.1705	0.047900	310.73	339.47	1.1536	0.040642	309.95	338.40	1.1389
110	0.059880	320.63	350.57	1.1971	0.049458	319.91	349.59	1.1803	0.042010	319.19	348.60	1.1658
120	0.061687	329.89	360.73	1.2233	0.050997	329.23	359.82	1.2067	0.043358	328.55	358.90	1.1924
130	0.063479	339.29	371.03	1.2491	0.052519	338.67	370.18	1.2327	0.044688	338.04	369.32	1.2186
140	0.065256	348.83	381.46	1.2747	0.054027	348.25	380.66	1.2584	0.046004	347.66	379.86	1.2444
150	0.067021	358.51	392.02	1.2999	0.055522	357.96	391.27	1.2838	0.047306	357.41	390.52	1.2699
160	0.068775	368.33	402.72	1.3249	0.057006	367.81	402.01	1.3088	0.048597	367.29	401.31	1.2951

T [°C]	P = 0.80 MPa (Tsat = 31.31°C)				P = 0.90 MPa (Tsat = 35.51°C)				P = 1.00 MPa (Tsat = 39.37°C)			
	v [m³/kg]	u [kJ/kg]	h [kJ/kg]	s [kJ/kg·K]	v [m³/kg]	u [kJ/kg]	h [kJ/kg]	s [kJ/kg·K]	v [m³/kg]	u [kJ/kg]	h [kJ/kg]	s [kJ/kg·K]
Sat.	0.025621	246.79	267.29	0.9183	0.022683	248.85	269.26	0.9169	0.020313	250.68	270.99	0.9156
40	0.027035	254.82	276.45	0.9480	0.023375	253.13	274.17	0.9327	0.020406	251.30	271.71	0.9179
50	0.028547	263.86	286.69	0.9802	0.024809	262.44	284.77	0.9660	0.021796	260.94	282.74	0.9525
60	0.029973	272.83	296.81	1.0110	0.026146	271.60	295.13	0.9976	0.023068	270.32	293.38	0.9850
70	0.031340	281.81	306.88	1.0408	0.027413	280.72	305.39	1.0280	0.024261	279.59	303.85	1.0160
80	0.032659	290.84	316.97	1.0698	0.028630	289.86	315.63	1.0574	0.025398	288.86	314.25	1.0458
90	0.033941	299.95	327.10	1.0981	0.029806	299.06	325.89	1.0860	0.026492	298.15	324.64	1.0748
100	0.035193	309.15	337.30	1.1258	0.030951	308.34	336.19	1.1140	0.027552	307.51	335.06	1.1031
110	0.036420	318.45	347.59	1.1530	0.032068	317.70	346.56	1.1414	0.028584	316.94	345.53	1.1308

Table B.11 Superheated Refrigerant-134a (Continued)

T [°C]	v [m³/kg]	u [kJ/kg]	h [kJ/kg]	s [kJ/kg·K]	v [m³/kg]	u [kJ/kg]	h [kJ/kg]	s [kJ/kg·K]	v [m³/kg]	u [kJ/kg]	h [kJ/kg]	s [kJ/kg·K]
120	0.037625	327.87	357.97	1.1798	0.033164	327.18	357.02	1.1684	0.029592	326.47	356.06	1.1580
130	0.038813	337.40	368.45	1.2061	0.034241	336.76	367.58	1.1949	0.030581	336.11	366.69	1.1846
140	0.039985	347.06	379.05	1.2321	0.035302	346.46	378.23	1.2210	0.031554	345.85	377.40	1.2109
150	0.041143	356.85	389.76	1.2577	0.036349	356.28	389.00	1.2467	0.032512	355.71	388.22	1.2368
160	0.042290	366.76	400.59	1.2830	0.037384	366.23	399.88	1.2721	0.033457	365.70	399.15	1.2623
170	0.043427	376.81	411.55	1.3080	0.038408	376.31	410.88	1.2972	0.034392	375.81	410.20	1.2875
180	0.044554	386.99	422.64	1.3327	0.039423	386.52	422.00	1.3221	0.035317	386.04	421.36	1.3124
	P = 1.20 MPa (T_{sat} = 46.29°C)				P = 1.40 MPa (T_{sat} = 52.40°C)				P = 1.60 MPa (T_{sat} = 57.88°C)			
Sat.	0.016715	253.81	273.87	0.9130	0.014107	256.37	276.12	0.9105	0.012123	258.47	277.86	0.9078
50	0.017201	257.63	278.27	0.9267								
60	0.018404	267.56	289.64	0.9614	0.015005	264.46	285.47	0.9389	0.012372	260.89	280.69	0.9163
70	0.019502	277.21	300.61	0.9938	0.016060	274.62	297.10	0.9733	0.013430	271.76	293.25	0.9535
80	0.020529	286.75	311.39	1.0248	0.017023	284.51	308.34	1.0056	0.014362	282.09	305.07	0.9875
90	0.021506	296.26	322.07	1.0546	0.017923	294.28	319.37	1.0364	0.015215	292.17	316.52	1.0194
100	0.022442	305.80	332.73	1.0836	0.018778	304.01	330.30	1.0661	0.016014	302.14	327.76	1.0500
110	0.023348	315.38	343.40	1.1118	0.019597	313.76	341.19	1.0949	0.016773	312.07	338.91	1.0795
120	0.024228	325.03	354.11	1.1394	0.020388	323.55	352.09	1.1230	0.017500	322.02	350.02	1.1081
130	0.025086	334.77	364.88	1.1664	0.021155	333.41	363.02	1.1504	0.018201	332.00	361.12	1.1360
140	0.025927	344.61	375.72	1.1930	0.021904	343.34	374.01	1.1773	0.018882	342.05	372.26	1.1632
150	0.026753	354.56	386.66	1.2192	0.022636	353.37	385.07	1.2038	0.019545	352.17	383.44	1.1900
160	0.027566	364.61	397.69	1.2449	0.023355	363.51	396.20	1.2298	0.020194	362.38	394.69	1.2163
170	0.028367	374.78	408.82	1.2703	0.024061	373.75	407.43	1.2554	0.020830	372.69	406.02	1.2421
180	0.029158	385.08	420.07	1.2954	0.024757	384.10	418.76	1.2807	0.021456	383.11	417.44	1.2676

TABLE B.11 Superheated Refrigerant-134a (Concluded)

Altitude [m]	Temperature [°C]	Pressure [kPa]	Gravity g [m/s²]	Speed of sound [m/s]	Density [kg/m³]	Viscosity μ [kg/m·s]	Conductivity [W/m·K]
0	15.00	101.33	9.807	340.3	1.225	1.789×10^{-5}	0.0253
200	13.70	98.95	9.806	339.5	1.202	1.783×10^{-5}	0.0252
400	12.40	96.61	9.805	338.8	1.179	1.777×10^{-5}	0.0252
600	11.10	94.32	9.805	338.0	1.156	1.771×10^{-5}	0.0251
800	9.80	92.08	9.804	337.2	1.134	1.764×10^{-5}	0.0250
1000	8.50	89.88	9.804	336.4	1.112	1.758×10^{-5}	0.0249
1200	7.20	87.72	9.803	335.7	1.090	1.752×10^{-5}	0.0248
1400	5.90	85.60	9.802	334.9	1.069	1.745×10^{-5}	0.0247
1600	4.60	83.53	9.802	334.1	1.048	1.739×10^{-5}	0.0245
1800	3.30	81.49	9.801	333.3	1.027	1.732×10^{-5}	0.0244
2000	2.00	79.50	9.800	332.5	1.007	1.726×10^{-5}	0.0243
2200	0.70	77.55	9.800	331.7	0.987	1.720×10^{-5}	0.0242
2400	−0.59	75.63	9.799	331.0	0.967	1.713×10^{-5}	0.0241
2600	−1.89	73.76	9.799	330.2	0.947	1.707×10^{-5}	0.0240
2800	−3.19	71.92	9.798	329.4	0.928	1.700×10^{-5}	0.0239
3000	−4.49	70.12	9.797	328.6	0.909	1.694×10^{-5}	0.0238
3200	−5.79	68.36	9.797	327.8	0.891	1.687×10^{-5}	0.0237
3400	−7.09	66.63	9.796	327.0	0.872	1.681×10^{-5}	0.0236
3600	−8.39	64.94	9.796	326.2	0.854	1.674×10^{-5}	0.0235
3800	−9.69	63.28	9.795	325.4	0.837	1.668×10^{-5}	0.0234
4000	−10.98	61.66	9.794	324.6	0.819	1.661×10^{-5}	0.0233
4200	−12.3	60.07	9.794	323.8	0.802	1.655×10^{-5}	0.0232
4400	−13.6	58.52	9.793	323.0	0.785	1.648×10^{-5}	0.0231
4600	−14.9	57.00	9.793	322.2	0.769	1.642×10^{-5}	0.0230

TABLE B.12 Properties of the Atmosphere at High Altitude

Altitude [m]	Temperature [°C]	Pressure [kPa]	Gravity g [m/s²]	Speed of sound [m/s]	Density [kg/m³]	Viscosity μ [kg/m·s]	Conductivity [W/m·K]
4800	−16.2	55.51	9.792	321.4	0.752	1.635×10^{-5}	0.0229
5000	−17.5	54.05	9.791	320.5	0.736	1.628×10^{-5}	0.0228
5200	−18.8	52.62	9.791	319.7	0.721	1.622×10^{-5}	0.0227
5400	−20.1	51.23	9.790	318.9	0.705	1.615×10^{-5}	0.0226
5600	−21.4	49.86	9.789	318.1	0.690	1.608×10^{-5}	0.0224
5800	−22.7	48.52	9.785	317.3	0.675	1.602×10^{-5}	0.0223
6000	−24.0	47.22	9.788	316.5	0.660	1.595×10^{-5}	0.0222
6200	−25.3	45.94	9.788	315.6	0.646	1.588×10^{-5}	0.0221
6400	−26.6	44.69	9.787	314.8	0.631	1.582×10^{-5}	0.0220
6600	−27.9	43.47	9.786	314.0	0.617	1.575×10^{-5}	0.0219
6800	−29.2	42.27	9.785	313.1	0.604	1.568×10^{-5}	0.0218
7000	−30.5	41.11	9.785	312.3	0.590	1.561×10^{-5}	0.0217
8000	−36.9	35.65	9.782	308.1	0.526	1.527×10^{-5}	0.0212
9000	−43.4	30.80	9.779	303.8	0.467	1.493×10^{-5}	0.0206
10,000	−49.9	26.50	9.776	299.5	0.414	1.458×10^{-5}	0.0201
12,000	−56.5	19.40	9.770	295.1	0.312	1.422×10^{-5}	0.0195
14,000	−56.5	14.17	9.764	295.1	0.228	1.422×10^{-5}	0.0195
16,000	−56.5	10.53	9.758	295.1	0.166	1.422×10^{-5}	0.0195
18,000	−56.5	7.57	9.751	295.1	0.122	1.422×10^{-5}	0.0195

Source: U.S. Standard Atmosphere Supplements, U.S. Government Printing Office, 1966. Based on year-round mean conditions at 45° latitude and varies with the time of the year and the weather patterns. The conditions at sea level (z = 0) are taken to be P = 101.325 kPa, T = 15°C, ρ = 1.2250 kg/m³, g = 9.80665 m²/s.

TABLE B.12   Properties of the Atmosphere at High Altitude (*Concluded*)

T [K]	h [kJ/kg]	P_r	u [kJ/kg]	V_r	s° [kJ/kg·K]	T [K]	h [kJ/kg]	P_r	u [kJ/kg]	V_r	s° [kJ/kg·K]
200	199.97	0.3363	142.56	1707.0	1.29559	580	586.04	14.38	419.55	115.7	2.37348
210	209.97	0.3987	149.69	1512.0	1.34444	590	596.52	15.31	427.15	110.6	2.39140
220	219.97	0.4690	156.82	1346.0	1.39105	600	607.02	16.28	434.78	105.8	2.40902
230	230.02	0.5477	164.00	1205.0	1.43557	610	617.53	17.30	442.42	101.2	2.42644
240	240.02	0.6355	171.13	1084.0	1.47824	620	628.07	18.36	450.09	96.92	2.44356
250	250.05	0.7329	178.28	979.0	1.51917	630	638.63	19.84	457.78	92.84	2.46048
260	260.09	0.8405	185.45	887.8	1.55848	640	649.22	20.64	465.50	88.99	2.47716
270	270.11	0.9590	192.60	808.0	1.59634	650	659.84	21.86	473.25	85.34	2.49364
280	280.13	1.0889	199.75	738.0	1.63279	660	670.47	23.13	481.01	81.89	2.50985
285	285.14	1.1584	203.33	706.1	1.65055	670	681.14	24.46	488.81	78.61	2.52589
290	290.16	1.2311	206.91	676.1	1.66802	680	691.82	25.85	496.62	75.50	2.54175
295	295.17	1.3068	210.49	647.9	1.68515	690	702.52	27.29	504.45	72.56	2.55731
298	298.18	1.3543	212.64	631.9	1.69528	700	713.27	28.80	512.33	69.76	2.57277
300	300.19	1.3860	214.07	621.2	1.70203	710	724.04	30.38	520.23	67.07	2.58810
305	305.22	1.4686	217.67	596.0	1.71865	720	734.82	32.02	528.14	64.53	2.60319
310	310.24	1.5546	221.25	572.3	1.73498	730	745.62	33.72	536.07	62.13	2.61803
315	315.27	1.6442	224.85	549.8	1.75106	740	756.44	35.50	544.02	59.82	2.63280
320	320.29	1.7375	228.42	528.6	1.76690	750	767.29	37.35	551.99	57.63	2.64737
325	325.31	1.8345	232.02	508.4	1.78249	760	778.18	39.27	560.01	55.54	2.66176
330	330.34	1.9352	235.61	489.4	1.79783	780	800.03	43.35	576.12	51.64	2.69013
340	340.42	2.149	242.82	454.1	1.82790	800	821.95	47.75	592.30	48.08	2.71787
350	350.49	2.379	250.02	422.2	1.85708	820	843.98	52.59	608.59	44.84	2.74504
360	360.58	2.626	257.24	393.4	1.88543	840	866.08	57.60	624.95	41.85	2.77170
370	370.67	2.892	264.46	367.2	1.91313	860	888.27	63.09	641.40	39.12	2.79783

TABLE B.13 Ideal-Gas Properties of Air

T [K]	h [kJ/kg]	P_r	u [kJ/kg]	V_r	s° [kJ/kg·K]	T [K]	h [kJ/kg]	P_r	u [kJ/kg]	V_r	s° [kJ/kg·K]
380	380.77	3.176	271.69	343.4	1.94001	880	910.56	68.98	657.95	36.61	2.82344
390	390.88	3.481	278.93	321.5	1.96633	900	932.93	75.29	674.58	34.31	2.84856
400	400.98	3.806	286.16	301.6	1.99194	920	955.38	82.05	691.28	32.18	2.87324
410	411.12	4.153	293.43	283.3	2.01699	940	977.92	89.28	708.08	30.22	2.89748
420	421.26	4.522	300.69	266.6	2.04142	960	1000.55	97.00	725.02	28.40	2.92128
430	431.43	4.915	307.99	251.1	2.06533	980	1023.25	105.2	741.98	26.73	2.94468
440	441.61	5.332	315.30	236.8	2.08870	1000	1046.04	114.0	758.94	25.17	2.96770
450	451.80	5.775	322.62	223.6	2.11161	1020	1068.89	123.4	776.10	23.72	2.99034
460	462.02	6.245	329.97	211.4	2.13407	1040	1091.85	133.3	793.36	23.29	3.01260
470	472.24	6.742	337.32	200.1	2.15604	1060	1114.86	143.9	810.62	21.14	3.03449
480	482.49	7.268	344.70	189.5	2.17760	1080	1137.89	155.2	827.88	19.98	3.05608
490	492.74	7.824	352.08	179.7	2.19876	1100	1161.07	167.1	845.33	18.896	3.07732
500	503.02	8.411	359.49	170.6	2.21952	1120	1184.28	179.7	862.79	17.886	3.09825
510	513.32	9.031	366.92	162.1	2.23993	1140	1207.57	193.1	880.35	16.946	3.11883
520	523.63	9.684	374.36	154.1	2.25997	1160	1230.92	207.2	897.91	16.064	3.13916
530	533.98	10.37	381.84	146.7	2.27967	1180	1254.34	222.2	915.57	15.241	3.15916
540	544.35	11.10	389.34	139.7	2.29906	1200	1277.79	238.0	933.33	14.470	3.17888
550	555.74	11.86	396.86	133.1	2.31809	1220	1301.31	254.7	951.09	13.747	3.19834
560	565.17	12.66	404.42	127.0	2.33685	1240	1324.93	272.3	968.95	13.069	3.21751
570	575.59	13.50	411.97	121.2	2.35531						
1260	1348.55	290.8	986.90	12.435	3.23638	1600	1757.57	791.2	1298.30	5.804	3.52364
1280	1372.24	310.4	1004.76	11.835	3.25510	1620	1782.00	834.1	1316.96	5.574	3.53879
1300	1395.97	330.9	1022.82	11.275	3.27345	1640	1806.46	878.9	1335.72	5.355	3.55381
1320	1419.76	352.5	1040.88	10.747	3.29160	1660	1830.96	925.6	1354.48	5.147	3.56867

TABLE B.13   Ideal-Gas Properties of Air (Continued)

T [K]	h [kJ/kg]	P_r	u [kJ/kg]	V_r	s° [kJ/kg·K]	T [K]	h [kJ/kg]	P_r	u [kJ/kg]	V_r	s° [kJ/kg·K]
1340	1443.60	375.3	1058.94	10.247	3.30959	1680	1855.50	974.2	1373.24	4.949	3.58335
1360	1467.49	399.1	1077.10	9.780	3.32724	1700	1880.1	1025	1392.7	4.761	3.5979
1380	1491.44	424.2	1095.26	9.337	3.34474	1750	1941.6	1161	1439.8	4.328	3.6336
1400	1515.42	450.5	1113.52	8.919	3.36200	1800	2003.3	1310	1487.2	3.994	3.6684
1420	1539.44	478.0	1131.77	8.526	3.37901	1850	2065.3	1475	1534.9	3.601	3.7023
1440	1563.51	506.9	1150.13	8.153	3.39586	1900	2127.4	1655	1582.6	3.295	3.7354
1460	1587.63	537.1	1168.49	7.801	3.41247	1950	2189.7	1852	1630.6	3.022	3.7677
1480	1611.79	568.8	1186.95	7.468	3.42892	2000	2252.1	2068	1678.7	2.776	3.7994
1500	1635.97	601.9	1205.41	7.152	3.44516	2050	2314.6	2303	1726.8	2.555	3.8303
1520	1660.23	636.5	1223.87	6.854	3.46120	2100	2377.7	2559	1775.3	2.356	3.8605
1540	1684.51	672.8	1242.43	6.569	3.47712	2150	2440.3	2837	1823.8	2.175	3.8901
1560	1708.82	710.5	1260.99	6.301	3.49276	2200	2503.2	3138	1872.4	2.012	3.9191
1580	1733.17	750.0	1279.65	6.046	3.50829	2250	2566.4	3464	1921.3	1.864	3.9474

Note: The properties P_r (relative pressure) and V_r (relative specific volume) are dimensionless quantities used in the analysis of isentropic processes and should not be confused with the properties pressure and specific volume.

Source: Kenneth Wark, *Thermodynamics*, 4th ed. (New York: McGraw-Hill, 1983), pp. 785–6, table A–5. Originally published in J. H. Keenan and J. Kaye, *Gas Tables* (New York: John Wiley & Sons, 1948).

TABLE B.13 Ideal-Gas Properties of Air (*Concluded*)

Table B.14 Ideal-Gas Properties of Nitrogen, N_2

T [K]	h [kJ/kmol]	u [kJ/kmol]	s° [kJ/kmol·K]	T [K]	h [kJ/kmol]	u [kJ/kmol]	s° [kJ/kmol·K]
0	0	0	0	460	13,399	9574	204.170
220	6391	4562	182.639	470	13,693	9786	204.803
230	6683	4770	183.938	480	13,988	9997	205.424
240	6975	4979	185.180	490	14,285	10,210	206.033
250	7266	5188	186.370	500	14,581	10,423	206.630
260	7558	5396	187.514	510	14,876	10,635	207.216
270	7849	5604	188.614	520	15,172	10,848	207.792
280	8141	5813	189.673	530	15,469	11,062	208.358
290	8432	6021	190.695	540	15,766	11,277	208.914
298	8669	6190	191.502	550	16,064	11,492	209.461
300	8723	6229	191.682	560	16,363	11,707	209.999
310	9014	6437	192.638	570	16,662	11,923	210.528
320	9306	6645	193.562	580	16,962	12,139	211.049
330	9597	6853	194.459	590	17,262	12,356	211.562
340	9888	7061	195.328	600	17,563	12,574	212.066
350	10,180	7270	196.173	610	17,864	12,792	212.564
360	10,471	7478	196.995	620	18,166	13,011	213.055
370	10,763	7687	197.794	630	18,468	13,230	213.541
380	11,055	7895	198.572	640	18,772	13,450	214.018
390	11,347	8104	199.331	650	19,075	13,671	214.489
400	11,640	8314	200.071	660	19,380	13,892	214.954
410	11,932	8523	200.794	670	19,685	14,114	215.413
420	12,225	8733	201.499	680	19,991	14,337	215.866
430	12,518	8943	202.189	690	20,297	14,560	216.314
440	12,811	9153	202.863	700	20,604	14,784	216.756
450	13,105	9363	203.523	710	20,912	15,008	217.192

T [K]	h [kJ/kmol]	u [kJ/kmol]	s° [kJ/kmol·K]	T [K]	h [kJ/kmol]	u [kJ/kmol]	s° [kJ/kmol·K]
720	21,220	15,234	217.624	980	29,476	21,328	227.398
730	21,529	15,460	218.059	990	29,803	21,571	227.728
740	21,839	15,686	218.472	1000	30,129	21,815	228.057
750	22,149	15,913	218.889	1020	30,784	22,304	228.706
760	22,460	16,141	219.301	1040	31,442	22,795	229.344
770	22,772	16,370	219.709	1060	32,101	23,288	229.973
780	23,085	16,599	220.113	1080	32,762	23,782	230.591
790	23,398	16,830	220.512	1100	33,426	24,280	231.199
800	23,714	17,061	220.907	1120	34,092	24,780	231.799
810	24,027	17,292	221.298	1140	34,760	25,282	232.391
820	24,342	17,524	221.684	1160	35,430	25,786	232.973
830	24,658	17,757	222.067	1180	36,104	26,291	233.549
840	24,974	17,990	222.447	1200	36,777	26,799	234.115
850	25,292	18,224	222.822	1220	37,452	27,308	234.673
860	25,610	18,459	223.194	1240	38,129	27,819	235.223
870	25,928	18,695	223.562	1260	38,807	28,331	235.766
880	26,248	18,931	223.927	1280	39,488	28,845	236.302
890	26,568	19,168	224.288	1300	40,170	29,361	236.831
900	26,890	19,407	224.647	1320	40,853	29,378	237.353
910	27,210	19,644	225.002	1340	41,539	30,398	237.867
920	27,532	19,883	225.353	1360	42,227	30,919	238.376
930	27,854	20,122	225.701	1380	42,915	31,441	238.878
940	28,178	20,362	226.047	1400	43,605	31,964	239.375
950	28,501	20,603	226.389	1420	44,295	32,489	239.865
960	28,826	20,844	226.728	1440	44,988	33,014	240.350
970	29,151	21,086	227.064	1460	45,682	33,543	240.827

TABLE B.14 Ideal-Gas Properties of Nitrogen, N_2 (Continued)

T [K]	h [kJ/kmol]	u [kJ/kmol]	s° [kJ/kmol·K]
1480	46,377	34,071	241.301
1500	47,073	34,601	241.768
1520	47,771	35,133	242.228
1540	48,470	35,665	242.685
1560	49,168	36,197	243.137
1580	49,869	36,732	243.585
1600	50,571	37,268	244.028
1620	51,275	37,806	244.464
1640	51,980	38,344	244.896
1660	52,686	38,884	245.324
1680	53,393	39,424	245.747
1700	54,099	39,965	246.166
1720	54,807	40,507	246.580
1740	55,516	41,049	246.990
1760	56,227	41,594	247.396
1780	56,938	42,139	247.798
1800	57,651	42,685	248.195
1820	58,363	43,231	248.589
1840	59,075	43,777	248.979
1860	59,790	44,324	249.365
1880	60,504	44,873	249.748
1900	61,220	45,423	250.128
1920	61,936	45,973	250.502
1940	62,654	46,524	250.874
1960	63,381	47,075	251.242
1980	64,090	47,627	251.607
2000	64,810	48,181	251.969
2050	66,612	49,567	252.858
2100	68,417	50,957	253.726
2150	70,226	52,351	254.578
2200	72,040	53,749	255.412
2250	73,856	55,149	256.227
2300	75,676	56,553	257.027
2350	77,496	57,958	257.810
2400	79,320	59,366	258.580
2450	81,149	60,779	259.332
2500	82,981	62,195	260.073
2550	84,814	63,613	260.799
2600	86,650	65,033	261.512
2650	88,488	66,455	262.213
2700	90,328	67,880	262.902
2750	92,171	69,306	263.577
2800	94,014	70,734	264.241
2850	95,859	72,163	264.895
2900	97,705	73,593	265.538
2950	99,556	75,028	266.170
3000	101,407	76,464	266.793
3050	103,260	77,902	267.404
3100	105,115	79,341	268.007
3150	106,972	80,782	268.601
3200	108,830	82,224	269.186
3250	110,690	83,668	269.763

Source: Tables B.14 through B.21 are adapted from Kenneth Wark, *Thermodynamics*, 4th ed. (New York: McGraw-Hill, 1983), pp. 787–98. Originally published in JANAF, *Thermochemical Tables*, NSRDS-NBS-37, 1971.

TABLE B.14 Ideal-Gas Properties of Nitrogen, N_2 *(Concluded)*

T [K]	h [kJ/kmol]	u [kJ/kmol]	s° [kJ/kmol·K]	T [K]	h [kJ/kmol]	u [kJ/kmol]	s° [kJ/kmol·K]
0	0	0	0	460	13,525	9710	218.016
220	6404	4575	196.171	470	13,842	9935	218.676
230	6694	4782	197.461	480	14,151	10,160	219.326
240	6984	4989	198.696	490	14,460	10,386	219.963
250	7275	5197	199.885	500	14,770	10,614	220.589
260	7566	5405	201.027	510	15,082	10,842	221.206
270	7858	5613	202.128	520	15,395	11,071	221.812
280	8150	5822	203.191	530	15,708	11,301	222.409
290	8443	6032	204.218	540	16,022	11,533	222.997
298	8682	6203	205.033	550	16,338	11,765	223.576
300	8736	6242	205.213	560	16,654	11,998	224.146
310	9030	6453	206.177	570	16,971	12,232	224.708
320	9325	6664	207.112	580	17,290	12,467	225.262
330	9620	6877	208.020	590	17,609	12,703	225.808
340	9916	7090	208.904	600	17,929	12,940	226.346
350	10,213	7303	209.765	610	18,250	13,178	226.877
360	10,511	7518	210.604	620	18,572	13,417	227.400
370	10,809	7733	211.423	630	18,895	13,657	227.918
380	11,109	7949	212.222	640	19,219	13,898	228.429
390	11,409	8166	213.002	650	19,544	14,140	228.932
400	11,711	8384	213.765	660	19,870	14,383	229.430
410	12,012	8603	214.510	670	20,197	14,626	229.920
420	12,314	8822	215.241	680	20,524	14,871	230.405
430	12,618	9043	215.955	690	20,854	15,116	230.885
440	12,923	9264	216.656	700	21,184	15,364	231.358
450	13,228	9487	217.342	710	21,514	15,611	231.827

TABLE B.15 Ideal-Gas Properties of Oxygen, O_2

T [K]	h [kJ/kmol]	u [kJ/kmol]	s° [kJ/kmol·K]	T [K]	h [kJ/kmol]	u [kJ/kmol]	s° [kJ/kmol·K]
720	21,845	15,859	232.291	980	30,692	22,544	242.768
730	22,177	16,107	232.748	990	31,041	22,809	243.120
740	22,510	16,357	233.201	1000	31,389	23,075	243.471
750	22,844	16,607	233.649	1020	32,088	23,607	244.164
760	23,178	16,859	234.091	1040	32,789	24,142	244.844
770	23,513	17,111	234.528	1060	33,490	24,677	245.513
780	23,850	17,364	234.960	1080	34,194	25,214	246.171
790	24,186	17,618	235.387	1100	34,899	25,753	246.818
800	24,523	17,872	235.810	1120	35,606	26,294	247.454
810	24,861	18,126	236.230	1140	36,314	26,836	248.081
820	25,199	18,382	236.644	1160	37,023	27,379	248.698
830	25,537	18,637	237.055	1180	37,734	27,923	249.307
840	25,877	18,893	237.462	1200	38,447	28,469	249.906
850	26,218	19,150	237.864	1220	39,162	29,018	250.497
860	26,559	19,408	238.264	1240	39,877	29,568	251.079
870	26,899	19,666	238.660	1260	40,594	30,118	251.653
880	27,242	19,925	239.051	1280	41,312	30,670	252.219
890	27,584	20,185	239.439	1300	42,033	31,224	252.776
900	27,928	20,445	239.823	1320	42,753	31,778	253.325
910	28,272	20,706	240.203	1340	43,475	32,334	253.868
920	28,616	20,967	240.580	1360	44,198	32,891	254.404
930	28,960	21,228	240.953	1380	44,923	33,449	254.932
940	29,306	21,491	241.323	1400	45,648	34,008	255.454
950	29,652	21,754	241.689	1420	46,374	34,567	255.968
960	29,999	22,017	242.052	1440	47,102	35,129	256.475
970	30,345	22,280	242.411	1460	47,831	35,692	256.978

TABLE B.15 Ideal-Gas Properties of Oxygen, O_2 (Continued)

T [K]	h [kJ/kmol]	u [kJ/kmol]	s° [kJ/kmol·K]	T [K]	h [kJ/kmol]	u [kJ/kmol]	s° [kJ/kmol·K]
1480	48,561	36,256	257.474	2000	67,881	51,253	268.655
1500	49,292	36,821	257.965	2050	69,772	52,727	269.588
1520	50,024	37,387	258.450	2100	71,668	54,208	270.504
1540	50,756	37,952	258.928	2150	73,573	55,697	271.399
1560	51,490	38,520	259.402	2200	75,484	57,192	272.278
1580	52,224	39,088	259.870	2250	77,397	58,690	273.136
1600	52,961	39,658	260.333	2300	79,316	60,193	273.891
1620	53,696	40,227	260.791	2350	81,243	61,704	274.809
1640	54,434	40,799	261.242	2400	83,174	63,219	275.625
1660	55,172	41,370	261.690	2450	85,112	64,742	276.424
1680	55,912	41,944	262.132	2500	87,057	66,271	277.207
1700	56,652	42,517	262.571	2550	89,004	67,802	277.979
1720	57,394	43,093	263.005	2600	90,956	69,339	278.738
1740	58,136	43,669	263.435	2650	92,916	70,883	279.485
1760	58,880	44,247	263.861	2700	94,881	72,433	280.219
1780	59,624	44,825	264.283	2750	96,852	73,987	280.942
1800	60,371	45,405	264.701	2800	98,826	75,546	281.654
1820	61,118	45,986	265.113	2850	100,808	77,112	282.357
1840	61,866	46,568	265.521	2900	102,793	78,682	283.048
1860	62,616	47,151	265.925	2950	104,785	80,258	283.728
1880	63,365	47,734	266.326	3000	106,780	81,837	284.399
1900	64,116	48,319	266.722	3050	108,778	83,419	285.060
1920	64,868	48,904	267.115	3100	110,784	85,009	285.713
1940	65,620	49,490	267.505	3150	112,795	86,601	286.355
1960	66,374	50,078	267.891	3200	114,809	88,203	286.989
1980	67,127	50,665	268.275	3250	116,827	89,804	287.614

TABLE B.15   Ideal-Gas Properties of Oxygen, O_2 (Concluded)

T [K]	h [kJ/kmol]	u [kJ/kmol]	s° [kJ/kmol·K]	T [K]	h [kJ/kmol]	u [kJ/kmol]	s° [kJ/kmol·K]
0	0	0	0	460	15,916	12,091	231.144
220	6601	4772	202.966	470	16,351	12,444	232.080
230	6938	5026	204.464	480	16,791	12,800	233.004
240	7280	5285	205.920	490	17,232	13,158	233.916
250	7627	5548	207.337	500	17,678	13,521	234.814
260	7979	5817	208.717	510	18,126	13,885	235.700
270	8335	6091	210.062	520	18,576	14,253	236.575
280	8697	6369	211.376	530	19,029	14,622	237.439
290	9063	6651	212.660	540	19,485	14,996	238.292
298	9364	6885	213.685	550	19,945	15,372	239.135
300	9431	6939	213.915	560	20,407	15,751	239.962
310	9807	7230	215.146	570	20,870	16,131	240.789
320	10,186	7526	216.351	580	21,337	16,515	241.602
330	10,570	7826	217.534	590	21,807	16,902	242.405
340	10,959	8131	218.694	600	22,280	17,291	243.199
350	11,351	8439	219.831	610	22,754	17,683	243.983
360	11,748	8752	220.948	620	23,231	18,076	244.758
370	12,148	9068	222.044	630	23,709	18,471	245.524
380	12,552	9392	223.122	640	24,190	18,869	246.282
390	12,960	9718	224.182	650	24,674	19,270	247.032
400	13,372	10,046	225.225	660	25,160	19,672	247.773
410	13,787	10,378	226.250	670	25,648	20,078	248.507
420	14,206	10,714	227.258	680	26,138	20,484	249.233
430	14,628	11,053	228.252	690	26,631	20,894	249.952
440	15,054	11,393	229.230	700	27,125	21,305	250.663
450	15,483	11,742	230.194	710	27,622	21,719	251.368

TABLE B.16   Ideal-Gas Properties of Carbon Dioxide, CO_2

T [K]	h [kJ/kmol]	u [kJ/kmol]	s° [kJ/kmol·K]	T [K]	h [kJ/kmol]	u [kJ/kmol]	s° [kJ/kmol·K]
720	28,121	22,134	252.065	980	41,685	33,537	268.119
730	28,622	22,522	252.755	990	42,226	33,995	268.670
740	29,124	22,972	253.439	1000	42,769	34,455	269.215
750	29,629	23,393	254.117	1020	43,859	35,378	270.293
760	30,135	23,817	254.787	1040	44,953	36,306	271.354
770	30,644	24,242	255.452	1060	46,051	37,238	272.400
780	31,154	24,669	256.110	1080	47,153	38,174	273.430
790	31,665	25,097	256.762	1100	48,258	39,112	274.445
800	32,179	25,527	257.408	1120	49,369	40,057	275.444
810	32,694	25,959	258.048	1140	50,484	41,006	276.430
820	33,212	26,394	258.682	1160	51,602	41,957	277.403
830	33,730	26,829	259.311	1180	52,724	42,913	278.361
840	34,251	27,267	259.934	1200	53,848	43,871	297.307
850	34,773	27,706	260.551	1220	54,977	44,834	280.238
860	35,296	28,125	261.164	1240	56,108	45,799	281.158
870	35,821	28,588	261.770	1260	57,244	46,768	282.066
880	36,347	29,031	262.371	1280	58,381	47,739	282.962
890	36,876	29,476	262.968	1300	59,522	48,713	283.847
900	37,405	29,922	263.559	1320	60,666	49,691	284.722
910	37,935	30,369	264.146	1340	61,813	50,672	285.586
920	38,467	30,818	264.728	1360	62,963	51,656	286.439
930	39,000	31,268	265.304	1380	64,116	52,643	287.283
940	39,535	31,719	265.877	1400	65,271	53,631	288.106
950	40,070	32,171	266.444	1420	66,427	54,621	288.934
960	40,607	32,625	267.007	1440	67,586	55,614	289.743
970	41,145	33,081	267.566	1460	68,748	56,609	290.542

TABLE B.16 Ideal-Gas Properties of Carbon Dioxide, CO_2 (Continued)

T [K]	h [kJ/kmol]	u [kJ/kmol]	s° [kJ/kmol·K]	T [K]	h [kJ/kmol]	u [kJ/kmol]	s° [kJ/kmol·K]
1480	66,911	57,606	291.333	2000	100,804	84,185	309.210
1500	71,078	58,606	292.114	2050	103,835	86,791	310.701
1520	72,246	59,609	292.888	2100	106,864	89,404	312.160
1540	73,417	60,613	292.654	2150	109,898	92,023	313.589
1560	74,590	61,620	294.411	2200	112,939	94,648	314.988
1580	76,767	62,630	295.161	2250	115,984	97,277	316.356
1600	76,944	63,741	295.901	2300	119,035	99,912	317.695
1620	78,123	64,653	296.632	2350	122,091	102,552	319.011
1640	79,303	65,668	297.356	2400	125,152	105,197	320.302
1660	80,486	66,592	298.072	2450	128,219	107,849	321.566
1680	81,670	67,702	298.781	2500	131,290	110,504	322.808
1700	82,856	68,721	299.482	2550	134,368	113,166	324.026
1720	84,043	69,742	300.177	2600	137,449	115,832	325.222
1740	85,231	70,764	300.863	2650	140,533	118,500	326.396
1760	86,420	71,787	301.543	2700	143,620	121,172	327.549
1780	87,612	72,812	302.217	2750	146,713	123,849	328.684
1800	88,806	73,840	302.884	2800	149,808	126,528	329.800
1820	90,000	74,868	303.544	2850	152,908	129,212	330.896
1840	91,196	75,897	304.198	2900	156,009	131,898	331.975
1860	92,394	76,929	304.845	2950	159,117	134,589	333.037
1880	93,593	77,962	305.487	3000	162,226	137,283	334.084
1900	94,793	78,996	306.122	3050	165,341	139,982	335.114
1920	95,995	80,031	306.751	3100	168,456	142,681	336.126
1940	97,197	81,067	307.374	3150	171,576	145,385	337.124
1960	98,401	82,105	307.992	3200	174,695	148,089	338.109
1980	99,606	83,144	308.604	3250	177,822	150,801	339.069

TABLE B.16 Ideal-Gas Properties of Carbon Dioxide, CO_2 (Concluded)

T [K]	h [kJ/kmol]	u [kJ/kmol]	s° [kJ/kmol·K]	T [K]	h [kJ/kmol]	u [kJ/kmol]	s° [kJ/kmol·K]
0	0	0	0	460	13,412	9587	210.243
220	6391	4562	188.683	470	13,708	9800	210.880
230	6683	4771	189.980	480	14,005	10,014	211.504
240	6975	4979	191.221	490	14,302	10,228	212.117
250	7266	5188	192.411	500	14,600	10,443	212.719
260	7558	5396	193.554	510	14,898	10,658	213.310
270	7849	5604	194.654	520	15,197	10,874	213.890
280	8140	5812	195.713	530	15,497	11,090	214.460
290	8432	6020	196.735	540	15,797	11,307	215.020
298	8669	6190	197.543	550	16,097	11,524	215.572
300	8723	6229	197.723	560	16,399	11,743	216.115
310	9014	6437	198.678	570	16,701	11,961	216.649
320	9306	6645	199.603	580	17,003	12,181	217.175
330	9597	6854	200.500	590	17,307	12,401	217.693
340	9889	7062	201.371	600	17,611	12,622	218.204
350	10,181	7271	202.217	610	17,915	12,843	218.708
360	10,473	7480	203.040	620	18,221	13,066	219.205
370	10,765	7689	203.842	630	18,527	13,289	219.695
380	11,058	7899	204.622	640	18,833	13,512	220.179
390	11,351	8108	205.383	650	19,141	13,736	220.656
400	11,644	8319	206.125	660	19,449	13,962	221.127
410	11,938	8529	206.850	670	19,758	14,187	221.592
420	12,232	8740	207.549	680	20,068	14,414	222.052
430	12,526	8951	208.252	690	20,378	14,641	222.505
440	12,821	9163	208.929	700	20,690	14,870	222.953
450	13,116	9375	209.593	710	21,002	15,099	223.396

TABLE B.17 Ideal-Gas Properties of Carbon Monoxide, CO

T [K]	h [kJ/kmol]	u [kJ/kmol]	s° [kJ/kmol·K]	T [K]	h [kJ/kmol]	u [kJ/kmol]	s° [kJ/kmol·K]
720	21,315	15,328	223.833	980	29,693	21,545	233.752
730	21,628	15,558	224.265	990	30,024	21,793	234.088
740	21,943	15,789	224.692	1000	30,355	22,041	234.421
750	22,258	16,022	225.115	1020	31,020	22,540	235.079
760	22,573	16,255	225.533	1040	31,688	23,041	235.728
770	22,890	16,488	225.947	1060	32,357	23,544	236.364
780	23,208	16,723	226.357	1080	33,029	24,049	236.992
790	23,526	16,957	226.762	1100	33,702	24,557	237.609
800	23,844	17,193	227.162	1120	34,377	25,065	238.217
810	24,164	17,429	227.559	1140	35,054	25,575	238.817
820	24,483	17,665	227.952	1160	35,733	26,088	239.407
830	24,803	17,902	228.339	1180	36,406	26,602	239.989
840	25,124	18,140	228.724	1200	37,095	27,118	240.663
850	25,446	18,379	229.106	1220	37,780	27,637	241.128
860	25,768	18,617	229.482	1240	38,466	28,426	241.686
870	26,091	18,858	229.856	1260	39,154	28,678	242.236
880	26,415	19,099	230.227	1280	39,844	29,201	242.780
890	26,740	19,341	230.593	1300	40,534	29,725	243.316
900	27,066	19,583	230.957	1320	41,226	30,251	243.844
910	27,392	19,826	231.317	1340	41,919	30,778	244.366
920	27,719	20,070	231.674	1360	42,613	31,306	244.880
930	28,046	20,314	232.028	1380	43,309	31,836	245.388
940	28,375	20,559	232.379	1400	44,007	32,367	245.889
950	28,703	20,805	232.727	1420	44,707	32,900	246.385
960	29,033	21,051	233.072	1440	45,408	33,434	246.876
970	29,362	21,298	233.413	1460	46,110	33,971	247.360

TABLE B.17 Ideal-Gas Properties of Carbon Monoxide, CO (Continued)

T [K]	h [kJ/kmol]	u [kJ/kmol]	s° [kJ/kmol·K]
1480	46,813	34,508	247.839
1500	47,517	35,046	248.312
1520	48,222	35,584	248.778
1540	48,928	36,124	249.240
1560	49,635	36,665	249.695
1580	50,344	37,207	250.147
1600	51,053	37,750	250.592
1620	51,763	38,293	251.033
1640	52,472	38,837	251.470
1660	53,184	39,382	251.901
1680	53,895	39,927	252.329
1700	54,609	40,474	252.751
1720	55,323	41,023	253.169
1740	56,039	41,572	253.582
1760	56,756	42,123	253.991
1780	57,473	42,673	254.398
1800	58,191	43,225	254.797
1820	58,910	43,778	255.194
1840	59,629	44,331	255.587
1860	60,351	44,886	255.976
1880	61,072	45,441	256.361
1900	61,794	45,997	256.743
1920	62,516	46,552	257.122
1940	63,238	47,108	257.497
1960	63,961	47,665	257.868
1980	64,684	48,221	258.236

T [K]	h [kJ/kmol]	u [kJ/kmol]	s° [kJ/kmol·K]
2000	65,408	48,780	258.600
2050	67,224	50,179	259.494
2100	69,044	51,584	260.370
2150	70,864	52,988	261.226
2200	72,688	54,396	262.065
2250	74,516	55,809	262.887
2300	76,345	57,222	263.692
2350	78,178	58,640	264.480
2400	80,015	60,060	265.253
2450	81,852	61,482	266.012
2500	83,692	62,906	266.755
2550	85,537	64,335	267.485
3200	87,383	65,766	268.202
3250	89,230	67,197	268.905
2700	91,077	68,628	269.596
2750	92,930	70,066	270.285
2800	94,784	71,504	270.943
2850	96,639	72,945	271.602
2900	98,495	74,383	272.249
2950	100,352	75,825	272.884
3000	102,210	77,267	273.508
3050	104,073	78,715	274.123
3100	105,939	80,164	274.730
3150	107,802	81,612	275.326
3200	109,667	83,061	275.914
3250	111,534	84,513	276.494

TABLE B.17 Ideal-Gas Properties of Carbon Monoxide, CO (Concluded)

T [K]	h [kJ/kmol]	u [kJ/kmol]	s° [kJ/kmol·K]	T [K]	h [kJ/kmol]	u [kJ/kmol]	s° [kJ/kmol·K]
0	0	0	0	840	24,359	17,375	160.891
260	7370	5209	126.636	880	25,551	18,235	162.277
270	7657	5412	127.719	920	26,747	19,098	163.607
280	7945	5617	128.765	960	27,948	19,966	164.884
290	8233	5822	129.775	1000	29,154	20,839	166.114
298	8468	5989	130.574	1040	30,364	21,717	167.300
300	8522	6027	130.754	1080	31,580	22,601	168.449
320	9100	6440	132.621	1120	32,802	23,490	169.560
340	9680	6853	134.378	1160	34,028	24,384	170.636
360	10,262	7268	136.039	1200	35,262	25,284	171.682
380	10,843	7684	137.612	1240	36,502	26,192	172.698
400	11,426	8100	139.106	1280	37,749	27,106	173.687
420	12,010	8518	140.529	1320	39,002	28,027	174.652
440	12,594	8936	141.888	1360	40,263	28,955	175.593
460	13,179	9355	143.187	1400	41,530	29,889	176.510
480	13,764	9773	144.432	1440	42,808	30,835	177.410
500	14,350	10,193	145.628	1480	44,091	31,786	178.291
520	14,935	10,611	146.775	1520	45,384	32,746	179.153
560	16,107	11,451	148.945	1560	46,683	33,713	179.995
600	17,280	12,291	150.968	1600	47,990	34,687	180.820
640	18,453	13,133	152.863	1640	49,303	35,668	181.632
680	19,630	13,976	154.645	1680	50,622	36,654	182.428
720	20,807	14,821	156.328	1720	51,947	37,646	183.208
760	21,988	15,669	157.923	1760	53,279	38,645	183.973
800	23,171	16,520	159.440	1800	54,618	39,652	184.724

TABLE B.18 Ideal-Gas Properties of Hydrogen, H_2

T [K]	h [kJ/kmol]	u [kJ/kmol]	s° [kJ/kmol·K]	T [K]	h [kJ/kmol]	u [kJ/kmol]	s° [kJ/kmol·K]
1840	55,962	40,663	185.463	2550	80,755	59,554	196.837
1880	57,311	41,680	186.190	2600	82,558	60,941	197.539
1920	58,668	42,705	186.904	2650	84,368	62,335	198.229
1960	60,031	43,735	187.607	2700	86,186	63,737	198.907
2000	61,400	44,771	188.297	2750	88,008	65,144	199.575
2050	63,119	46,074	189.148	2800	89,838	66,558	200.234
2100	64,847	47,386	189.979	2850	91,671	67,976	200.885
2150	66,584	48,708	190.796	2900	93,512	69,401	201.527
2200	68,328	50,037	191.598	2950	95,358	70,831	202.157
2250	70,080	51,373	192.385	3000	97,211	72,268	202.778
2300	71,839	52,716	193.159	3050	99,065	73,707	203.391
2350	73,608	54,069	193.921	3100	100,926	75,152	203.995
2400	75,383	55,429	194.669	3150	102,793	76,604	204.592
2450	77,168	56,798	195.403	3200	104,667	78,061	205.181
2500	78,960	58,175	196.125	3250	106,545	79,523	205.765

TABLE B.18 Ideal-Gas Properties of Hydrogen, H_2 (Concluded)

T [K]	h [kJ/kmol]	u [kJ/kmol]	s° [kJ/kmol·K]	T [K]	h [kJ/kmol]	u [kJ/kmol]	s° [kJ/kmol·K]
0	0	0	0	460	15,428	11,603	203.497
220	7295	5466	178.576	470	15,777	11,869	204.247
230	7628	5715	180.054	480	16,126	12,135	204.982
240	7961	5965	181.471	490	16,477	12,403	205.705
250	8294	6215	182.831	500	16,828	12,671	206.413
260	8627	6466	184.139	510	17,181	12,940	207.112
270	8961	6716	185.399	520	17,534	13,211	207.799
280	9296	6968	186.616	530	17,889	13,482	208.475
290	9631	7219	187.791	540	18,245	13,755	209.139
298	9904	7425	188.720	550	18,601	14,028	209.795
300	9966	7472	188.928	560	18,959	14,303	210.440
310	10,302	7725	190.030	570	19,318	14,579	211.075
320	10,639	7978	191.098	580	19,678	14,856	211.702
330	10,976	8232	192.136	590	20,039	15,134	212.320
340	11,314	8487	193.144	600	20,402	15,413	212.920
350	11,652	8742	194.125	610	20,765	15,693	213.529
360	11,992	8998	195.081	620	21,130	15,975	214.122
370	12,331	9255	196.012	630	21,495	16,257	214.707
380	12,672	9513	196.920	640	21,862	16,541	215.285
390	13,014	9771	197.807	650	22,230	16,826	215.856
400	13,356	10,030	198.673	660	22,600	17,112	216.419
410	13,699	10,290	199.521	670	22,970	17,399	216.976
420	14,043	10,551	200.350	680	23,342	17,688	217.527
430	14,388	10,813	201.160	690	23,714	17,978	218.071
440	14,734	11,075	201.955	700	24,088	18,268	218.610
450	15,080	11,339	202.734	710	24,464	18,561	219.142

TABLE B.19 Ideal-Gas Properties of Water Vapor, H_2O

T [K]	h [kJ/kmol]	u [kJ/kmol]	s° [kJ/kmol·K]	T [K]	h [kJ/kmol]	u [kJ/kmol]	s° [kJ/kmol·K]
720	24,840	18,854	219.668	980	35,061	26,913	231.767
730	25,218	19,148	220.189	990	35,472	27,240	232.184
740	25,597	19,444	220.707	1000	35,882	27,568	232.597
750	25,977	19,741	221.215	1020	36,709	28,228	233.415
760	26,358	20,039	221.720	1040	37,542	28,895	234.223
770	26,741	20,339	222.221	1060	38,380	29,567	235.020
780	27,125	20,639	222.717	1080	39,223	30,243	235.806
790	27,510	20,941	223.207	1100	40,071	30,925	236.584
800	27,896	21,245	223.693	1120	40,923	31,611	237.352
810	28,284	21,549	224.174	1140	41,780	32,301	238.110
820	28,672	21,855	224.651	1160	42,642	32,997	238.859
830	29,062	22,162	225.123	1180	43,509	33,698	239.600
840	29,454	22,470	225.592	1200	44,380	34,403	240.333
850	29,846	22,779	226.057	1220	45,256	35,112	241.057
860	30,240	23,090	226.517	1240	46,137	35,827	241.773
870	30,635	23,402	226.973	1260	47,022	36,546	242.482
880	31,032	23,715	227.426	1280	47,912	37,270	243.183
890	31,429	24,029	227.875	1300	48,807	38,000	243.877
900	31,828	24,345	228.321	1320	49,707	38,732	244.564
910	32,228	24,662	228.763	1340	50,612	39,470	245.243
920	32,629	24,980	229.202	1360	51,521	40,213	245.915
930	33,032	25,300	229.637	1380	52,434	40,960	246.582
940	33,436	25,621	230.070	1400	53,351	41,711	247.241
950	33,841	25,943	230.499	1420	54,273	42,466	247.895
960	34,247	26,265	230.924	1440	55,198	43,226	248.543
970	34,653	26,588	231.347	1460	56,128	43,989	249.185

TABLE B.19 Ideal-Gas Properties of Water Vapor, H_2O (Continued)

T [K]	h [kJ/kmol]	u [kJ/kmol]	s° [kJ/kmol·K]	T [K]	h [kJ/kmol]	u [kJ/kmol]	s° [kJ/kmol·K]
1480	57,062	44,756	249.820	2000	82,593	65,965	264.571
1500	57,999	45,528	250.450	2050	85,156	68,111	265.838
1520	58,942	46,304	251.074	2100	87,735	70,275	267.081
1540	59,888	47,084	251.693	2150	90,330	72,454	268.301
1560	60,838	47,868	252.305	2200	92,940	74,649	269.500
1580	61,792	48,655	252.912	2250	95,562	76,855	270.679
1600	62,748	49,445	253.513	2300	98,199	79,076	271.839
1620	63,709	50,240	254.111	2350	100,846	81,308	272.978
1640	64,675	51,039	254.703	2400	103,508	83,553	274.098
1660	65,643	51,841	255.290	2450	106,183	85,811	275.201
1680	66,614	52,646	255.873	2500	108,868	88,082	276.286
1700	67,589	53,455	256.450	2550	111,565	90,364	277.354
1720	68,567	54,267	257.022	2600	114,273	92,656	278.407
1740	69,550	55,083	257.589	2650	116,991	94,958	279.441
1760	70,535	55,902	258.151	2700	119,717	97,269	280.462
1780	71,523	56,723	258.708	2750	122,453	99,588	281.464
1800	72,513	57,547	259.262	2800	125,198	101,917	282.453
1820	73,507	58,375	259.811	2850	127,952	104,256	283.429
1840	74,506	59,207	260.357	2900	130,717	106,605	284.390
1860	75,506	60,042	260.898	2950	133,486	108,959	285.338
1880	76,511	60,880	261.436	3000	136,264	111,321	286.273
1900	77,517	61,720	261.969	3050	139,051	113,692	287.194
1920	78,527	62,564	262.497	3100	141,846	116,072	288.102
1940	79,540	63,411	263.022	3150	144,648	118,458	288.999
1960	80,555	64,259	263.542	3200	147,457	120,851	289.884
1980	81,573	65,111	264.059	3250	150,272	123,250	290.756

TABLE B.19 Ideal-Gas Properties of Water Vapor, H_2O (Concluded)

T [K]	h [kJ/kmol]	u [kJ/kmol]	s° [kJ/kmol·K]	T [K]	h [kJ/kmol]	u [kJ/kmol]	s° [kJ/kmol·K]
0	0	0	0	2400	50,894	30,940	204.932
298	6852	4373	160.944	2450	51,936	31,566	205.362
300	6892	4398	161.079	2500	52,979	32,193	205.783
500	11,197	7040	172.088	2550	54,021	32,820	206.196
1000	21,713	13,398	186.678	2600	55,064	33,447	206.601
1500	32,150	19,679	195.143	2650	56,108	34,075	206.999
1600	34,234	20,931	196.488	2700	57,152	34,703	207.389
1700	36,317	22,183	197.751	2750	58,196	35,332	207.772
1800	38,400	23,434	198.941	2800	59,241	35,961	208.148
1900	40,482	24,685	200.067	2850	60,286	36,590	208.518
2000	42,564	25,935	201.135	2900	61,332	37,220	208.882
2050	43,605	26,560	201.649	2950	62,378	37,851	209.240
2100	44,646	27,186	202.151	3000	63,425	38,482	209.592
2150	45,687	27,811	202.641	3100	65,520	39,746	210.279
2200	46,728	28,436	203.119	3200	67,619	41,013	210.945
2250	47,769	29,062	203.588	3300	69,720	42,283	211.592
2300	48,811	29,688	204.045	3400	71,824	43,556	212.220
2350	49,852	30,314	204.493	3500	73,932	44,832	212.831

TABLE B.20   Ideal-Gas Properties of Monatomic Oxygen, O

T [K]	h [kJ/kmol]	u [kJ/kmol]	s° [kJ/kmol·K]	T [K]	h [kJ/kmol]	u [kJ/kmol]	s° [kJ/kmol·K]
0	0	0	0	2400	77,015	57,061	248.628
298	9188	6709	183.594	2450	78,801	58,431	249.364
300	9244	6749	183.779	2500	80,592	59,806	250.088
500	15,181	11,024	198.955	2550	82,388	61,186	250.799
1000	30,123	21,809	219.624	2600	84,189	62,572	251.499
1500	46,046	33,575	232.506	2650	85,995	63,962	252.187
1600	49,358	36,055	234.642	2700	87,806	65,358	252.864
1700	52,706	38,571	236.672	2750	89,622	66,757	253.530
1800	56,089	41,123	238.606	2800	91,442	68,162	254.186
1900	59,505	43,708	240.453	2850	93,266	69,570	254.832
2000	62,952	46,323	242.221	2900	95,095	70,983	255.468
2050	64,687	47,642	243.077	2950	96,927	72,400	256.094
2100	66,428	48,968	243.917	3000	98,763	73,820	256.712
2150	68,177	50,301	244.740	3100	102,447	76,673	257.919
2200	69,932	51,641	245.547	3200	106,145	79,539	259.093
2250	71,694	52,987	246.338	3300	109,855	82,418	260.235
2300	73,462	54,339	247.116	3400	113,578	85,309	261.347
2350	75,236	55,697	247.879	3500	117,312	88,212	262.429

TABLE B.21 Ideal-Gas Properties of Hydroxyl, OH

Substance	Formula	$h_f°$ [kJ/kmol]	$g_f°$ [kJ/kmol]	$s°$ [kJ/kmol·K]
Carbon	C(s)	0	0	5.74
Hydrogen	$H_2(g)$	0	0	130.68
Nitrogen	$N_2(g)$	0	0	191.61
Oxygen	$O_2(g)$	0	0	205.04
Carbon monoxide	CO(g)	−110,530	−137,150	197.65
Carbon dioxide	$CO_2(g)$	−393,520	−394,360	213.80
Water vapor	$H_2O(g)$	−241,820	−228,590	188.83
Water	$H_2O(l)$	−285,830	−237,180	69.92
Hydrogen peroxide	$H_2O_2(g)$	−136,310	−105,600	232.63
Ammonia	$NH_3(g)$	−46,190	−16,590	192.33
Methane	$CH_4(g)$	−74,850	−50,790	186.16
Acetylene	$C_2H_2(g)$	+226,730	+209,170	200.85
Ethylene	$C_2H_4(g)$	+52,280	+68,120	219.83
Ethane	$C_2H_6(g)$	−84,680	−32,890	229.49
Propylene	$C_3H_6(g)$	+20,410	+62,720	266.94
Propane	$C_3H_8(g)$	−103,850	−23,490	269.91
n-Butane	$C_4H_{10}(g)$	−126,150	−15,710	310.12
n-Octane	$C_8H_{18}(g)$	−208,450	+16,530	466.73
n-Octane	$C_8H_{18}(l)$	−249,950	+6,610	360.79
n-Dodecane	$C_{12}H_{26}(g)$	−291,010	+50,150	622.83
Benzene	$C_6H_6(g)$	+82,930	+129,660	269.20
Methyl alcohol	$CH_3OH(g)$	−200,670	−162,000	239.70
Methyl alcohol	$CH_3OH(l)$	−238,660	−166,360	126.80
Ethyl alcohol	$C_2H_5OH(g)$	−235,310	−168,570	282.59
Ethyl alcohol	$C_2H_5OH(l)$	−277,690	−174,890	160.70
Oxygen	O(g)	+249,190	+231,770	161.06
Hydrogen	H(g)	+218,000	+203,290	114.72
Nitrogen	N(g)	+472,650	+455,510	153.30
Hydroxyl	OH(g)	+39,460	+34,280	183.70

Source: From JANAF, *Thermochemical Tables* (Midland, MI: Dow Chemical Co., 1971); Selected Values of Chemical Thermodynamic Properties, NBS Technical Note 270-3, 1968; and *API Research Project 44* (Carnegie Press, 1953).

TABLE B.22 Enthalpy of Formation, Gibbs Function of Formation, and Absolute Entropy at 25°C, 1 atm

Fuel (phase)	Formula	Molar mass [kg/kmol]	Density[1] [kg/L]	Enthalpy of vaporization[2] [kJ/kg]	Specific heat[1] c_p [kJ/kg]	Higher heating value[3] [kJ/kg]	Lower heating value[3] [kJ/kg]
Carbon (s)	C	12.011	2	—	0.708	32,800	32,800
Hydrogen (g)	H_2	2.016	—	—	14.4	141,800	120,000
Carbon monoxide (g)	CO	28.013	—	—	1.05	10,100	10,100
Methane (g)	CH_4	16.043	—	509	2.20	55,530	50,050
Methanol (l)	CH_4O	32.042	0.790	1168	2.53	22,660	19,920
Acetylene (g)	C_2H_2	26.038	—	—	1.69	49,970	48,280
Ethane (g)	C_2H_6	30.070	—	172	1.75	51,900	47,520
Ethanol (l)	C_2H_6O	46.069	0.790	919	2.44	29,670	26,810
Propane (l)	C_3H_8	44.097	0.500	335	2.77	50,330	46,340
Butane (l)	C_4H_{10}	58.123	0.579	362	2.42	49,150	45,370
1-Pentene (l)	C_5H_{10}	70.134	0.641	363	2.20	47,760	44,630
Isopentane (l)	C_5H_{12}	72.150	0.626	—	2.32	48,570	44,910
Benzene (l)	C_6H_6	78.114	0.877	433	1.72	41,800	40,100
Hexene (l)	C_6H_{12}	84.161	0.673	392	1.84	47,500	44,400
Hexane (l)	C_6H_{14}	86.177	0.660	366	2.27	48,310	44,740
Toluene (l)	C_7H_8	92.141	0.867	412	1.71	42,400	40,500
Heptane (l)	C_7H_{16}	100.204	0.684	365	2.24	48,100	44,600
Octane (l)	C_8H_{18}	114.231	0.703	363	2.23	47,890	44,430
Decane (l)	$C_{10}H_{22}$	142.285	0.730	361	2.21	47,640	44,240
Gasoline (l)	$C_nH_{1.87n}$	100–110	0.72–0.78	350	2.4	47,300	44,000
Light diesel (l)	$C_nH_{1.8n}$	170	0.78–0.84	270	2.2	46,100	43,200
Heavy diesel (l)	$C_nH_{1.7n}$	200	0.82–0.88	230	1.9	45,500	42,800
Natural gas (g)	$C_nH_{3.8n}N_{0.1n}$	18	—	—	2	50,000	45,000

[1] At 1 atm and 20°C.
[2] At 25°C for liquid fuels, and 1 atm and normal boiling temperature for gaseous fuels.
[3] At 25°C. Multiply by molar mass to obtain heating values in kJ/kmol.

TABLE B.23 Properties of Some Common Fuels and Hydrocarbons

Temp. [K]	$H_2 \rightleftharpoons 2H$	$O_2 \rightleftharpoons 2O$	$N_2 \rightleftharpoons 2N$	$H_2O \rightleftharpoons H_2 + \frac{1}{2}O_2$	$H_2O \rightleftharpoons \frac{1}{2}H_2 + OH$	$CO_2 \rightleftharpoons CO + \frac{1}{2}O_2$	$\frac{1}{2}N_2 + \frac{1}{2}O_2 \rightleftharpoons NO$
298	−164.005	−186.975	−367.480	−92.208	−106.208	−103.762	−35.052
500	−92.827	−105.630	−213.372	−52.691	−60.281	−57.616	−20.295
1000	−39.803	−45.150	−99.127	−23.163	−26.034	−23.529	−9.388
1200	−30.874	−35.005	−80.011	−18.182	−20.283	−17.871	−7.569
1400	−24.463	−27.742	−66.329	−14.609	−16.099	−13.842	−6.270
1600	−19.637	−22.285	−56.055	−11.921	−13.066	−10.830	−5.294
1800	−15.866	−18.030	−48.051	−9.826	−10.657	−8.497	−4.536
2000	−12.840	−14.622	−41.645	−8.145	−8.728	−6.635	−3.931
2200	−10.353	−11.827	−36.391	−6.768	−7.148	−5.120	−3.433
2400	−8.276	−9.497	−32.011	−5.619	−5.832	−3.860	−3.019
2600	−6.517	−7.521	−28.304	−4.648	−4.719	−2.801	−2.671
2800	−5.002	−5.826	−25.117	−3.812	−3.763	−1.894	−2.372
3000	−3.685	−4.357	−22.359	−3.086	−2.937	−1.111	−2.114
3200	−2.534	−3.072	−19.937	−2.451	−2.212	−0.429	−1.888
3400	−1.516	−1.935	−17.800	−1.891	−1.576	0.169	−1.690
3600	−0.609	−0.926	−15.898	−1.392	−1.088	0.701	−1.513
3800	0.202	−0.019	−14.199	−0.945	−0.501	1.176	−1.356
4000	0.934	0.796	−12.660	−0.542	−0.044	1.599	−1.216
4500	2.486	2.513	−9.414	0.312	0.920	2.490	−0.921
5000	3.725	3.895	−6.807	0.996	1.689	3.197	−0.686
5500	4.743	5.023	−4.666	1.560	2.318	3.771	−0.497
6000	5.590	5.963	−2.865	2.032	2.843	4.245	−0.341

The equilibrium constant K_p for the reaction $v_A A + n_B B \rightleftharpoons v_C C + v_D D$ is defined as $K_p = \dfrac{p_C^{v_C} p_D^{v_D}}{p_A^{v_A} p_B^{v_B}}$

Source: Gordon J. Van Wylen and Richard E. Sonntag, *Fundamentals of Classical Thermodynamics*, English/SI Version, 3rd ed. (New York: John Wiley & Sons, 1986), p. 723, table A.14. Based on thermodynamic data given in JANAF, *Thermochemical Tables* (Midland, MI: Thermal Research Laboratory, The Dow Chemical Company, 1971).

TABLE B.24   Natural Logarithms of the Equilibrium Constant K_p

Ma	Ma*	A/A*	P/P_0	ρ/ρ_0	T/T_0
0	0	∞	1.0000	1.0000	1.0000
0.1	0.1094	5.8218	0.9930	0.9950	0.9980
0.2	0.2182	2.9635	0.9725	0.9803	0.9921
0.3	0.3257	2.0351	0.9395	0.9564	0.9823
0.4	0.4313	1.5901	0.8956	0.9243	0.9690
0.5	0.5345	1.3398	0.8430	0.8852	0.9524
0.6	0.6348	1.1882	0.7840	0.8405	0.9328
0.7	0.7318	1.0944	0.7209	0.7916	0.9107
0.8	0.8251	1.0382	0.6560	0.7400	0.8865
0.9	0.9146	1.0089	0.5913	0.6870	0.8606
1.0	1.0000	1.0000	0.5283	0.6339	0.8333
1.2	1.1583	1.0304	0.4124	0.5311	0.7764
1.4	1.2999	1.1149	0.3142	0.4374	0.7184
1.6	1.4254	1.2502	0.2353	0.3557	0.6614
1.8	1.5360	1.4390	0.1740	0.2868	0.6068
2.0	1.6330	1.6875	0.1278	0.2300	0.5556
2.2	1.7179	2.0050	0.0935	0.1841	0.5081
2.4	1.7922	2.4031	0.0684	0.1472	0.4647
2.6	1.8571	2.8960	0.0501	0.1179	0.4252
2.8	1.9140	3.5001	0.0368	0.0946	0.3894
3.0	1.9640	4.2346	0.0272	0.0760	0.3571
5.0	2.2361	25.000	0.0019	0.0113	0.1667
∞	2.2495	∞	0	0	0

TABLE B.25 One-Dimensional Isentropic Compressible-Flow Functions for an Ideal Gas with $k = 1.4$

Ma_1	Ma_2	P_2/P_1	ρ_2/ρ_1	T_2/T_1	P_{02}/P_{01}	P_{02}/P_1
1.0	1.0000	1.0000	1.0000	1.0000	1.0000	1.8929
1.1	0.9118	1.2450	1.1691	1.0649	0.9989	2.1328
1.2	0.8422	1.5133	1.3416	1.1280	0.9928	2.4075
1.3	0.7860	1.8050	1.5157	1.1909	0.9794	2.7136
1.4	0.7397	2.1200	1.6897	1.2547	0.9582	3.0492
1.5	0.7011	2.4583	1.8621	1.3202	0.9298	3.4133
1.6	0.6684	2.8200	2.0317	1.3880	0.8952	3.8050
1.7	0.6405	3.2050	2.1977	1.4583	0.8557	4.2238
1.8	0.6165	3.6133	2.3592	1.5316	0.8127	4.6695
1.9	0.5956	4.0450	2.5157	1.6079	0.7674	5.1418
2.0	0.5774	4.5000	2.6667	1.6875	0.7209	5.6404
2.1	0.5613	4.9783	2.8119	1.7705	0.6742	6.1654
2.2	0.5471	5.4800	2.9512	1.8569	0.6281	6.7165
2.3	0.5344	6.0050	3.0845	1.9468	0.5833	7.2937
2.4	0.5231	6.5533	3.2119	2.0403	0.5401	7.8969
2.5	0.5130	7.1250	3.3333	2.1375	0.4990	8.5261
2.6	0.5039	7.7200	3.4490	2.2383	0.4601	9.1813
2.7	0.4956	8.3383	3.5590	2.3429	0.4236	9.8624
2.8	0.4882	8.9800	3.6636	2.4512	0.3895	10.5694
2.9	0.4814	9.6450	3.7629	2.5632	0.3577	11.3022
3.0	0.4752	10.3333	3.8571	2.6790	0.3283	12.0610
4.0	0.4350	18.5000	4.5714	4.0469	0.1388	21.0681
5.0	0.4152	29.000	5.0000	5.8000	0.0617	32.6335
∞	0.3780	∞	6.0000	∞	0	∞

TABLE B.26 One-Dimensional Normal-Shock Functions for an Ideal Gas with $k = 1.4$

Ma	T_0/T_0^*	P_0/P_0^*	T/T^*	P/P^*	V/V^*
0.0	0.0000	1.2679	0.0000	2.4000	0.0000
0.1	0.0468	1.2591	0.0560	2.3669	0.0237
0.2	0.1736	1.2346	0.2066	2.2727	0.0909
0.3	0.3469	1.1985	0.4089	2.1314	0.1918
0.4	0.5290	1.1566	0.6151	1.9608	0.3137
0.5	0.6914	1.1141	0.7901	1.7778	0.4444
0.6	0.8189	1.0753	0.9157	1.5957	0.5745
0.7	0.9085	1.0431	0.9929	1.4235	0.6975
0.8	0.9639	1.0193	1.0255	1.2658	0.8101
0.9	0.9921	1.0049	1.0245	1.1246	0.9110
1.0	1.0000	1.0000	1.0000	1.0000	1.0000
1.2	0.9787	1.0194	0.9118	0.7958	1.1459
1.4	0.9343	1.0777	0.8054	0.6410	1.2564
1.6	0.8842	1.1756	0.7017	0.5236	1.3403
1.8	0.8363	1.3159	0.6089	0.4335	1.4046
2.0	0.7934	1.5031	0.5289	0.3636	1.4545
2.2	0.7561	1.7434	0.4611	0.3086	1.4938
2.4	0.7242	2.0451	0.4038	0.2648	1.5252
2.6	0.6970	2.4177	0.3556	0.2294	1.5505
2.8	0.6738	2.8731	0.3149	0.2004	1.5711
3.0	0.6540	3.4245	0.2803	0.1765	1.5882

TABLE B.27 Rayleigh Flow Functions for an Ideal Gas with $k = 1.4$

Dimension	Metric	Metric/English
Acceleration	$1\ m/s^2 = 100\ cm/s^2$	$1\ m/s^2 = 3.2808\ ft/s^2$ $1\ ft/s^2 = 0.3048^*\ m/s^2$
Area	$1\ m^2 = 10^4\ cm^2 = 10^6\ mm^2 = 10^{-6}\ km^2$	$1\ m^2 = 1550\ in^2 = 10.764\ ft^2$ $1\ ft^2 = 144\ in^2 = 0.09290304^*\ m^2$
Density	$1\ g/cm^3 = 1\ kg/L = 1000\ kg/m^3$	$1\ g/cm^3 = 62.428\ lbm/ft^3 = 0.036127\ lbm/in^3$ $1\ lbm/in^3 = 1728\ lbm/ft^3$ $1\ kg/m^3 = 0.062428\ lbm/ft^3$
Energy, heat, work, internal energy, enthalpy	$1\ kJ = 1000\ J = 1000\ N \cdot m = 1\ kPa \cdot m^3$ $1\ kJ/kg = 1000\ m^2/s^2$ $1\ kWh = 3000\ kJ$ $1\ cal^{\dagger} = 4.184\ J$ $1\ IT\ cal^{\dagger} = 4.1868\ J$	$1\ kJ = 0.94782\ Btu$ $1\ Btu = 1.055056\ kJ$ $= 5.40395\ psia \cdot ft^3 = 778.169\ lbf \cdot ft$ $1\ Btu/lbm = 25,037\ ft^2/s^2 = 2.326^*\ kJ/kg$ $1\ kJ/kg = 0.430\ Btu/lbm$ $1\ kWh = 3412.14\ Btu$ $1\ therm = 10^5\ Btu = 1.055 \times 10^5\ kJ\ (natural\ gas)$
Force	$1\ N = 1\ kg \cdot m/s^2 = 10^5\ dyne$ $1\ kgf = 9.80665\ N$	$1\ N = 0.22481\ lbf$ $1\ lbf = 32.174\ lbm \cdot ft/s^2 = 4.44822\ N$
Heat flux	$1\ W/cm^2 = 10^4\ W/m^2$	$1\ W/m^2 = 0.3171\ Btu/h \cdot ft^2$
Heat transfer coefficient	$1\ W/m^2 \cdot {}^{\circ}C = 1\ W/m^2 \cdot K$	$1\ W/m^2 \cdot {}^{\circ}C = 0.17612\ Btu/h \cdot ft^2 \cdot {}^{\circ}F$
Length	$1\ m = 100\ cm = 1000\ mm = 10^6\ \mu m$ $1\ km = 1000\ m$	$1\ m = 39.370\ in = 3.2808\ ft = 1.0926\ yd$ $1\ ft = 12\ in = 0.3048^*\ m$ $1\ mile = 5280\ ft = 1.6093\ km$ $1\ in = 2.54^*\ cm$
Mass	$1\ kg = 1000\ g$ $1\ metric\ ton = 1000\ kg$	$1\ kg = 2.2046226\ lbm$ $1\ lbm = 0.45359237^*\ kg$ $1\ ounce = 28.3495\ g$ $1\ slug = 32.174\ lbm = 14.5939\ kg$ $1\ short\ ton = 2000\ lbm = 907.1847\ kg$

Conversion Factors

Dimension	Metric	Metric/English
Power, heat transfer rate	1 W = 1 J/s 1 kW = 1000 W = 1341 hp 1 hp‡ = 745.7 W	1 kW = 3412.14 Btu/h = 737.56 lbf·ft/s 1 hp = 550 lbf·ft/s = 0.7068 Btu/s = 42.41 Btu/min = 2544.5 Btu/h = 0.74570 kW 1 boiler hp = 33,475 Btu 1 Btu/h = 1.055056 kJ/h 1 ton of refrigeration = 200 Btu/min
Pressure	1 Pa = 1 N/m² 1 kPa = 10³ Pa = 10⁻² MPa 1 atm = 101.325 kPa = 1.01325 bars = 760 mm Hg at 0°C = 1.03323 kgf/cm² 1 mm Hg = 0.1333 kPa	1 Pa = 1.4504 × 10⁻⁴ psia = 0.020886 lbf/ft² 1 psi = 144 lbf/ft² = 6.894757 kPa 1 atm = 14.696 psia = 29.92 in Hg at 30°F 1 in Hg = 3.387 kPa
Specific heat	1 kJ/kg·°C = 1 kJ/kg·K = 1 J/g·°C	1 Btu/lbm·°F = 4.1868 kJ/kg·°C 1 Btu/lbmol·R = 4.1868 kJ/kmol·K 1 kJ/kg·°C = 0.23885 Btu/lbm·°F = 0.23885 Btu/lbm·R
Specific volume	1 m³/kg = 1000 L/kg = 1000 cm³/g	1 m³/kg = 16.02 ft³/lbm 1 ft³/lbm = 0.062428 m³/kg
Temperature	T (K) = T (°C) + 273.15 ΔT (K) = ΔT (°C)	T (R) = T (°F) + 459.67 = $1.8T$ (K) T (°F) = $1.8T$ (°C) + 32 ΔT (°F) = ΔT (R) = $1.8\Delta T$ (K)
Thermal conductivity	1 W/m·°C = 1 W/m·K	1 W/m·°C = 0.57782 Btu/h·ft·°F
Velocity	1 m/s = 3.60 km/h	1 m/s = 3.2808 ft/s = 2.237 mi/h 1 mi/h = 1.46667 ft/s 1 mi/h = 1.6093 km/h

Conversion Factors (*Continued*)

Dimension	Metric	Metric/English
Volume	1 m³ = 1000 L = 10^6 cm³ (cc)	1 m³ = 6.1024×10^4 in³ = 35.315 ft³ = 264.17 gal (U.S.) 1 U.S. gallon = 231 in³ = 3.7854 L 1 fl ounce = 29.5735 cm³ = 0.0295735 L 1 U.S. gallon = 128 fl ounces
Volume flow rate	1 m³/s = 60,000 L/min = 10^6 cm³/s	1 m³/s = 15,850 gal/min (gpm) = 35.315 ft³/s = 2118.9 ft³/min (cfm)

*Exact conversion factor between metric and English units.

†Calorie is originally defined as the amount of heat needed to raise the temperature or 1 g of water by 1°C, but it varies with temperature. The international steam table (IT) calorie (generally preferred by engineers) is exactly 4.1868 J by definition and corresponds to the specific heat of water at 15°C. The thermochemical calorie (generally preferred by physicists) is exactly 4.184 J by definition and corresponds to the specific heat of water at room temperature. The difference between the two is about 0.06 percent, which is negligible. The capitalized Calorie used by nutritionists is actually a kilocalorie (1000 IT calories).

‡Mechanical horsepower. The electrical horsepower is taken to be exactly 746 W.

Conversion Factors (*Concluded*)

Universal gas constant	$R_u = 8.31447$ kJ/kmol·K
	$= 8.31447$ kPa·m^3/kmol·K
	$= 0.0831447$ bar·m^3/kmol·K
	$= 82.05$ L·atm/kmol·K
	$= 1.9858$ Btu/lbmol·R
	$= 1545.37$ ft·lbf/lbmol·R
	$= 10.73$ psia·ft^3/lbmol·R
Standard acceleration of gravity	$g = 9.80665$ m/s^2
	$= 32.174$ ft/s^2
Standard atmospheric pressure	1 atm $= 101.325$ kPa
	$= 1.01325$ bars
	$= 14.696$ psia
	$= 760$ mm Hg (0°C)
	$= 29.9213$ in Hg (32°F)
	$= 10.3323$ m H$_2$O (4°C)
Stefan–Boltzmann constant	$\sigma = 5.6704 \times 10^{-8}$ W/m^2·K^4
	$= 0.1714 \times 10^{-8}$ Btu/h·ft^2·R^4
Boltzmann constant	$k = 1.380650 \times 10^{-23}$ J/K
Speed of light in vacuum	$c_o = 2.9979 \times 10^8$ m/s $= 9.836 \times 108$ ft/s
Speed of sound in dry air at 0°C and 1 atm	$c = 331.36$ m/s
	$= 1089$ ft/s
Heat of fusion of water at 1 atm	$h_{if} = 333.7$ kJ/kg
	$= 143.5$ Btu/lbm
Enthalpy of vaporization of water at 1 atm	$h_{fg} = 2256.5$ kJ/kg
	$= 970.12$ Btu/lbm

Some Physical Constants

APPENDIX C

Guidelines for Report Writing

C.1   Course Projects

Upfront a reminder to the student: A one-page proposal is required for course-project (C-P) topics not listed below. Recycling previous reports or copying reports is unethical. Recall that a hard copy of the course-project report (C-PR) is due on final exam day (at the latest), and that the associated pdf file is needed if the C-PR contains video links. Note that the C-P presentation should be professional and tutorial so that students learn something new. Finally, keep in mind that *a C-PR is not just an elaborate homework-problem solution.*

C.2   Guidelines for Writing a C-PR

As mentioned, a course project is much more in quality and quantity than a homework assignment. As a starter, it typically requires a brief literature review to document the state-of-the-art for the given task. The antagonistic interplay between a rather accurate mathematical description of an assigned problem and finding a tractable solution method puts a premium on making the right assumptions and selecting a suitable equation solver, often in terms of available software and hardware. The resulting work has to be documented in an appropriate format, i.e., a *project report* that should have the following features.

- *Cover Page:* List project title, author, date, instructor, course number, and department/university.
- *TOC:* Table of contents with page numbers (if needed for lengthy reports).
- *Nomenclature:* Optional but recommended for complex problem solutions.
- *Abstract:* Summary of the problem analyzed, solution method employed, and novel results with physical insight presented.
- *Introduction:* Project objectives and task justification with state-of-the-art system description; basic approach or concept; references of the *literature review* in terms of "author (year);" discussion of the succinct differences between published system/problem solutions and anticipated results.
- *Theory:* Comprehensive system sketch with coordinate system, etc.; basic assumptions, governing equations, inlet/boundary conditions, property values

in form of a data table, ranges of dimensionless groups, and closure models; also, list references as well as information sources.

- *Solution Method and Model Validations:* **It is always recommended to devise an interesting thermodynamics system for which an analytic/approximate solution can be obtained. Still, linear PDEs and most ODEs may be solved with the popular MatLab software. In many cases, numerical equation solvers (e.g., OpenFOAM or commercial software) are needed to obtain realistic results.** For *model validation* the two basic approaches are: (a) quantitative proof, e.g., comparison with an exact solution of a simpler system and/or experimental data comparisons, or (at least) qualitative justification that the solution is correct; (b) numerical accuracy checks of computer simulation models, i.e., mesh-independence of the results as well as mass and momentum residuals less than, say, 10^{-4}.

- *Results and Discussion:* Graphs with interpretations, i.e., explain physical insight (note: number the figures and discuss them concisely); draw conclusions and provide applications.

- *Conclusions and Future Work:* Conclusions are an extended form of the Abstract, listing what novel contributions have been made and what their significance is, discussing limitations of the study as well as future work is important as well.

- *References:* List all literature cited in alphabetical order; for single and multiple authors use Weiss (1898), Pott & Deckel (2013), Unsinn et al. (2020).

- *Appendices:* Lengthy derivations, computer programs, etc.

Expanding on the stated *guidelines for writing a course-project report,* there are several basic tasks to be accomplished:

(a) As mentioned, a thorough literature review is necessary, taking up the major part of the *Introduction.* Clearly, all topics are well documented in the open literature, and videos illuminating the unique flow phenomena plus nifty applications are readily available.

(b) A challenge is to set up a specific system/problem with which the fundamentals are explained, and thereby natural as well as industrial applications can be illustrated. The necessary *math model* builds on and expands the material learned so far. Some problem solutions can be gained via analytical means, while others require computational tools. In any case, as all reports, the course project should exhibit a fine balance between math and physics of thermodynamics topics.

(c) The *Theory* section connects the chosen thermodynamic system/device with its solution method by describing the process mathematically and thereby indicating a suitable solution method. The three preliminaries, i.e., system *sketch,* list of *assumptions,* and best *method* to be employed, should be the starting point. If the system-describing equations appear to be too difficult, the project objectives have to be curtailed and hence a revised, well-posed math model developed. Next, necessary input data sets have to be collected and the expected value-ranges of key operational groups have to be determined.

(d) *A prime course-project choice* should be an interesting thermodynamic device/system for which an *analytic/approximate solution* is available. The popular MatLab software can be part of this approach when linear PDEs or ODEs have to be solved. In many cases, however, *numerical equation solvers* (e.g., the free OpenFOAM library or commercial software) are needed to obtain realistic results.

(e) *Results,* in terms of figures, graphs, tables, and/or correlations, should be well displayed, ideally in dimensionless form. With parametric sensitivity analyses (PSA), i.e., varying the device input data and/or cycle's operational conditions, the full range of system behavior can be explored. Figures with their captions and legends should be largely self-explanatory. The related *discussions* explain the reasons for unusual trends and summarize new physical insight, expanding on scientific implications and/or on practical applications.

(f) *Conclusions* are an extended form of the Abstract, listing what novel contributions have been made and what their significance is, while adding limitations of the study as well as future work.

C.3   C-PR Evaluation Criteria

While the instructor's goal is that the students doing C-PRs plus PowerPoint slide presentations significantly improve their knowledge base and skill level in thermodynamics, students are typically only concerned with better final course grades. So, catering to the students' objective, the C-PR grades (or extra course points) are based on the following criteria.

- *Impressive format:* Follow guidelines, provide typed text/equations, draw figures/sketches, import images, list active video-links, etc.

- Project uniqueness and depth

- Balance between theory (physics and math and numerical examples), ideas for device/process improvements, as well as applications with realistic cost/benefit analysis and insightful discussions

- Relative competitiveness

C.4   C-PR Presentation

Presenting report results is a standard professional requirement for an employee or a graduate student. Therefore, a presentation with PowerPoint slides in front of a supportive audience is a first great learning step. The performance criteria are as follows:

- Effectiveness and appearance of PowerPoint slides/videos

- Comprehensiveness and consistency, i.e., being in line with the course material and your C-P report

- Value of the results and degree of physical insight gained

- Level of oral presentation in terms of clarity, logic, entertainment factor

APPENDIX **D**

Samples of Tests and Final Exams

The two tests and final exams illustrate the novel approach of how to probe the depth of the students' *knowledge base (Part A)* and *skill level (Part B)*. Clearly, the same type of questions for Part A and problems for Part B should be required for all homework sets. The solutions of the Part A insight questions should feature basic system drawings, some math, i.e., derivations and/or process functions, as well as explanations and possible applications. Part B problem solutions should feature the three prelims (Sketch, Assumptions, and Method) before launching into the actual solution in well-organized form.

Test I

(90 minutes; only Property Tables and Equation Sheets are allowed)

Part A: Insight (10 points each)

1 List different forms of energy (with mode of transfer) and show their applications in two different devices/systems, including energy and mass balances.

2 A P-C device with stops contains 0.6 kg of steam at 1.0 MPa and 400°C. After cooling to 250°C (still at 1.0 MPa), find the intermediate work done. What is T_{final} when $P_{final} = 500$ kPa as the piston touches the stops, which are located at 40% of the initial volume?

3 Considering the saturation curve for H_2O (see Equation Sheets), discuss how it is generated, its meaning, and practical applications.

Part B: Problems

4 (40 points) A cylinder containing R-134a at 800 kPa and 80°C features a frictionless piston that just touches a pair of stops at the top. The piston mass is such that a 500 kPa pressure is required to move it. Now, a valve at the bottom is opened until half of the refrigerant is withdrawn and the temperature has dropped to 20°C. Determine the heat transferred and plot the states/process in a phase-diagram.

5 (30 points) A steam turbine ($\dot{m} = 22$ kg/s and P = 12,350 kW) operates with 1.6 MPa and 350°C at its inlet and saturated vapor at 30°C at its exit. What is the heat loss rate? Depict the states/process in a phase-diagram and comment!

Solutions to Test I

Part A: Insight

1a Forms of Energy and Transfer Mode

- Heat, Q, via conduction (or convection)
- Work, W, via mechanical or electric actions
- Enthalpy $H = U + p\forall$, *via* fluid flow (i.e., mass transfer)

1b Devices with M-balances and E-balances

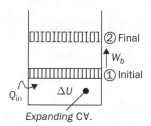

- p-c device
  $$m_{system} = \cent;\quad Q_{in} - W_b = \Delta U|_{C\forall}$$
  <Closed system>

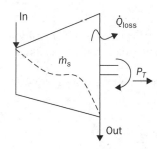

- Turbine
  $$\dot{m}_{steam} = \cent;\quad \dot{m}_s(h_{in} - h_{out}) = P_T + \dot{Q}_b$$

  <Steady open system>

2 Steam in p-c device 2 with stops.

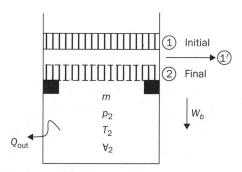

Find property data:

- For this isochoric process:

  $W_b = mp(v_{1'} - v_1)$ where $v_{1'}$ is intermediate

- Property Tables (see App. B)

p_1 & $T_1 \rightarrow v_1 = 0.30661$ m³; $p_{1'} = p_1$ & $T_{1'} \rightarrow v_{1'} = 0.23275$ m³

Hence,

$W_b = 0.6$ kg $*$ 1000 kPa $*$ (0.23275 − 0.30661) m³/kg = 44.3 kJ

- Final state (2)

$p_2 = 0.5$ MPa and $v_2 = 0.4 * v_1 = 0.1226$ m³/kg

∴ from Property Table B.5: $T_2 = 263.94°C$

3   Liquid–vapor saturation curve for H₂O

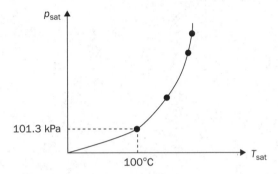

Given a liquid pressure, phase-change occurs at T_{sat} and vice versa. For example, $T_{sat} = 100°C$ at $p_{sat} = 101.3$ kPa for water. So, running a number of experiments, the $p_{sat}(T_{sat})$ – curve can be plotted. Clearly, water at low ambient pressure (Mount Everest) boils at lower temperature. In contrast, higher boiling temperatures (see pressure cooker) mean shorter cooking times.

Part B: Problems

4   R-134a in a p-c device with stops and a discharge pipe

Sketch

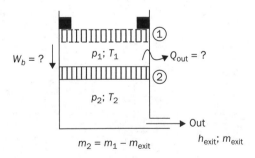

Assumptions

- Transient, open system
- R-134a $\cong$ 2-phase fluid
- $h_{exit} \approx \frac{1}{2}(h_1 + h_2)$, as more accurately $h = h(\text{time})$

- ΔpE & ΔkE are negligible
- Quasi-equilibrium process

Method

- M-B: $-m_{exit} = m_2 - m_1$
- E-B: $W_b - Q_{out} = -(mh)_{exit} = \Delta U |_{CV}$

  $\forall_i = m_i v_i; i = 1, 2$
- Property Tables (see App. B)

Solution   As the expansion process occurs in the superheated vapor region, the p-v (or T-v) diagram can be readily drawn.

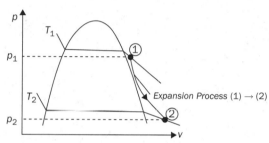

Note $T_1 > T_{sat@p_1}$ and $T_2 > T_{sat@p_2}$

- M-B: $m_{exit} = m_1 - m_2 = 1$ kg
- E-B: $W_b - Q_{out} = -(mh)_{exit} = (m_2 u_2 - m_1 u_1) |_{CV}$

  where,

  $$h_{exit} = \frac{1}{2}(h_1 + h_2) \text{ and } W_b = p_2(\forall_1 - \forall_2)$$

  with,

  $\forall_1 = m_1 v_1$ and $\forall_2 = m_2 v_2$
- Property values:

$$p_1 \text{ \& } T_1 \rightarrow \begin{cases} v_1 = 0.032659 \text{ m}^3/\text{kg} \\ u_1 = 290.86 \text{ kJ/kg} \\ h_1 = 316.99 \text{ kJ/kg} \end{cases}$$

$$p_2 \text{ \& } T_2 \rightarrow \begin{cases} v_2 = 0.042115 \text{ m}^3/\text{kg} \\ u_1 = 242.42 \text{ kJ/kg} \\ h_1 = 263.48 \text{ kJ/kg} \end{cases}$$

Thus, $h_{exit} = 290.23$ kJ/k and $W_b = 11.6$ kJ

as $\forall_1 = 0.06532$ m³ and $\forall_2 = 0.04213$ m³ as $m_2 = \frac{1}{2}m_1$

Solving for $Q = Q_{out} = W_b - (mh)_{exit} - \Delta U |_{CV}$

$Q = 60.7$ kJ

5   Steam turbine with heat loss

Sketch

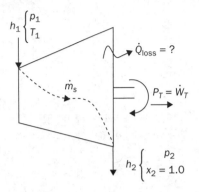

Assumptions

- Steady open system
- $H_2O \cong 2-$ phase fluid
- Quasi-equilibrium expansion process

Method

- $\dot{m}_1 = \dot{m}_2 = \dot{m} = \not\subset$

- $-(P_T + \dot{Q}_{loss}) = \dot{m}\Delta h$
- Property Tables (see App. B)

Solution

- $\dot{m} = 22$ kg

- $\sum \dot{E}_{in} - \sum \dot{E}_{out} = 0 > \dot{m}_1 h_1 = \dot{m}_2 h_2 + \dot{W}_T + \dot{Q}_{loss}$

  or

- $\dot{Q}_{loss} \equiv \dot{Q} = \dot{m}(h_1 - h_2) - \dot{W}_T$
- Property values

  Inlet (1): p_1 & $T_1 \rightarrow h_1 = 3146.0$ kJ/kg

  Outlet (2): T_2 & $x_1 \rightarrow h_2 = 2555.6$ kJ/kg

  $\therefore \dot{Q} = 640$ kW

T-v Diagram

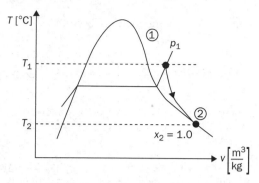

Comments

✓ Relatively high $\dot{Q}_{loss}$, i.e., turbine insulation is needed

✓ Exhaust of "wet steam" may be desirable

Test II

(90 minutes; only Property Tables and Equation Sheets are allowed)

Part A: Insight

1 (5 points) Given T_{in} for air in an idealized throttling device, determine T_{out} and prove your answer.

2 (5 points) Entropy being a state function implies that for a closed system $\Delta S_{real} = \Delta S_{ideal} = \Delta S_{system} = S_2 - S_1$, i.e., for both real and (*very different*) isentropic processes. Provide detailed math/physics explanations.

3 (10 points) Derive an expression for Δh of an ideal gas, considering a polytropic process, being adiabatic and reversible, with given temperatures T_1 and T_2 as well as constant heat capacity c_p.

Part B: Problems (40 points each)

4 Via a supply-line, air at 500 kPa and 70°C is pressed into an insulated p-c device, initially with 0.40 m³ containing 1.3 kg of air at 30°C. The process stops when the cylinder volume has increased by 50%. Assuming constant specific heats [i.e., $c_v = 0.718$ kJ/(kg·K) and $c_p = 1.005$ kJ/(kg·K)], develop an equation for T_{final}, solve for T_{final}, and find the entropy generated (use 315 K, if needed).

5 An adiabatic air-compressor (In: 98 kPa, 295 K, and 10 kg/s of air; Out: 1 MPa and 620 K) is powered by an adiabatic turbine (In: 12.5 MPa, 500°C, and 25 kg/s of steam; Out: 10 kPa and a quality of 92%). Determine the net power delivered to the generator as well as the entropy generated within the compressor (employing averaged c-values) and the turbine.

Provide a *T-s* diagram for the turbine and a $p(\forall)$ plot for the compressor work.

Solutions to Test II

Part A: Insight

1   Ideal throttle

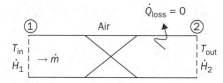

Notes

- Steady open device
- Air $\cong$ I.G.
- $\dot{W} = 0$; $\dot{Q} \approx 0$
- c_p is constant
    Hence, $c_p T_1 = c_p T_2$ or $T_{out} = T_{in}$

E-B:

$\dot{H}_1 = \dot{H}_2$ or $h_1 = h_2$ where for an I.G. $h = u + RT$ w. $|u(T)$ only

2 $\Delta S_{real} = \Delta S_{ideal}$?

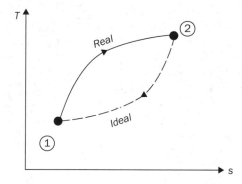

Notes

- Indeed, $\Delta S_{system} = S_2 - S_1$ is the same for the two very different processes because entropy is a state function.

- $\Delta S|_{ideal} = S_2 - S_1 = \int_1^2 \dfrac{\delta Q}{T}|_{internally\ reversible}$; however,

- $\Delta S|_{real} = S_2 - S_1 = \int_1^2 \dfrac{\delta Q}{T} + S_{gen.}$

- ΔS_{system}^{real} implies that in addition to the entropy transfer with heat, some extra entropy is generated due to irreversibilities.

- On a microscopic level, for $\Delta S_{\text{system}}^{\text{ideal}}$ only internally reversible heat transfer is needed to cause an increase in molecular disorder. For real processes while $Q(T)$ is being transferred, actual molecular dynamics (e.g., collision/deformation, vibration, spin, etc.) create additional disorder, i.e., S_{gen}.

3   Find Δh-expression for an isentropic, polytropic process of an ideal gas.

Recall from Gibbs II for $s = c$ process: $dh = v\,dp \cong \delta w$ where $v(p) = C^{1/k} - p^{-1/k}$; $\kappa \equiv \dfrac{c_p}{c_v}$

Hence, $dh = \left(\dfrac{c}{p}\right)^{1/\kappa}$ with $C^{1/k} = vp^{1/k}$ and $pv = RT$. Integration yields

$$\Delta h = C^{1/\kappa}\left.\frac{p^{(1-1/\kappa)}}{1-1/\kappa}\right|_1^2 = \left.\frac{\kappa}{\kappa-1} * RT\right|_1^2 = \frac{\kappa}{\kappa-1} * R(T_2 - T_1); R = c_p - c_v$$

$$\therefore \Delta h = c_p(T_2 - T_1)$$

Part B: Problems

4   Filling of a p-c device with air

Sketch

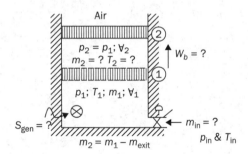

Air

$p_2 = p_1; \forall_2$
$m_2 = ? \, T_2 = ?$

$W_b = ?$

$p_1; T_1; m_1; \forall_1$

$S_{\text{gen}} = ?$

$m_2 = m_1 - m_{\text{exit}}$

$m_{\text{in}} = ?$
$p_{\text{in}} \,\&\, T_{\text{in}}$

Assumptions

- Transient, open system
- Air $\cong$ I.G.
- $Q = 0; W = W_b$
- Quasi-equilibrium expansion process at $p|_{\text{CV}} \approx c$
- Constant c_v & c_p
- $T_2 = 315.3$ K to be proven

Method

- Reduced M-B, E-B, and S-B
- $p\forall = mRT$
- Gibbs II for ΔS
- $W_b = p\Delta\forall$, as $p = \text{constant}$

- Expect $T_1 < T_2 < T_{in}$
- $h = c_p T$ and $u = c_v T$ for IGs

Solution

- M-B: $m_{in} = m_2 - m_1$; $m_2 = \dfrac{p_2 \forall_2}{RT_2}$; T_2 in K! $p_2 = p_1$

$$p_2 = p_1 = \frac{m_1 R T_1}{\forall_1} := 282.6 \text{ kPa}$$

w. $| \forall_2 = 1.5 \forall_1$: $m_{in} = m_2 - m_1 = \dfrac{590.8}{T_2} - 1.3 \text{ kg}$

- E-B: $\rightarrow (mh)_{in} - W_b = \Delta U|_{system} = m_2 u_2 - m_1 u_1$

where $W_b = p_1 \Delta \forall = 56.52 \text{ kJ}$
Hence, we balance

$$m_i(T_2) c_p T_{in} - W_b = m_2(T_2) c_v T_2 - m_1 c_v T_1$$

Numerically,

$$\left(\frac{590.8}{T_2} - 1.3\right) * 1.005(70 + 273) - 56.52 = \left(\frac{590.8}{T_2}\right) * 0.718 T_2 - 1.3 * 0.718(30 + 273)$$

Solving for T_2 yields $T_{final} = T_2 = 315 \text{ K}$ or $42°\text{C} = \begin{cases} > T_1 \\ < T_{in} \end{cases}$
Now,

$$m_2 = \frac{590.8}{315} = 1.87 \text{ kg} > m_{in} = 0.574 \text{ kg}$$

- S-B: $\rightarrow S_{gen} = \Delta S_{system} + \sum S_{out} - \sum S_{in} = (ms)_2 - (ms)_1 + 0 - (ms)_{in}$
Or

$$S_{gen} = m_2(s_2 - s_{in}) - m_1(s_1 - s_{in})$$

Note With Gibbs II for an I.G.:

$$S_{gen} = m_2 \left(c_p \ln \frac{T_2}{T_{in}} - R \ln \frac{p_2}{p_{in}}\right) - m_1 \left(c_p \ln \frac{T_1}{T_{in}} - R \ln \frac{p_1}{p_{in}}\right)$$

$$S_{gen} = 0.0971 \text{ kJ/K}$$

Comments

✓ For the supply line $h_{in} = ¢$ as p_{in} & T_{in} are constant.
✓ The E-B delivers T_2 implicitly in conjunction with the M-B using an iterative program; or, one can solve for T_2 explicitly.

✓ The S_{gen} – equation is revealing which processes contribute to and which decrease S_{gen}.

✓ Again, physically more interactive is the $\Delta S_{system} = S_{final} - S_{initial}$ equation to track entropy increase vs. decrease.

5

Sketch

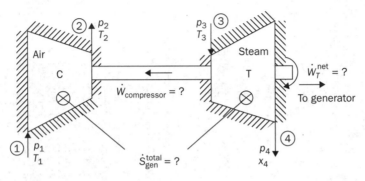

Assumptions

• Steady open systems

• $Air\big|_C \cong I.G.; H_2O\big|_T = 2\text{-phase}$

• $\dot{Q} = 0; \dot{W}_T^{net} = P_T - \dot{W}_C$

• Quasi-equilibrium expansion/compression processes

• Averaged specific heats

Method

• E-Bs: done separately for each device: M-B: $m_{in} = ¢; m_{steam} = ¢$

• S-Bs: $\dot{S}_{gen}^{total} = \dot{S}_{gen}^T + \dot{S}_{gen}^C$, using Gibbs II for $\dot{S}_{gen}^C$

$$\bar{c} = c_{avg} = c\left(\frac{1}{2}\left(T_{high} + T_{low}\right)\right); \text{ Property Tables in App. B used for both air and steam}$$

Solution

• Balances → $\dot{m}_{air}$ and $\dot{m}_{steam}$ are constant

$$\sum \dot{E}_{in} = \sum \dot{E}_{out} \text{ for compressor and turbine}$$

Thus, $\dot{W}_C + \dot{m}_{air}h_1 = \dot{m}_{air}h_2 > \dot{W}_C = \dot{m}_{air}\left(h_2 - h_1\right)$

$$\dot{m}_{steam}h_3 = P_T + \dot{m}_{steam}h_4 > \dot{W}_T^{total} = \dot{m}_{steam}\left(h_3 - h_4\right)$$

• $\dot{S}$-B:

$$\dot{S}_{gen}^{total} = \dot{S}_{gen}^C = \dot{S}_{gen}^T = \dot{m}_{air}\left(s_2 - s_1\right) + \dot{m}_{steam}\left(s_4 - s_3\right)$$

- Properties
    - (i) Compressor:

$$@ \ T_1 \ \& \ T_2 \begin{cases} \text{State (1)} \rightarrow h_1 = 285.17 \dfrac{\text{kJ}}{\text{kg}} @ 295 \text{ K} \\[3mm] \text{State (2)} \rightarrow h_2 = 628.07 \dfrac{\text{kJ}}{\text{kg}} @ 620 \text{ K} \end{cases}$$

    - (ii) Turbine:

        State (3) $\rightarrow p_3 \ \& \ T_3 > h_2 = 3343.6$ kJ/kg; $s_3 = 6.4651$ kJ/kg·K

        $p_4 \ \& \ x_4 = 0.92 > h_4 = h_f + x_4 h_{fg} = 2392.5$ kJ/kg; $s_4 = 7.5489$ kJ/kg·K

        Now, $\dot{W}_C = 10 * (628.07 - 295.17) = 3329$ kW and $P_T = \dot{W}_T^{\text{total}} = 23{,}777$ kW

        $\therefore \dot{W}_T^{\text{net}} = 20{,}448$ kW

        $\dot{S}_{\text{gen}}^{\text{total}} = \dot{m}_{\text{air}} \left(s_2 - s_1 \right) + \dot{m}_{\text{steam}} \left(s_4 - s_3 \right)$

        With Gibbs II, $\dot{S}_{\text{gen}}^C = \dot{m}_{\text{air}} \left(s_2 - s_1 \right) = \dot{m}_{\text{air}} \left(\displaystyle\int_1^2 c_p \left(T \right) \dfrac{dT}{T} - R_{\text{air}} \ ln \ \dfrac{p_2}{p_1} \right)$

        where $\displaystyle\int_1^2 c_p \left(T \right) \dfrac{dT}{T}$ uses $c_p \left(T \right) \approx \overline{c}_p = c_p \left(\dfrac{T_1 + T_2}{2} \right) = c_p (457.5 \text{ K}) \approx 1.021$ kJ/kg·K

        Hence, $\dot{S}_{\text{gen}}^C = 0.92$ kW/K

        $\dot{S}_{\text{gen}}^T = 25(7.5489 - 6.4651) = 27.1$ kW/K

        Thus, $S_{\text{gen}}^{\text{total}} = 28$ kW/K

Graphs

- Turbine

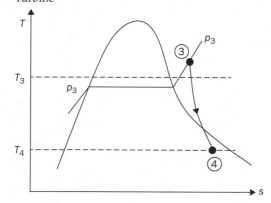

- Compressor

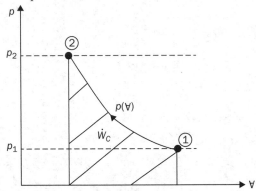

Comments

✓ $\dfrac{\dot{W}_T}{\dot{W}_C} \gg 1$ and $\dfrac{\dot{S}_T}{\dot{S}_C} \gg 1$, which makes sense as ΔT_T & $\Delta h_T \gg \Delta T_C$ & Δh_C.

✓ Although $\Delta S_T = s_4 - s_3$ is relatively small, $\dot{S}_{gen}^T$ is large because of $\dot{m}_{steam} \gg 1$.

Final Exam I

(3 hours; only Property Tables and Equation Sheets are allowed)

Part A: Insight

1 (5 points) Draw the C-HE cycle in both the T-s diagram and p-v diagram, indicating the heat transferred and work performed.

2 (10 points) An initially empty vessel ($\forall$ and T_0) is filled adiabatically with an ideal gas (p_{in} and T_{in}) of constant properties (c_v, c_p, R). Given the vessel volume, the gas properties as well as the inlet, final, and dead states, develop an expression for the rate of entropy generation. *Hint*: start with an appropriate exergy balance in differential form.

3 (5 points) You as a patent officer must examine a new refrigerator with an apparent $COP_R = 12$, which maintains the refrigerated space at 5°C while operating in a room at 25°C. What is your verdict?

Part B: Problems

4 (20 points) Consider a p-c device with 1.2 kg of air at 700 kPa and 200°C, where the piston is pressed against a pair of stops (the piston alone causes 600 kPa of pressure). Now a bottom valve is opened until the initial volume has decrease by 20%, while 40 kJ of heat is being lost. Find the final air temperature.

5 (20 points) A heat pump with R-134a keeps a space at 25°C. Its heat source is water that enters the evaporator in cross flow at 60°C with a mass flow rate of 0.065 kg/s and exits at 40°C. The refrigerant leaves the throttle at 12°C with a quality of 15% and enters (with the same pressure) the 1.6 kW compressor as saturated vapor. Find: (i) the R-134a mass flow rate; (ii) the rate of heat supply; (iii) the COP; and (iv) theoretically the minimum compressor power for the same rate of heat supply.

6 (20 points) Two rigid tanks are connected by a valve, where the ambient is at 25°C. Tank A contains 0.2 m³ of water at 400 kPa and 80% quality. Tank B contains 0.5 m³ of water at 200 kPa and 250°C. Now, the valve is opened until thermal equilibrium has been reached. Determine the final pressure and total heat transferred. Show the three states in a p-v diagram and show Q as a function of $T_{ambient}$ for $0 < T_{ambient} < 50°C$. Comment!

7 (10 points) An insulated p-c device contains 0.8 L of a saturated H_2O mixture at 120 kPa, where the surrounding is at 25°C and 100 kPa. An electric resistance heater supplies 1400 kJ. Determine the reversible work input during this process as well as the amount of exergy destruction. Show the two states in a T-s diagram and comment on $EX_{destroyed}$ as a function of $W_{electric}$.

8 (10 points) A house loses heat at a rate of 3800 kJ/h per °C difference between the indoors and outdoors. Using a heat pump with a required 4 kW to keep the house at 24°C, find the theoretically lowest outdoor temperature.

Solutions to Final Exam I

Part A: Insight

1 *p-v* diagram and *T-s* diagram

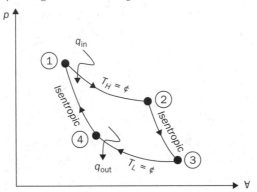

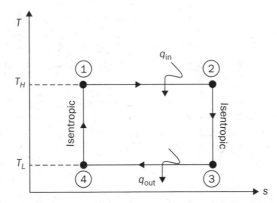

As the *T-s* diagram indicates, heat addition proceeds in the direction of increasing entropy, while heat rejection proceeds in the direction of decreasing entropy. The two isentropic processes (i.e., internally reversible and adiabatic) proceed at constant entropy. The area under the heat addition process in a *T-s* diagram is a geometric measure of the total heat supplied during the cycle, q_{in}, and the area under the heat rejection process is a measure of the total heat rejected, q_{out}. The difference between these two (the area enclosed by the cyclic curve) is the net heat transfer, which is also the network produced during the cycle. Therefore, in a *T-s* diagram, the ratio of the area enclosed by the cyclic curve to the area under the heat addition process curve represents the thermal efficiency of the cycle. Any modification that increases the ratio of these two areas will also increase the thermal efficiency of the cycle.

2 Air tank filling and rate of entropy generation

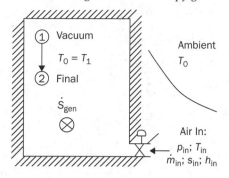

With $T_{\text{vacuum}} \cong T_0$ and $u_0 = h_0 = s_0 = 0$, for a transient open system in differential form.

$$\frac{dEX}{dt}\bigg|_{\text{vessel}} = \frac{dEX}{dt}\bigg|_{\text{in}} - \frac{dEX}{dt}\bigg|_{\text{out}} - \frac{dEX}{dt}\bigg|_{\text{destroyed}}$$

or

$$\frac{dEX}{dt}\bigg|_{\text{vessel}} = (h_{\text{in}} - T_0 s_{\text{in}})\underbrace{\frac{dm}{dt}}_{\dot{m}_{\text{in}}} - T_0 \dot{S}_{\text{gen}} = \frac{d(U - T_0 S)}{dt}$$

Hence,

$$T_0 \dot{S}_{\text{gen}} = -\frac{d}{dt}(U - T_0 S) + (h - T_0 s)_{\text{in}}\frac{dm}{dt}$$

or

$$\dot{S}_{\text{gen}}dt = -\frac{1}{T_0}(u_2 - T_0 s_2)dm + (h_{\text{in}} - T_0 s_{\text{in}})dm$$

Integration yields

$$\dot{S}_{\text{gen}} = \frac{m_2}{T_0}\bigg[h_{\text{in}} - u_2 + \underbrace{T_0 s_2 - T_0 s_{\text{in}}}_{-T_0(s_{\text{in}} - s_2)}\bigg]$$

where:

- $m_{\text{fuel}} = m_2 = \dfrac{p_2 \forall}{R_{\text{air}} T_2}$

- $s_{\text{in}} - s_2 = c_p \ln\dfrac{T_{\text{in}}}{T_2} - R \ln\dfrac{p_{\text{in}}}{p_0}$ (Gibbs II)

- $h_{\text{in}} - u_2 = c_p T_{\text{in}} - c_v T_2$

Finally, $\dot{S}_{gen} = \dfrac{p_2 \forall}{R_{I.G.} T_0 T_2} \left[c_p T_{in} - c_v T_2 - T_0 \left(c_p \ln \dfrac{T_{in}}{T_2} - R \ln \dfrac{p_{in}}{p_2} \right) \right]$

3 A COP_R of 12! Is it a valid claim?

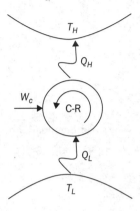

Clearly, $COP_{R,real} < COP_{R,reversible} = \dfrac{1}{T_H / T_L - 1}$

Here, the Carnot COP_R is $\dfrac{1}{\dfrac{298}{278} - 1} = 13.9$

Hence, the claim is valid.

Part B: Problems

4 Air leaving a p-c device

Sketch

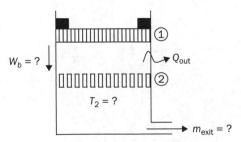

Assumptions

- Transient, open system
- Air $\cong$ I.G.
- $T_{exit} \approx \frac{1}{2}(T_1 + T_2)$ as in reality $T_{exit} = T(t)$
- Constant property values

Method

- $W_b - Q - H_e = \Delta U |_{CV}$ where $m_e = m_1 - m_2$
- $W_b = p_2 \, \Delta \forall$; and $H_{exit} = (mc_p T)_{exit}$
- $p\forall = mRT$

Solution

- M-B: $m_{exit} = m_1 - m_2$; $m_2 = m_{exit} T_2$
- E-B: $\Delta U |_{CV} = m_2 u_2 - m_1 u_1 = W_b - Q_{out} - (mh)_{exit}$
  Or

$$W_b^{\cdot} - Q - m_{exit} c_p T_{exit} = m_2 c_v T_2 - m_1 c_v T_1; \; T_{exit} = \frac{1}{2}(T_1 + T_2)$$

Note $T_2 = T_{final}$ is the principle (implicit) unknown
with

- $W_b = p_2(\forall_2 - \forall_1) = 600\left[\left(\dfrac{mRT}{p}\right)_1 + 0.8\forall_1\right] = 27.9 \text{ kJ}$

- $m_2 = \dfrac{p_2 \forall_2}{RT_2} = \dfrac{389.18}{T_2} \text{ kg}; \; m_{exit} = \left(1.2 - \dfrac{389.18}{T_2}\right) \text{ kg}$

one must solve for T_2:

$$27.9 - 40 - \left(1.2 - \frac{389.18}{T_2}\right) 1.02 * \frac{1}{2}(473 + T_2) = \frac{389.18}{T_2} * 0.733 T_2 - 1.2 * 0.733 * 273$$

$\therefore T_2 = 415$ K after, say, two iterations based on an educated initial guess

Hence, $m_{exit} = 1.2 - \dfrac{389.18}{415} = 0.262 \text{ kg}$

5 Heat pump for heating with a focus on the evaporator

Sketch

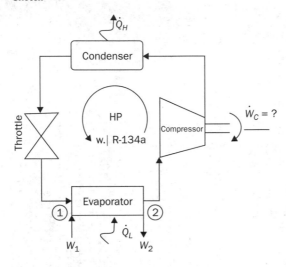

Assumptions

- Steady HP cycle
- Evaporator is a steady open system
- Water props @ $x_w = 0$
- Neglect ΔkE & ΔpE
- $\dot{Q}_{water} \approx \dot{Q}_{evaporator} \cong \dot{Q}_L$

Method

- $\dot{m} = \mathcal{C}$

- $\dot{Q} = \dot{m}\Delta h$ of $\dot{W}_C = \dot{Q}_H - \dot{Q}_L$

- $COP_{HP} = \dfrac{\dot{Q}_H}{\dot{W}_C}$

- Property Tables (see App. B)

Solutions

Evaporator Properties

- $T_1 = 12°C$ & $x_1 = 0.12 \rightarrow \begin{cases} p_1 = p_{sat} = 443.3 \text{ kPa} \\ h_1 = h_f + x_1 h_{fg} = 96.54 \text{ kJ/kg} \end{cases}$

- $p_2 = p_1$ & $x_2 = 1.0 \rightarrow h_2 = 257.33 \text{ kJ/kg} = h_g$

- Water in: $T_{W1} = 60°C$ & $x_{W1} = 0.0 \rightarrow h_{W1} = 251.18 \text{ kJ/kg}$

- Water out: $T_{W2} = 40°C$ & $x_{W2} = 0.0 \rightarrow h_{W2} = 167.53 \text{ kJ/kg}$

   E-B: $\dot{Q}_{water} = \dot{m}_W (h_{W1} - h_{W2}) = \dot{Q}_L = \dot{m}_R (h_2 - h_1)$

   $\dot{Q}_{water} = 5.437 \text{ kW} = \dot{m}_R \Delta h$

Hence,

(i) $\dot{m}_R = \dfrac{\dot{Q}_w}{\Delta h} = 0.0338 \text{ kg/s}$

(ii) Heating load $\dot{Q}_H = \dot{Q}_L + \dot{W}_C = 5.437 + 1.6 = 7.04 \text{ kW}$

(iii) $\text{COP}_{\text{HP}} = \dfrac{\dot{Q}_H}{\dot{W}_C} = \dfrac{7.04}{1.6} = 4.4$ and $\text{COP}_{\text{C-HP}} = \dfrac{1}{1 - \dfrac{T_L}{T_H}} = 9.51$

Hence,

(iv) $\dot{W}_{C,\text{min}} = \dfrac{\dot{Q}_H}{\text{COP}_{\text{max}}} = \dfrac{7.04}{9.51} = 0.74 \text{ kW} \approx 10\% \text{ of } \dot{W}_{C,\text{real}}$

6 Two connected "water" tanks

Sketch

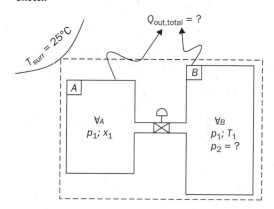

Notes

- $C\forall \cong \text{Tank A} + \text{Tank B}$
- Final state: $T_2 = T_{\text{surr}}$

Assumptions

- Transient closed system
- $H_2O \cong$ 2-phase
- $W = 0; \Delta pE$ & $\Delta kE \approx 0$

Method

- $\Delta U \,|_{C\forall} = \Delta U_A + \Delta U_B = 0 - Q_{\text{out}}$

- At initial state: $m_{\text{total}} = m_A + m_B$

  Where $m = \dfrac{\forall}{v}$

- Property Tables (App. B)

Solution

- E-B: $-Q_{\text{out}} = \left[U_2^{A+B} - \left(U_1^A + U_1^B \right) \right]$

  or

  $Q_{\text{out}} = -\left[m_{\text{total}} u_2 - (m_1 u_1)\big|_A - (m_1 u_1)\big|_B \right]; \quad m_{A,B} = \left(\dfrac{\forall}{v} \right)_{A,B}$

- Properties:

$$\text{Tank A: } p_1 \ \& \ x_1 \rightarrow \begin{cases} v_{11A} = v_f + x_1 v_{fg} = 0.37015 \ \dfrac{m^3}{kg} \\[2mm] u_{11A} = u_f + x_f u_{fg} = 2163.3 \ \dfrac{kJ}{kg} \end{cases}$$

$$\text{Tank B: } p_1 \ \& \ T_1 \rightarrow \begin{cases} v_{11B} = 1.1989 \ \dfrac{m^3}{kg} \\[2mm] u_{11B} = 2731.4 \ \dfrac{kJ}{kg} \end{cases} \therefore \ m_{1,B} = \dfrac{\forall_B}{v_{1,B}} = 0.417 \ kg$$

Final (equilibrium) State 2:

$$v_2 = \frac{\forall_{A+B}}{m_{total}} = \frac{0.7}{0.9573} = 0.73117 \ \frac{m^3}{kg}$$

$$\text{Now, with } T_2 \quad T_{surr} = 25°C \ \& \ v_2 \rightarrow \begin{cases} v_f = 0.001003 \ \dfrac{m^3}{kg}; \ v_g = 43.34 \ \dfrac{m^3}{kg} \\[2mm] u_f = 104.83 \ \dfrac{kJ}{kg}; \ u_{fg} = 2304.3 \dfrac{kJ}{kg} \end{cases}$$

Note, with $v_f < v_2 < v_g$, $p_2 = p_{sat@25°C} = 3.17 \ kPa$
where

$$x_2 = \frac{v_2 - v_g}{v_{fg}} = 0.01685 \text{ and } u_2 = u_f + x_2 u_{fg} = 143.65 \ kJ/kg$$

Hence,
$Q_{out} = 2170 \ kJ$

p-v Diagram

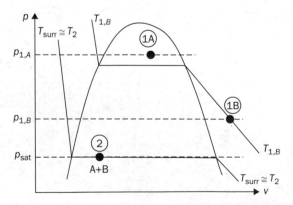

Graph $Q_{out} = fct.(T_2)$

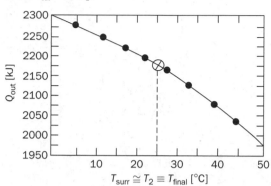

Comments

✓ The very high heat loss, Q_{out}, i.e., after $T_{final} = T_{surr}$, almost liquifies the vapors in Tanks A and B.

✓ Clearly as $T_{final} = T_2 = T_{surr}$ decreases, Q_{out} increases.

7 Exergy calculations for an insulated p-c device

Sketch   See basic insulated piston-cylinder system

Assumptions

• Transient closed system

• $H_2O \cong$ 2-phase

• $W_{in} = \begin{cases} W_{electric,in} \\ W_{reversible,in} \end{cases}$; $Q = 0$

• ΔpE & $\Delta kE \approx 0$

• $W_{boundary} = p\Delta\forall$

• Quasi-equilibrium process @ $p = ¢$

Method

• $m = \dfrac{\forall}{v}$

• $\Delta U|_{C\forall} = W_{electric} - W_{boundary}$
  Where
  $$\Delta U + W_b = \Delta H = m\Delta h$$

• $\Delta EX|_{C\forall} = EX_{in} - EX_{out} - EX_{destroyed}$ for $W_{rev,in}$ & $EX_{destroyed} = 0$

Solution

- Properties:
  - State 1:

$$p_1, x_1 = 0 \rightarrow \begin{cases} u_1 = 439.27 \text{ kJ/kg} \\ v_1 = 0.001047 \text{ m}^3\text{/kg} \\ h_1 = 439.36 \text{ kJ/kg} \\ s_1 = 1.3609 \text{ kJ/kg·K} \end{cases}$$

$$\therefore m = \frac{\forall}{v} = 7.639 \text{ kg and from E-B with } W_{el} = \Delta H = m(h_2 - h_1)$$

$$h_2 = h_1 + \frac{W_{el}}{m} = 439.36 + \frac{1.400}{7.369} = 622.63 \text{ kJ/kg}$$

- State 2:
  $$p_2 = p_1 \text{ 120 kPa \& } h_2 = 622.63 \text{ kJ/kg} \rightarrow x_2 = \frac{h_2 - h_f}{h_{fg}}; \quad \therefore x_2 = 0.08168$$

  so that:
  $$v_2 = v_f + x_2 v_{fg} = 0.1176 \text{ m}^3\text{/kg}; \, u_2 = 608.52 \text{ kJ/kg}; \, s_2 = 1.845 \text{ kJ/kg·K}$$

- With the EX-B: $W_{rev,in} = (EX_2 - EX_1)_{Cv} = m[(u_2 - u_1) - T_0(s_2 - s_1) + p_0(v_2 - v_1)]$

  Numerically, $W_{rev} = 278 \text{ kJ} \cong W_{min}$

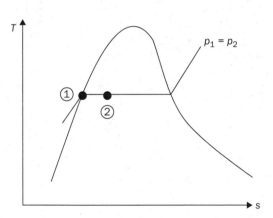

$$EX_{destroyed} = T_0 S_{gen} = T_0 m(s_2 - s_1)$$
$$= 1104 \text{ kJ}$$

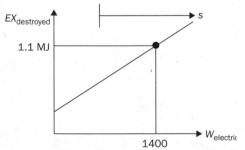

8 House Heating

Sketch

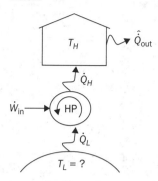

Assumptions

- $T_L \cong T_{outdoors}$
- Steady cyclic system
- Carnot HP to work with $T_L = T_{min}$

Method

- $$\text{COP}_{HP} = \frac{\dot{Q}_H}{\dot{W}_{in}} \rightarrow \text{COP}_{C\text{-}HP} = \frac{1}{1 - \dfrac{T_L}{T_H}}$$

- $\dot{Q}_H = \dot{Q}_{out}(T_H - T_{outdoor})$

Solution

- $\dot{Q}_H = 3800 \text{ kJ/h·K}\,(297 - T_2) = 1.056 \text{ kW/K}(297 - T_2)\text{ K}$

  With

  $\text{COP}_{HP} = \text{COP}_{C\text{-}HP}$ for $T_L = T_{min}$

  $$\text{COP} = \frac{\dot{Q}_H}{\dot{W}_{in}} = \frac{1.056(297 - T_L)}{4} = \frac{1}{1 - \dfrac{T_L}{297}}$$

  $\therefore T_L^{min} = 263.5 \text{ K} \cong -9.5°C$

Final Exam II

(3 hours; only Property Tables and Equation Sheets are allowed)

Part A: Insight (5 points each)

1   Is an isothermal process necessarily internally reversible? Provide an example!

(No, as an isothermal process can be irreversible. For example, a system with $W_{stir}^{in} = Q_{out}$ is left with S_{gen}.)

2   Does a refrigerator with a higher COP necessarily have a higher 2nd-Law efficiency than one with a lower COP?

(No, the fridge with a lower COP may have a higher $\eta_{II} = \dfrac{COP}{COP_{rev}}$.)

3   Plot in both *T-s* and *p-v* diagrams polytropic process paths for gases with $1 \leq n \leq k = c_p/c_v$. Comment on the use of the $n = 1$ vs. $n = k$ graphs.

Notes   The areas under the $p(v)$ or $T(s)$ graphs are $W_{compression} = m\int_1^2 v\, dp$ or $Q = m\int_1^2 T\, ds$.

4   A thin metal wire $(m, c, T_{initial})$ is subjected to heat-transfer rates $\dot{Q}_{in} = \mathcal{C}$ and $\dot{Q}_{out}(t)$. Develop an equation for the initial temperature rise, plot $T(t)$, and discuss the graph.

(For this closed 1-D system with constant props, the E-B reads

$$\dot{Q}_{net} = \dot{Q}_{in\backslash out} = \frac{dU}{dt} = mc\frac{dT}{dt}$$

Thus $\dfrac{dT}{dt} = \dfrac{\dot{Q}_{net}}{mc}$ s.t. $T(t = 0) = T_{initial}$

Note that when $\dot{Q}_{out} = \dot{Q}_{in}$, i.e., $\dot{Q}_{net} = 0$, a steady state is reached with $T = T_{final}$.)

5   Consider a reversible heat engine receiving heat from a finite body $(m, c;$ initially at $T_1)$ and rejecting waste-heat to a second body (same m and c; but initially at T_2). Work is produced until the two bodies are at $T_{equilibrium}$. Develop an equation for $T_{equilibrium}.(T_1, T_2)$. Comment!

(For $\dot{W}_{max}^{HE}$, $\dot{S}_{gen} = 0$! Also $\Delta S_{HE} = 0$ being cyclic. Hence, $\dot{S}_{gen} = \Delta\dot{S}|_{surroundings} + 0 + \Delta\dot{S}|_{sink} = 0$.

Or

$$mc\ln\frac{T_{eq}}{T_1} + 0 = mc\ln\frac{T_{eq}}{T_2} = 0$$

$\therefore \ln\dfrac{T_{eq}}{T_1} + \ln\dfrac{T_{eq}}{T_2} = 0 > \ln\dfrac{T_{eq}}{T_1}\dfrac{T_{eq}}{T_2} = 0 > T_{eq}^2 = T_1 T_2$, i.e., $T_{eq} = \sqrt{T_1 T_2}$.)

Comments   For max. power production, $T_{equilibrium}$ is equal to the geometric mean of T_1 & T_2.

Part B: Problems

6   (20 points) Steam enters a 5-MW two-stage adiabatic turbine system at 8 MPa and 500°C. The steam, leaving the 1st stage with 2 MPa and 350°C, is isobarically reheated to 500°C to enter

the 2nd stage where it exits at 30 kPa with a quality of 97%. The surrounding being at 25°C, determine the reversible power output and the rate of exergy destruction. Plot the process in a *T-s* diagram.

(Reversible power output, $\dot{W}_{rev} = 5457$ kW; Rate of exergy destruction, $\dot{EX}_{destroyed} = 457$ kW)

7 (20 points) Consider a well-insulated room (4 m × 5 m × 7 m; 100 kPa, 22°C). Now a container with one ton of water at 80°C is installed. Of interest are the final temperature and the total entropy change [kJ/K] during this process. Use properties at room temperature; ignore the container volume.

(Final temperature, $T_{final} = 78.4$°C; Total entropy change, $\Delta S_{total} = 1.79$ kJ/K > 0%)

8 (20 points) A p-c device in a 100-kPa and 25°C surrounding initially contains 2 L of air at 100 kPa and 25°C. With a useful work input of 1.2 kJ, the air is now compressed to 600 kPa and 25°C. Neglecting heat losses, find the exergy at both the initial and final states, the minimal compression work, and the 2nd-Law efficiency. Take $\forall_0 = \forall_1$; averaged c-values.

(Exergy at the initial state, $EX_1 = 0$ as $p_1 = p_0$ & $T_1 = T_0$ (dead state); Exergy at the final state, $EX_2 = 2$ L; Minimal compression work, $W_{rev} = 0.171$ kJ; 2nd-Law efficiency, $\eta_{II} = 14.3\%$)

9 (15 points) Steam is to be condensed in a tube-and-shell heat exchanger at 50°C with water entering the meandering tube at 12°C with 240 kg/s and leaving at 20°C. Neglecting any losses, determine the rate of steam condensation, i.e., from g to f, and the exergy-destruction rate in the heat exchanger.

(Rate of steam condensation, $\dot{m} = 3.369$ kg/s; Exergy-destruction rate in the heat exchanger, $\dot{EX}_{destroyed} = 837$ kW)

(A 20-point bonus problem) A 4-L tank, filled with 2 L of water and 2 L of sat. vapor at 175 kPa, is equipped with a 175-kPa relief valve. The ambient conditions are 25°C and 100 kPa. Now 750 W are supplied for 20 min from a 180°C source. How much H_2O does remain, and what is the exergy destruction?

(Amount of H_2O remaining, m = 1.507 kg; Exergy destruction $EX_{destroyed} = 96.8$ kJ)

References

Bejan, A. (1996). Method of entropy generation minimization, or modeling and optimization based on combined heat transfer and thermodynamics. *Revue Générale De Thermique*, *35*(418–419), 637–646. https://doi.org/10.1016/s0035-3159(96)80059-6

Borgnakke, C., Sonntag, R. E., Van, W. G. J., & Sonntag, R. E. (2009). *Fundamentals of Thermodynamics*. Hoboken, NJ: Wiley.

Çengel, Y. A., & Boles, M. A. (2015). *Thermodynamics: An Engineering Approach,* 8th ed. Boston: McGraw-Hill.

Jaynes, E. T. (1965). Gibbs vs. Boltzmann Entropies. *American Journal of Physics*, *33*(5), 391–398. https://doi.org/10.1119/1.1971557

Moran, M. J., & Shapiro, H. N. (2011). *Fundamentals of Engineering Thermodynamics,* 7th ed. Chichester, England: Wiley.

Turns, S. R. (2006). *Thermodynamics: Concepts and Applications*. New York: Cambridge University Press.

Index

F